Hiqmet Kamberaj
Classical Mechanics

Also of Interest

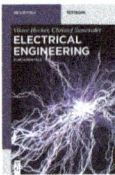

Electrical Engineering
Fundamentals
Viktor Hacker, Christof Sumereder, 2020
ISBN 978-3-11-052102-3, e-ISBN (PDF) 978-3-11-052111-5,
e-ISBN (EPUB) 978-3-11-052113-9

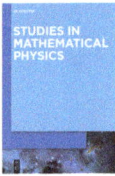

De Gruyter Studies in Mathematical Physics
Edited by Michael Efroimsky, Leonard Gamberg, Dmitry Gitman,
Alexander Lazarian, Boris M. Smirnov
ISSN 2194-3532, e-ISSN 2194-3540

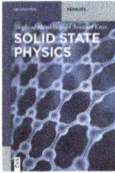

Solid State Physics
Siegfried Hunklinger, Christian Enss, planned for 2021
ISBN 978-3-11-066645-8, e-ISBN (PDF) 978-3-11-066650-2,
e-ISBN (EPUB) 978-3-11-066708-0

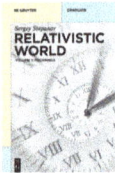

Relativistic World
Volume 1 Mechanics
Serhii Stepanov, 2018
ISBN 978-3-11-051587-9, e-ISBN (PDF) 978-3-11-051588-6,
e-ISBN (EPUB) 978-3-11-051600-5

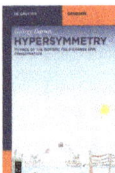

Hypersymmetry
Physics of the Isotopic Field-Charge Spin Conservation
György Darvas, 2020
ISBN 978-3-11-071317-6, e-ISBN (PDF) 978-3-11-071318-3,
e-ISBN (EPUB) 978-3-11-071348-0

Hiqmet Kamberaj

Classical Mechanics

——

DE GRUYTER

Mathematics Subject Classification 2010
Primary: 00A79, 51P05, 00A69; Secondary: 70F15, 76A02

Author
Dr. Hiqmet Kamberaj
Metodija Mitevski Nr. 12/4-3
1000 Skopje
North Macedonia
h.kamberaj@gmail.com

ISBN 978-3-11-075581-7
e-ISBN (PDF) 978-3-11-075582-4
e-ISBN (EPUB) 978-3-11-075593-0

Library of Congress Control Number: 2021940883

Bibliographic information published by the Deutsche Nationalbibliothek
The Deutsche Nationalbibliothek lists this publication in the Deutsche Nationalbibliografie;
detailed bibliographic data are available on the Internet at http://dnb.dnb.de.

© 2021 Walter de Gruyter GmbH, Berlin/Boston
Cover image: Jaredd Craig / unsplash
Typesetting: VTeX UAB, Lithuania
Printing and binding: CPI books GmbH, Leck

www.degruyter.com

Preface

Physics is part of any curriculum in science and engineering. This course's main objective is to help engineering and other science students in more advanced courses in these fields. A solid foundation in fundamental theories of physics is a must for accomplishing an engineering or science degree. In this textbook, the emphasis will be on introducing the students to physics' fundamental concepts and how different theories are developed from physical observations and phenomena.

The textbook gives an introduction to the fundamental concepts of classical mechanics.

This textbook starts with a brief introduction to the essential calculus for classical mechanics (Chapter 1). Then, Chapter 2 introduces the basic units and measurements in physics. Chapter 3 gives the definition and relationship of the essential quantities that describe a general object's kinematics in one dimension. Then, Chapter 4 generalizes the same concepts for two- and three-dimensional motion. The free-fall and projectile motion are described in detail as examples of one- and two-dimensional motion, respectively.

Chapter 5 introduces the laws that describe the dynamics of the motion; also, the circular motion is discussed as a particular case of applying Newton's laws in Chapter 6. Chapter 7 introduces the definitions of work and kinetic energy. Besides, the work–kinetic energy theorem is given. The potential energy and conservation laws of mechanics are described in Chapter 8. In Chapter 9, the concept of momentum and impulse are introduced. Also, collisions are introduced.

Chapter 10 describes the kinematics and dynamics of the rotational motion for a general rigid object. Chapter 11 introduces the rolling motion and represents the angular momentum for a rigid body and a system of particles. Chapter 12 introduces the static equilibrium conditions and elasticity of solids. Oscillatory motion is presented in Chapter 13. The last chapter (Chapter 14) introduces gravity and Kepler's laws.

This textbook is geared more towards examples and problem-solving techniques. The students will get a firsthand experience of how physics theories are applied to everyday problems in engineering and science. The learning outcome will be broad knowledge and know-how for problem-solving techniques, crucial in training engineers and scientists for a successful career in these fields. At the end of each chapter, several exercises are shown, and the solutions to those exercises are presented at the end of the textbook as an extra chapter.

The book is mainly aimed at undergraduate students in engineering and science. Chapter 1 reviews some elementary mathematics that can be useful to the students in the transition from high school to university.

May 2021   *Hiqmet Kamberaj*

https://doi.org/10.1515/9783110755824-201

Acknowledgment

I thank my family for their continuous support: Nera (my wife), Jon (my son) and Lina (my daughter).

https://doi.org/10.1515/9783110755824-202

Contents

1 A brief review of calculus for physics

1.1 Coordinate systems

Often in physics, as we will see, we deal with the location in space, which is done with the use of *coordinates*. The ones most often used are the *Cartesian coordinates*, (x, y). These are used in a Cartesian coordinate system, where the horizontal and vertical axes intersect at a single point, called the *origin* (see also Fig. 1.1). Cartesian coordinates are also called *rectangular coordinates*.

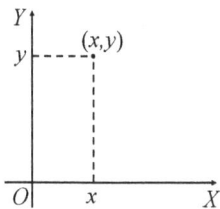

Figure 1.1: Points in a Cartesian coordinate system labeled with coordinates (x, y).

For convenience reasons sometimes we represent a point in a plane by its plane *polar coordinates* (r, θ), shown in Fig. 1.2. In the polar coordinate system, r is the distance from the origin to the point having Cartesian coordinates (x, y), and θ is the angle between r- and x-axis, taken as the positive x-axis. The angle θ is usually measured counterclockwise from the positive x-axis.

From the right triangle in Fig. 1.2, we can obtain the Cartesian coordinates as a function of the plane polar coordinates of any point using the following equations:

$$x = r \cos \theta \tag{1.1}$$
$$y = r \sin \theta.$$

Using the trigonometric relations, $\tan \theta = \sin \theta / \cos \theta$ and $\sin^2 \theta + \cos^2 \theta = 1$, we can find that

$$\tan \theta = \frac{y}{x} \tag{1.2}$$
$$r = \sqrt{x^2 + y^2}.$$

Equation (1.2) can be used to relate the Cartesian coordinated (x, y) with the polar coordinates (r, θ) only when θ is defined as in Fig. 1.2, with positive θ angle measured counterclockwise from the positive x-axis.

These standard conventions are often used from some scientific calculators to perform conversions between Cartesian and polar coordinates. Moreover, if we choose the reference axis for the polar angle θ to be different from the positive x-axis or if the sense of increasing θ is different, then the expressions in eq. (1.1) or in eq. (1.2) will change.

https://doi.org/10.1515/9783110755824-001

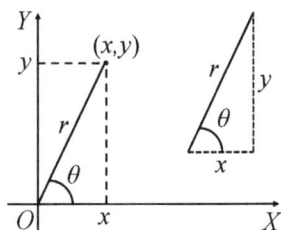

Figure 1.2: (a) The plane polar coordinates of a point are represented by the distance r and the angle θ, where θ is measured counterclockwise from the positive x-axis. (b) The right triangle used to relate (x, y) to (r, θ).

1.2 Scalar and vector quantities

In physics, there exist both scalar quantities and vector quantities. For example, for expressing the temperature, we need its value (that is a number) and the units, either in "degrees Celsius" (°C) or "degrees Fahrenheit" (°F). Therefore, we say that temperature is a scalar quantity, entirely defined by a number and appropriate units. Other scalar quantities include volume, mass, and time intervals.

Definition 1.1 (Scalar quantity). By definition, a scalar quantity is specified by a single value with an appropriate unit. A scalar quantity has no direction.

The scalar quantities obey the same rules of ordinary arithmetic when used for calculations.

Often, when we determine an object's position (say a car) that starts the motion from some reference frame's origin, it is necessary to know both its speed and its direction. Both the speed and the direction are named as a single quantity, *velocity*. Therefore, velocity is a vector quantity because it depends on the direction. As we will see, it is defined as a physical quantity that is completely specified by a value (speed) with appropriate units and direction.

Definition 1.2 (Vector quantity). By definition, a vector quantity is determined by both magnitude and direction.

The displacement is another example of the vector quantity. Suppose a particle moves from point A to point B along a straight path, as shown in Fig. 1.3. The displacement is graphically represented by an arrow from A to B, with the tip of the arrow pointing toward the final point B. The arrowhead's direction indicates the direction of the displacement, and the length of this arrow characterizes its magnitude. Note that, if the particle travels along some other path, but from A to B, as indicated by the dashed line in Fig. 1.3, its displacement is still the arrow drawn from A to B.

Here, we will use a boldface letter ($\mathbf{A}$) to represent a vector quantity. Besides, the vector can be notated with an arrow over a letter. The magnitude of the vector $\mathbf{A}$ will be written either A or $|\mathbf{A}|$. The magnitude of a vector also has physical units; for example, the displacement has SI units of meters and the velocity meters per second.

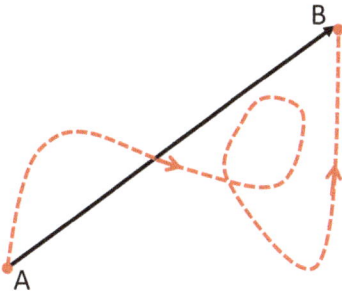

Figure 1.3: The trajectory of a particle moving from some point A to some point B.

1.3 Some properties of vectors

1.3.1 Equality of two vectors

Definition 1.3 (Equal vectors). Two vectors **A** and **B** may be defined to be equal if they have the same magnitude and the same direction. That is, **A** = **B** only if $A = B$ and if **A** and **B** point in the same direction along parallel lines.

In Fig. 1.4 we show several equal vectors parallel to each other even though they all have different starting points. Using this property in a diagram, we can move a vector to a position parallel to itself without affecting the vector. That is often used to draw the free diagram of forces.

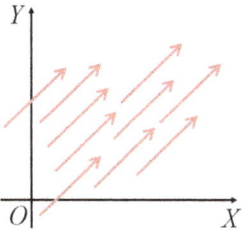

Figure 1.4: Illustration of equal vectors.

1.3.2 Adding vectors

Triangle rule of addition
To add vector **B** to vector **A**, we will use the geometric description. That is, first, draw vector **A**, with its magnitude represented by a convenient scale, on graph paper and then draw vector **B** to the same scale with its tail starting from the tip of **A**, as shown in Fig. 1.5.

The final resultant vector **R** is written as

$$\mathbf{R} = \mathbf{A} + \mathbf{B}. \tag{1.3}$$

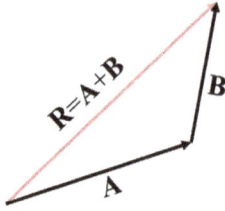

Figure 1.5: The sum of two vectors **A** and **B** is the vector **R** = **A** + **B**.

R is the vector drawn from the tail of **A** to the tip of **B**. This procedure is known as the *triangle rule of addition*.

We can use the same geometrical description to add more than two vectors, as shown in Fig. 1.6, for the case of four different vectors. The resultant vector **R** is the vector drawn from the tail of the first vector to the tip of the last vector completing in this way a polygon:

$$\mathbf{R} = \mathbf{A} + \mathbf{B} + \mathbf{C} + \mathbf{D}. \tag{1.4}$$

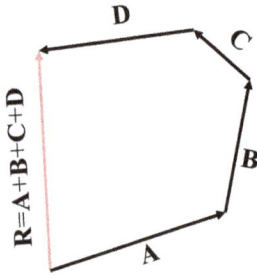

Figure 1.6: The sum of four vectors, namely **A**, **B**, **C** and **D**, is the vector **R** = **A** + **B** + **C** + **D**.

Parallelogram rule of addition
The parallelogram rule of addition is another alternative, a graphical procedure used to add two vectors, shown in Fig. 1.7. According to the parallelogram rule, the tails of the two vectors **A** and **B** are joined together, and the resultant vector **R** is the diagonal of a parallelogram formed with **A** and **B** as two of its four sides.

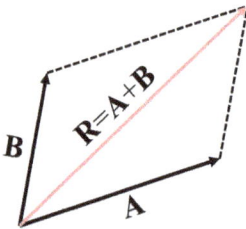

Figure 1.7: The parallelogram rule of addition of two vectors.

Commutative law of addition

Commutative law of addition says that the sum is independent of the order of the addition when two vectors are added, which is known as the commutative law of addition (see also Fig. 1.8):

$$\mathbf{A} + \mathbf{B} = \mathbf{B} + \mathbf{A}. \tag{1.5}$$

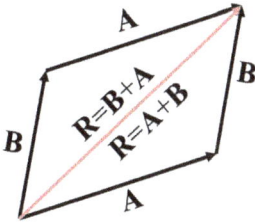

Figure 1.8: The commutative law of addition of two vectors.

Associative law of addition

Similarly, if three or more vectors are added, their sum is independent of how the individual vectors are grouped. A geometric proof of this rule for three vectors is given in Fig. 1.9. That is called the associative law of addition:

$$(\mathbf{A} + \mathbf{B}) + \mathbf{C} = \mathbf{A} + (\mathbf{B} + \mathbf{C}). \tag{1.6}$$

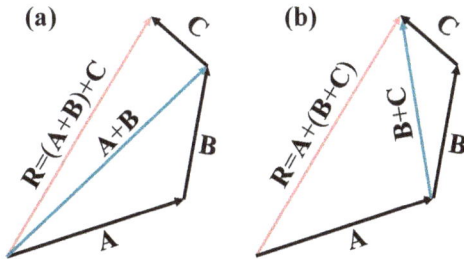

Figure 1.9: The associative law of addition of three vectors. (a) $\mathbf{R} = (\mathbf{A} + \mathbf{B}) + \mathbf{C}$ and (b) $\mathbf{R} = \mathbf{A} + (\mathbf{B} + \mathbf{C})$.

Note that a vector quantity is characterized by both magnitude and direction, and also, it obeys the rules of vector addition, as described above.

> Besides, the vectors involved in the summation must have the same units. It would not be meaningful to add a velocity vector (for example, 10 km/h to the west) to a displacement vector (for example, 30 km to the south) because they represent different physical quantities. The same rule also applies to scalars. For instance, it would practically be inappropriate, to sum up the time intervals and temperatures.

1.3.3 Negative of a vector

Definition 1.4 (Negative of a vector). The negative of the vector **A** is defined as the vector that when added to **A** gives zero for the vector sum. That is,

$$\mathbf{A} + (-\mathbf{A}) = 0. \tag{1.7}$$

The vectors **A** and −**A** have the same magnitude but point in opposite directions, as shown in Fig. 1.10.

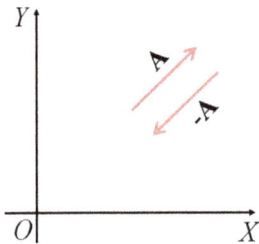

Figure 1.10: The negative of vector **A** is the vector −**A**.

1.3.4 Subtracting vectors

The vector subtraction uses the definition of the negative of a vector.

Definition 1.5 (Vector subtraction). We define the subtraction **A** − **B** as the vector −**B** added to the vector **A**:

$$\mathbf{A} - \mathbf{B} = \mathbf{A} + (-\mathbf{B}). \tag{1.8}$$

The geometric construction for subtracting two vectors in this way is illustrated in Fig. 1.11.

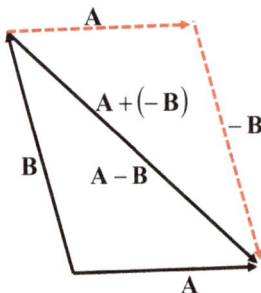

Figure 1.11: Illustration of the geometric construction for subtracting two vectors.

Alternatively, the vector subtraction, which is the difference **A** − **B** between two vectors **A** and **B**, can be obtained by adding the second vector to the tip of the first vector.

1.3.5 Multiplying a vector by a scalar

If vector **A** is multiplied by a positive scalar quantity m, then the product $m\mathbf{A}$ is a vector that has the same direction as **A** and magnitude $m|\mathbf{A}|$. If vector **A** is multiplied by a negative scalar quantity $-m$, then the product $-m\mathbf{A}$ is directed opposite **A**. For example, the vector $2\mathbf{A}$ is two times as long as **A**, and it has the same direction as **A**; the vector $-(1/3)\mathbf{A}$ is one-third the length of **A** and points in the direction opposite **A**.

1.3.6 Components of a vector and unit vectors

Whenever a high accuracy is required or in three-dimensional problems, the vectors' projections along the coordinate axes method are suggested rather than the geometric way of adding vectors.

Components of a vector

The projections are called the *components* of the vector, and the vector can, in general, be described by these components.

Consider a vector **A** positioned in the xy plane and making an arbitrary angle θ with the positive x-axis, as shown in Fig. 1.12, which can be expressed as the sum of the vectors $\mathbf{A}_x$ and $\mathbf{A}_y$ with which it can form a right triangle:

$$\mathbf{A} = \mathbf{A}_x + \mathbf{A}_y. \qquad (1.9)$$

We will often refer to the "components of a vector **A**," written A_y, and A_x (without the boldface notation).

The component A_x represents the projection of **A** along the x-axis, and the component A_y represents the projection of **A** along the y-axis. These components can be positive or negative. The component A_x is positive if **A** points in the positive x direction and is negative if **A** points in the negative x direction. The same is true for the component A_y.

Using the definitions of sine and cosine functions, we write

$$\cos\theta = A_x/A \qquad (1.10)$$
$$\sin\theta = A_y/A.$$

Hence,

$$A_x = A\cos\theta \qquad (1.11)$$
$$A_y = A\sin\theta.$$

The triangle formed by the components A_x and A_y and vector **A** is a right triangle with a length of the hypotenuse of A. Thus, it follows that the magnitude and direction of **A** are related to its components through the expressions

$$A = \sqrt{A_x^2 + A_y^2} \tag{1.12}$$

$$\theta = \tan^{-1}\left(\frac{A_y}{A_x}\right). \tag{1.13}$$

Note that the signs of the components A_x and A_y depend on the angle θ. For example, if $90° < \theta \leq 180°$, then A_x is negative and A_y is positive. If $180° \leq \theta < 270°$, then both A_x and A_y are negative.

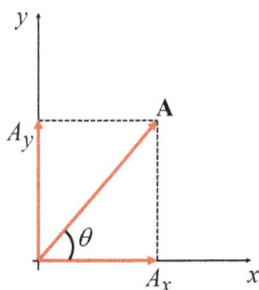

Figure 1.12: Representation of a vector **A** in xy-plane.

Suppose we are working with a problem in physics that requires resolving a vector into its components. For convenience, they are often expressed in a coordinate system rotated with respect to the system having axes that are horizontal and vertical. If we choose reference axes or an angle other than the axes and angle showed in Fig. 1.12, we must modify the components accordingly.

Suppose a vector **B** makes an angle θ' with the x'-axis defined in Fig. 1.13. The components of **B** along the x'- and y'-axes are

$$B_{x'} = B\cos\theta' \tag{1.14}$$
$$B_{y'} = B\sin\theta'.$$

The magnitude and direction of **B** are obtained from expressions equivalent to those given above. That indicates that the components of a vector can be expressed in any convenient arbitrary coordinate system.

Unit vectors

The unit vector is a dimensionless vector having a magnitude equal to one. This vector is used to specify a given direction and has no other physical meaning. That is, it is just used as a convenience in describing a direction in space. The symbols **i**, **j** and **k** are

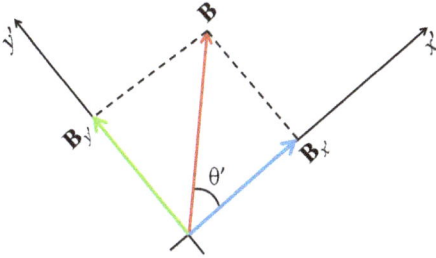

Figure 1.13: Rotation of the *xy* plane.

often used to indicate unit vectors along the positive x, y, and z axes, respectively. The unit vectors **i**, **j** and **k** form a set of mutually perpendicular vectors in a right-handed coordinate system, as shown in Fig. 1.14.

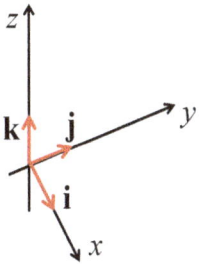

Figure 1.14: A right-handed coordinate system.

The magnitude of each unit vector equals 1:

$$|\mathbf{i}| = |\mathbf{j}| = |\mathbf{k}| = 1. \tag{1.15}$$

For a vector **A** in the *xy* plane we can write

$$\mathbf{A}_x = A_x \mathbf{i} \tag{1.16}$$
$$\mathbf{A}_y = A_y \mathbf{j}.$$

Thus, the vector **A** can be written as

$$\mathbf{A} = A_x \mathbf{i} + A_y \mathbf{j}. \tag{1.17}$$

Furthermore, consider a point in the *xy* plane with Cartesian coordinates (x, y). The location of that point can be specified by the position vector **r**:

$$\mathbf{r} = x\mathbf{i} + y\mathbf{j}. \tag{1.18}$$

Hence, the components of **r** are the lengths x and y.

Consider addition of two vectors **A** and **B** in two-dimensional space, where

$$\mathbf{A} = A_x\mathbf{i} + A_y\mathbf{j} \tag{1.19}$$
$$\mathbf{B} = B_x\mathbf{i} + B_y\mathbf{j}.$$

To do so, we add the x and y components separately, then the resultant vector $\mathbf{R} = \mathbf{A}+\mathbf{B}$ is

$$\mathbf{R} = (A_x\mathbf{i} + A_y\mathbf{j}) + (B_x\mathbf{i} + B_y\mathbf{j}) \tag{1.20}$$
$$= (A_x + B_x)\mathbf{i} + (A_y + B_y)\mathbf{j}$$
$$= R_x\mathbf{i} + R_y\mathbf{j}$$

where

$$R_x = A_x + B_x \tag{1.21}$$
$$R_y = A_y + B_y.$$

The magnitude and the angle the vector **R** forms with x-axis are, respectively,

$$R = \sqrt{R_x^2 + R_y^2} = \sqrt{(A_x + B_x)^2 + (A_y + B_y)^2} \tag{1.22}$$

and

$$\tan\theta = \frac{R_y}{R_x} = \frac{A_y + B_y}{A_x + B_x}. \tag{1.23}$$

1.4 Vector multiplication

We can perform vector multiplication in two ways. Depending on the physical quantity, one may require the *dot* or *scalar* product of two vectors and the *cross* or the *vector* product of two vectors.

1.4.1 Dot product of two vectors

Let **F** and **d** be two vectors making an angle θ. These vectors can be multiplied to produce a scalar through a vector multiplication operation called the *dot product, scalar product* or *inner product*. Fig. 1.15 shows the graphical representation of the above statement.

The scalar product W of the two vectors **F** and **d** is given by

$$W = \mathbf{F} \cdot \mathbf{d} \tag{1.24}$$

or

$$W = Fd \cos \theta. \tag{1.25}$$

Here, θ is the angle formed by the vectors **F** and **d** (see also Fig. 1.15).

Note, later on in this book, the physical quantity called work (W) is the dot product of two vectors, the force (**F**) and the displacement (**d**).

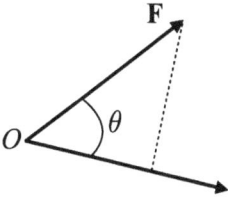

d   **Figure 1.15:** The scalar product of two vectors **F** and **d**.

Scalar product of vectors can also be performed using the **i**, **j**, and **k** vector representation. Let **A** and **B** be two vectors defined in terms of (**i**, **j**, **k**) components as

$$\mathbf{A} = A_x\mathbf{i} + A_y\mathbf{j} + A_z\mathbf{k} \tag{1.26}$$
$$\mathbf{B} = B_x\mathbf{i} + B_y\mathbf{j} + B_z\mathbf{k}.$$

The dot or scalar product of **A** and **B** is defined as

$$\mathbf{A} \cdot \mathbf{B} = (A_x\mathbf{i} + A_y\mathbf{j} + A_z\mathbf{k}) \cdot (B_x\mathbf{i} + B_y\mathbf{j} + B_z\mathbf{k}) \tag{1.27}$$
$$= A_xB_x + A_yB_y + B_zB_z$$

where the following relations were used:

$$\mathbf{i} \cdot \mathbf{i} = \mathbf{j} \cdot \mathbf{j} = \mathbf{k} \cdot \mathbf{k} = 1 \tag{1.28}$$
$$\mathbf{i} \cdot \mathbf{j} = \mathbf{i} \cdot \mathbf{k} = \mathbf{j} \cdot \mathbf{k} = 0. \tag{1.29}$$

1.4.2 Cross product of two vectors

The cross product or vector product of the vectors **A** and **B** gives another vector, namely **R**:

$$\mathbf{R} = \mathbf{A} \times \mathbf{B}. \tag{1.30}$$

Since the product is a vector, we must not only calculate its magnitude but also find its direction. The magnitude of the cross product vector **R** is

$$|\mathbf{R}| = |\mathbf{A} \times \mathbf{B}| = |\mathbf{A}||\mathbf{B}| \sin \theta. \tag{1.31}$$

Here, θ is the angle formed by the vectors **A** and **B**.

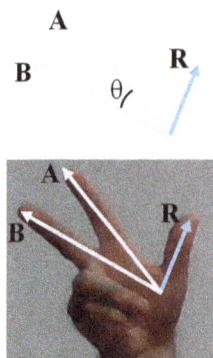

Figure 1.16: The cross product of two vectors **A** and **B**.

The direction of the vector **R** is along the line perpendicular to the plane created by the two vectors **A** and **B**, as shown in Fig. 1.16. The sign of the vector will change if the order of the cross-multiplication changes:

$$\mathbf{A} \times \mathbf{B} = -\mathbf{B} \times \mathbf{A}. \tag{1.32}$$

The direction and the sign can easily be found by the so-called *right-hand-rule*. As shown in Fig. 1.16, point your index finger in the direction of the first vector, in this case, vector **A**, and your middle finger in the direction of the second vector **B**, then your thumb will point to the direction of the vector product **R**. Note your index finger and middle finger make an angle θ together. If you point your index finger in the direction of the vector **B** and take it to be the first vector, then you have to turn your **A** with your middle finger, therefore, your thumb will point downward, and that would be the direction of the vector **R**.

Note that the cross product of a vector **A** with itself is zero, because its magnitude is determined in terms of the $\sin \theta$, and since $\theta = 0°$, then the magnitude is zero. Therefore,

$$\mathbf{A} \times \mathbf{A} = 0. \tag{1.33}$$

We can also determine the cross product of vectors using the $(\mathbf{i}, \mathbf{j}, \mathbf{k})$ vector representation. Let us define two vectors, **A** and **B**, and their $(\mathbf{i}, \mathbf{j}, \mathbf{k})$ representations:

$$\mathbf{A} = A_x\mathbf{i} + A_y\mathbf{j} + A_z\mathbf{k} \tag{1.34}$$

$$\mathbf{B} = B_x\mathbf{i} + B_y\mathbf{j} + B_z\mathbf{k}. \tag{1.35}$$

The cross product is

$$\mathbf{A} \times \mathbf{B} = (A_x\mathbf{i} + A_y\mathbf{j} + A_z\mathbf{k}) \times (B_x\mathbf{i} + B_y\mathbf{j} + B_z\mathbf{k}) \tag{1.36}$$
$$= (A_xB_y - A_yB_x)\mathbf{i} \times \mathbf{j}$$
$$+ (A_xB_z - A_zB_x)\mathbf{i} \times \mathbf{k}$$

$$+ (A_y B_z - A_z B_y)\mathbf{j} \times \mathbf{k}$$
$$= (A_y B_z - A_z B_y)\mathbf{i} - (A_x B_z - A_z B_x)\mathbf{j} + (A_x B_y - A_y B_x)\mathbf{k}$$

where the following relations were used:

$$\mathbf{i} \times \mathbf{j} = \mathbf{k}; \quad \mathbf{i} \times \mathbf{k} = -\mathbf{j}; \quad \mathbf{j} \times \mathbf{k} = \mathbf{i} \tag{1.37}$$
$$\mathbf{i} \times \mathbf{i} = \mathbf{j} \times \mathbf{j} = \mathbf{k} \times \mathbf{k} = 0. \tag{1.38}$$

Therefore, the components of the vector $\mathbf{A} \times \mathbf{B}$ are given as

$$(\mathbf{A} \times \mathbf{B})_x = (A_y B_z - A_z B_y) \tag{1.39}$$
$$(\mathbf{A} \times \mathbf{B})_y = -(A_x B_z - A_z B_x) \tag{1.40}$$
$$(\mathbf{A} \times \mathbf{B})_z = (A_x B_y - A_y B_x). \tag{1.41}$$

We can also determine the cross product as the determinant of the following matrix:

$$\mathbf{A} \times \mathbf{B} = \det \begin{pmatrix} \mathbf{i} & \mathbf{j} & \mathbf{k} \\ A_x & A_y & A_z \\ B_x & B_y & B_z \end{pmatrix} \tag{1.42}$$

or

$$\mathbf{A} \times \mathbf{B} = \det \begin{pmatrix} A_y & A_z \\ B_y & B_z \end{pmatrix} \mathbf{i} - \det \begin{pmatrix} A_x & A_z \\ B_x & B_z \end{pmatrix} \mathbf{j} + \det \begin{pmatrix} A_x & A_y \\ B_x & B_y \end{pmatrix} \mathbf{k}. \tag{1.43}$$

The magnitude of the cross product vector is determined as follows:

$$|\mathbf{A} \times \mathbf{B}| = \sqrt{(\mathbf{A} \times \mathbf{B})_x^2 + (\mathbf{A} \times \mathbf{B})_y^2 + (\mathbf{A} \times \mathbf{B})_z^2}. \tag{1.44}$$

1.5 Some scientific notations

Many quantities that we deal often are very large or very small. For example, the speed of light in vacuum is about 300000000 m/s. It can be seen that it is cumbersome to read or even write and keep track of zeros in such numbers. To avoid this problem we use powers of the number 10:

$$10^0 = 1 \tag{1.45}$$
$$10^1 = 10$$
$$10^2 = 10 \times 10 = 100$$
$$10^3 = 10 \times 10 \times 10 = 1000$$
$$10^4 = 10 \times 10 \times 10 \times 10 = 10000$$

$$10^5 = 10 \times 10 \times 10 \times 10 \times 10 = 100000$$

and so on. Here, the power to which 10 is raised corresponds to the number of zeros. This number is also called the *exponent* of 10.

Using the same method, the number smaller than 1 can be represented as

$$10^{-1} = \frac{1}{10} = 0.1 \tag{1.46}$$

$$10^{-2} = \frac{1}{10 \times 10} = 0.01$$

$$10^{-3} = \frac{1}{10 \times 10 \times 10} = 0.001$$

$$10^{-4} = \frac{1}{10 \times 10 \times 10 \times 10} = 0.0001$$

$$10^{-5} = \frac{1}{10 \times 10 \times 10 \times 10 \times 10} = 0.00001$$

and so on. Here, the value of the (negative) exponent equals the number of places the decimal point is to the left of the digit 1.

In scientific notation, the numbers are expressed as a power of 10 multiplied by another number between 1 and 10. The scientific notation for 6134000000 is 6.134×10^9 and that for 0.000135 is 1.35×10^{-4}.

The following rules apply when numbers expressed in scientific notation are being multiplied:

$$10^m \times 10^n = 10^{m+n}. \tag{1.47}$$

In eq. (1.47), n and m can be any real or integer number. For example,

$$10^2 \times 10^3 = 10^{2+3} = 10^5 = 100000. \tag{1.48}$$

The same rule applies when the exponents are negative (either one or both):

$$10^3 \times 10^{-2} = 10^{3+(-2)} = 10^1 = 10 \tag{1.49}$$

$$10^{-4} \times 10^{-5} = 10^{(-4)+(-5)} = 10^{-9}.$$

When dividing numbers expressed in scientific notation, we write

$$\frac{10^m}{10^n} = 10^{m-n} \tag{1.50}$$

where, similarly, n and m can be any numbers (reals or integers).

1.6 Some basic rules of algebra

In general, the symbols, such as x, y and z, are considered as unknowns when algebraic operations are performed. For example,

$$8x = 32. \tag{1.51}$$

Equation (1.51) is an equation, which is solved for x. To solve it, we can multiply or divide both sides with the same factor (different from zero) without destroying the equality. In our example, if we divide both sides by 8, we get

$$\frac{8x}{8} = \frac{32}{8} \tag{1.52}$$

or

$$x = 4. \tag{1.53}$$

Now, let us consider the equation

$$x + a = b. \tag{1.54}$$

To solve this type of equation, we can add or subtract the same quantity from each side:

$$(x + a) - a = b - a \tag{1.55}$$

or

$$x = b - a. \tag{1.56}$$

Now, consider the following equation:

$$\frac{x}{a} = b \tag{1.57}$$

where $a \neq 0$. Multiplying each side by a, we get

$$\frac{x}{a}(a) = b(a) \tag{1.58}$$

or

$$x = ba. \tag{1.59}$$

As a general rule, whatever operation is performed on the left side of the equation must also be performed on the right side.

The rules for multiplying, dividing, adding, and subtracting fractions are summarized in Table 1.1, where a, b, c, and d are any four real numbers.

Table 1.1: Some rules for multiplying, dividing, adding, and subtracting fractions.

Operation	Rule
Multiplication	$\left(\frac{a}{b}\right)\left(\frac{c}{d}\right) = \frac{ac}{bd}$
Division	$\frac{a/b}{c/d} = \frac{ad}{bc}$
Addition	$\frac{a}{b} \pm \frac{c}{d} = \frac{ad \pm bc}{bd}$

1.6.1 Power

The following rule holds when multiplying two powers with the same base x:

$$x^n x^m = x^{n+m} \tag{1.60}$$

where n and m are called exponents. When dividing two powers the following rule holds:

$$\frac{x^n}{x^m} = x^{n-m}. \tag{1.61}$$

The fraction powers can be written as

$$x^{1/m} = \sqrt[m]{x}. \tag{1.62}$$

For any power raised to a certain power n, the following rule holds:

$$\left(x^n\right)^m = x^{nm}. \tag{1.63}$$

Some rules of the powers are summarized in Table 1.2.

Table 1.2: Some rules for powers.

Rules of exponents
$x^0 = 1$
$x^1 = x$
$x^n x^m = x^{n+m}$
$x^n / x^m = x^{n-m}$
$x^{1/m} = \sqrt[m]{x}$
$\left(x^n\right)^m = x^{nm}$

1.6.2 Factoring

Some practical formulas for factoring an equation are

$$ax + ay + az = a(x + y + z) \quad \text{common factor} \tag{1.64}$$
$$a^2 \pm 2ab + b^2 = (a \pm b)^2 \quad \text{perfect square}$$
$$a^2 - b^2 = (a - b)(a + b) \quad \text{differences of squares.}$$

1.6.3 Quadratic equations

In general, the following form is given for a quadratic equation:

$$ax^2 + bx + c = 0. \tag{1.65}$$

Here, x is an unknown quantity and a, b, and c are real numerical factors known as the coefficients of the equation. This equation has two roots, given by

$$x = \frac{-b \pm \sqrt{b^2 - 4ac}}{2a}. \tag{1.66}$$

If $b^2 \geq 4ac$, the roots are reals.

1.6.4 Linear equations

The linear equation has the general form

$$y = mx + b \tag{1.67}$$

where m and b are constants.

This equation is called linear because the graph of y versus x is a straight line (see Fig. 1.17).

The constant b, called the y-intercept, denotes the value of y at which the straight line intersects the y-axis. The constant m equals the slope of the straight line. Besides, m equals the tangent of the angle between the line and x-axis. If any two points on the straight line, given by the coordinates (x_1, y_1) and (x_2, y_2) (see Fig. 1.17), are determined, then the slope of the straight line can be expressed as

$$\text{Slope} \equiv m = \frac{y_2 - y_1}{x_2 - x_1} = \frac{\Delta y}{\Delta x} = \tan \theta. \tag{1.68}$$

Note that m and b can have either positive or negative values. If $m > 0$, the straight line has a positive slope, as in Fig. 1.18. If $m < 0$, the straight line has a negative slope. In Fig. 1.17, both m and b are positive. Three other possible situations are shown in Fig. 1.18.

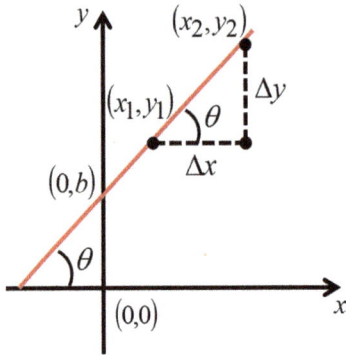

Figure 1.17: A straight line graph.

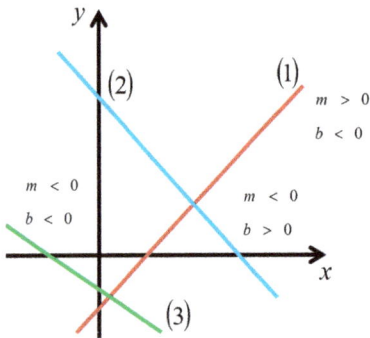

Figure 1.18: Different straight line graphs.

1.6.5 Logarithms

Let x be a quantity expressed as a power of a as

$$x = a^y. \tag{1.69}$$

a is called the base number and y is called the logarithm of x with base a:

$$y = \log_a x. \tag{1.70}$$

In practice, the most used bases are the base 10, also called *common logarithm base*, and base $e = 2.718\ldots$, called Euler's constant or the *natural logarithm base*. In the case of the common logarithm:

$$y = \log_{10} x \tag{1.71}$$

and in the case of the natural logarithm

$$y = \ln x. \tag{1.72}$$

Some properties of the logarithms are summarized by

$$\log(ab) = \log a + \log b \tag{1.73}$$
$$\log(a/b) = \log a - \log b$$
$$\log(a^n) = n \log a$$
$$\ln e = 1$$
$$\ln e^n = n$$
$$\ln(1/a) = -\ln a.$$

1.7 Geometry

The distance between two points in three-dimensions with coordinates (x_1, y_1, z_1) and (x_2, y_2, z_2) is defined as

$$d = \sqrt{(x_2 - x_1)^2 + (y_2 - y_1)^2 + (z_2 - z_1)^2}. \tag{1.74}$$

In two dimensions, the distance between any two points with coordinates (x_1, y_1) and (x_2, y_2) is defined as

$$d = \sqrt{(x_2 - x_1)^2 + (y_2 - y_1)^2}. \tag{1.75}$$

The arc length s of a circular arc (Fig. 1.19) is proportional to the radius r for a fixed value of θ (in radians):

$$s = r\theta. \tag{1.76}$$

Thus,

$$\theta = \frac{s}{r}. \tag{1.77}$$

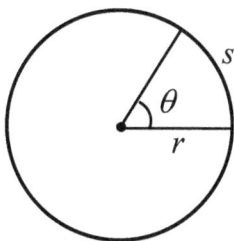

Figure 1.19: The circular arc.

The equation of a circle with radius R and center at (x_0, y_0) is

$$(x - x_0)^2 + (y - y_0)^2 = R^2. \tag{1.78}$$

The equation of an ellipse with origin at (x_0, y_0) is

$$\frac{(x - x_0)^2}{a^2} + \frac{(y - y_0)^2}{b^2} = 1 \tag{1.79}$$

where a is the length of semi-major axis (x-axis) and b the length of semi-minor axis (y-axis).

The equation of a parabola with vertex at $y = b$ is

$$y = ax^2 + b. \tag{1.80}$$

The equation of a rectangular hyperbola is

$$xy = \text{constant}. \tag{1.81}$$

1.8 Trigonometry

By definition, a right triangle is one containing a 90° angle. Consider the right triangle shown in Fig. 1.20, where side a is opposite the angle θ, side b is adjacent to the angle θ, and side c is the hypotenuse of the triangle.

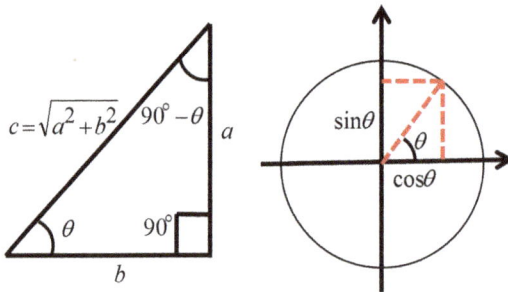

Figure 1.20: The right triangle.

In such a triangle, there are three basic trigonometric functions defined, namely sine (sin), cosine (cos), and tangent (tan) functions. In terms of the angle θ, these functions are defined by

$$\sin \theta = \frac{a}{c} \tag{1.82}$$

$$\cos \theta = \frac{b}{c} \tag{1.83}$$

$$\tan \theta = \frac{a}{b}. \tag{1.84}$$

Using the Pythagorean theorem, we get

$$c^2 = a^2 + b^2. \tag{1.85}$$

Form this definition it follows that

$$\sin^2 \theta + \cos^2 \theta = 1 \tag{1.86}$$

$$\tan \theta = \frac{\sin \theta}{\cos \theta}. \tag{1.87}$$

The following relations can also be derived:

$$\sin \theta = \cos(90° - \theta) \tag{1.88}$$

$$\cos \theta = \sin(90° - \theta) \tag{1.89}$$

$$\cot \theta = \tan(90° - \theta) \tag{1.90}$$

where

$$\cot \theta = \frac{1}{\tan \theta}. \tag{1.91}$$

Some properties of the trigonometric functions are

$$\sin(-\theta) = -\sin \theta \tag{1.92}$$

$$\cos(-\theta) = \cos \theta \tag{1.93}$$

$$\tan(-\theta) = -\tan \theta. \tag{1.94}$$

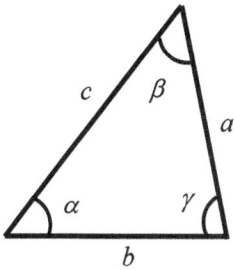

Figure 1.21: A general triangle.

For any triangle, see Fig. 1.21, we have

$$\alpha + \beta + \gamma = 180°. \tag{1.95}$$

The law of cosines says that

$$a^2 = b^2 + c^2 - 2bc \cos \alpha \tag{1.96}$$

$$b^2 = a^2 + c^2 - 2ac \cos \beta \tag{1.97}$$

$$c^2 = a^2 + b^2 - 2ab \cos \gamma \tag{1.98}$$

and from the law of sines

$$\frac{a}{\sin \alpha} = \frac{b}{\sin \beta} = \frac{c}{\sin \gamma}. \tag{1.99}$$

1.9 Derivative

By definition, the derivative of y for x is the limit, as Δx approaches zero, of the slopes of chords that are drawn between two points on the y versus x curve, as shown in Fig. 1.22.

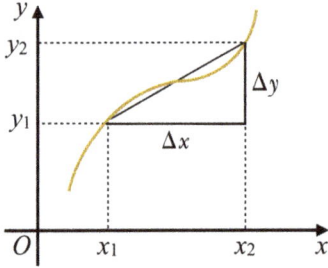

Figure 1.22: The derivative dy/dx.

Mathematically, we write this definition as

$$\frac{dy}{dx} = \lim_{\Delta x \to 0} \frac{\Delta y}{\Delta x} = \lim_{\Delta x \to 0} \frac{y(x + \Delta x) - y(x)}{\Delta x} \tag{1.100}$$

where Δy and Δx are defined as

$$\Delta x = x_2 - x_1; \quad \Delta y = y_2 - y_1. \tag{1.101}$$

A useful expression to know is the derivative with respect to x of

$$y(x) = ax^n \tag{1.102}$$

where a is a constant and n number, given as

$$\frac{dy}{dx} = nax^{n-1}. \tag{1.103}$$

1.9.1 Derivative of the product of two functions

If $f(x)$ is a function given as product of two other functions, say $g(x)$ and $h(x)$, then

$$\frac{df}{dx} = \frac{d}{dx}(g(x)h(x)) = g(x)\frac{dh}{dx} + h(x)\frac{dg}{dx}. \tag{1.104}$$

1.9.2 Derivative of the sum of two functions

If $f(x)$ is a function given as sum of two other functions, say $g(x)$ and $h(x)$, then

$$\frac{df}{dx} = \frac{d}{dx}(g(x) + h(x)) = \frac{dg}{dx} + \frac{dh}{dx}. \tag{1.105}$$

1.9.3 The chain rule of differential calculus

If $y = f(x)$ and $x = g(z)$, then dy/dz can be written as the product of two derivatives:

$$\frac{dy}{dz} = \frac{dy}{dx}\frac{dx}{dz}.$$
(1.106)

1.9.4 The second derivative

By definition, the second derivative of a function $y(x)$ with respect to x is the derivative of the function dy/dx with respect to x. It is usually written

$$\frac{d^2y}{dx^2} = \frac{d}{dx}\left(\frac{dy}{dx}\right).$$
(1.107)

1.10 Exercises

Exercise 1.1. If you walk 3.0 m toward the east and then 4.0 m toward the north, as shown in Fig. 1.23. Determine the final position.

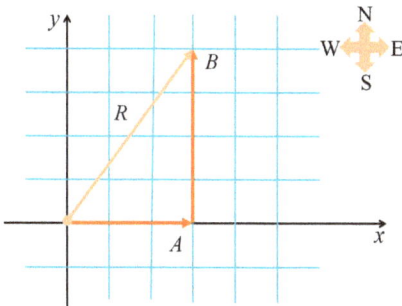

Figure 1.23: Displacement of the walker.

Exercise 1.2. Let **A** and **B** be two vectors (see Fig. 1.24) given as

$$\mathbf{A} = (3.5\mathbf{i} - 2.0\mathbf{j}); \quad \mathbf{B} = (4.0\mathbf{i} + 3.0\mathbf{j}).$$
(1.108)

Find (a) in which plane these two vectors are positioned; (b) the sum $\mathbf{A} + \mathbf{B}$ and (c) the magnitude.

Exercise 1.3. Consider a right triangle with hypotenuse of length 3.0 m, and one of its angles is 30°. What is the length of (a) the side opposite the 30° angle and (b) the side adjacent to the 30° angle?

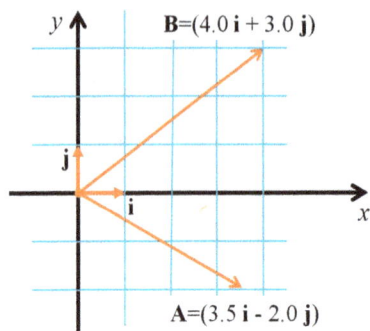

Figure 1.24: The vectors **A** and **B** in the *xy* plane.

Exercise 1.4. Suppose $y(x)$ (that is, y as a function of x) is given by

$$y(x) = ax^5 + bx^3 + c \tag{1.109}$$

where a, b and c are constants. Determine dy/dx.

Exercise 1.5. Suppose displacement y as a function of the time t is given by

$$y(t) = y_0 + v_0 t + at^2 \tag{1.110}$$

where y_0, v_0 and a are constants. Determine dy/dt.

Exercise 1.6. Find the resultant vector of these two vectors:

$$\mathbf{A} = 2\mathbf{i} + 5\mathbf{j} + 6\mathbf{k}; \quad \mathbf{B} = -3\mathbf{i} - 4\mathbf{j} + 4\mathbf{k}. \tag{1.111}$$

Exercise 1.7. Let **A** and **B** be two vectors making an angle of $\theta = 60°$. Find the dot product of these two vectors.

Exercise 1.8. Calculate the dot product of the two vectors **A** and **B** given as

$$\mathbf{A} = 3\mathbf{i} + 2\mathbf{j} + 4\mathbf{k}; \quad \mathbf{B} = -2\mathbf{i} + 2\mathbf{j} - 4\mathbf{k}. \tag{1.112}$$

Exercise 1.9. An airplane is traveling due north-east at 800 km/h. Find the components of its velocity along the east and the north.

Exercise 1.10. A person walks to the east for 2 km and then toward the north for 1 km. Calculate the resultant displacement vector and its angle with the east.

Exercise 1.11. A driver is driving toward the south with speed 60 km/h and then toward the west with speed 50 km/h. Calculate the resultant velocity vector and its angle with the east.

Exercise 1.12. Find the sum and the difference of the two vectors

$$\mathbf{A} = 3\mathbf{i} + 7\mathbf{j} - 2\mathbf{k} \tag{1.113}$$

$$\mathbf{B} = \mathbf{i} + 7\mathbf{j} - 6\mathbf{k}. \tag{1.114}$$

2 Physics and measurement

2.1 Standards of length, mass, and time

There are three basic quantities to express the laws of physics, namely:
1. length (L),
2. mass (M),
3. and time (T).

In mechanics, the other quantities can be expressed in terms of only these three quantities. When we report the results of a measurement, a standard must be defined for the results to be reproduced. For instance, if we report that a wall is 7 meters high and 1 meter is determined to the unit of length, we say that the wall's height is seven times the length unit. Similarly, if we report that a person has a mass of m, and the mass unit is 1 kilogram, the person is m times bigger than the unit of mass.

When we choose a standard, it must be readily accessible and possess some property that must give the same results when measured by different people in different places. In 1960, an international committee defined standards for length, mass, and other fundamental quantities. That is today known as the SI system of units. In SI, the units of length, mass, and time are the *meter, kilogram*, and *second*, respectively. Other SI standards include temperature (the *kelvin*), electric current (the *ampère*), luminous intensity (the *candela*), and the amount of substance (the *mole*). Here, we will be concerned only with the units of length, mass, and time.

Note that very recently, in May 2019, the basic units, such as the ampère, the kilogram, the kelvin, and the mole, were redefined based on the relationship to fundamental constants, rather than arbitrary or abstract definitions. That redefinition of the basic units will not affect the outcome of the experimental measurements. Still, it will allow us to perform analyses with higher precision and do them in multiple ways, at any place and time and to any scale, with the same accuracy.

2.1.1 Length

Since in A. D. 1120 was decided (by the king of England) that the standard of length would be named the *yard* (in his country).

Definition 2.1 (Yard). By definition, one *yard* is equal to the distance from the tip of the nose to the end of the outstretched arm.

Definition 2.2 (Foot). Similarly, the distance of 1 *foot*, adopted by the French King Louis XIV, is the length of the foot.

https://doi.org/10.1515/9783110755824-002

Definition 2.3 (Meter). In 1799, the legal standard of length became *meter*, which was defined as one ten-millionth the distance from the equator to the North Pole calculated along a longitudinal line passing through Paris.

In 1960, the length of the meter was redefined to be the separation, under controlled conditions, between two lines on the platinum-iridium bar stored in France. However, this standard was abandoned for the reason of being with limited accuracy.

In the 1960s and 1970s, one meter equaled 1650763.73 wavelengths of orange-red light emitted from a krypton-86 lamp, which was redefined, in October 1983, as the distance traveled by light in vacuum during a time of $\frac{1}{299792458}$ second, where 299792458 represents the light's speed in vacuum in m/s.

Table 2.1 gives approximate values of some measured lengths.

Table 2.1: Approximate values of some measured lengths.

Measurement	Length (m)
Distance from Earth to most remote known quasar	1.4×10^{26}
Distance from Earth to most remote known normal galaxies	9×10^{25}
Distance from Earth to nearest large galaxy	2×10^{22}
Distance from Sun to the nearest star	4×10^{16}
One light year	9.46×10^{15}
The average radius of the Earth's orbit about the Sun	1.50×10^{11}
The average distance of the Earth from Moon	3.84×10^{8}
Distance between the equator to the North Pole	1.00×10^{7}
Mean radius of the Earth	6.37×10^{6}
Length of a football field	9.4×10^{1}
Length of a housefly	5×10^{-3}
Size of the smallest dust particles	$\sim 10^{-4}$
Size of cells of most living organisms	$\sim 10^{-5}$
Diameter of a hydrogen atom	$\sim 10^{-10}$
Diameter of an atomic nucleus	$\sim 10^{-14}$
Diameter of a proton	$\sim 10^{-15}$

2.1.2 Mass

Since 1887, the kilogram (kg) is the basic SI unit of mass, which is defined to be equal to the mass of a specific platinum-iridium alloy cylinder. Today, this is kept at the International Bureau of Weights and Measures at Sévres, France.

Definition 2.4 (Mass). In May 2019, a new definition of the kilogram is introduced, such as one kilogram equals the Planck constant h divided by $6.62607015 \times 10^{-34}$ m^{-2}s:

$$1\,\text{kg} = \frac{h}{6.62607015 \times 10^{-34}\text{m}^{-2}\text{s}}. \tag{2.1}$$

Table 2.2 gives approximate values of the masses of various objects.

Table 2.2: Approximate values of some measured masses.

Body	Mass (kg)
Visible Universe	$\sim 10^{52}$
Milky Way galaxy	7×10^{41}
Sun	1.99×10^{30}
Earth	5.98×10^{24}
Moon	7.36×10^{22}
Horse	$\sim 10^{3}$
Human	$\sim 10^{2}$
Frog	$\sim 10^{-1}$
Mosquito	$\sim 10^{-5}$
Bacterium	$\sim 10^{-15}$
Hydrogen atom	1.67×10^{-27}
Electron	9.11×10^{-31}

2.1.3 Time

The basic unit of time is second (s), which until 1960 was considered in terms of the mean solar day for the year 1900.

It is well known that the rotation of the Earth varies slightly with time, and therefore, it cannot be considered a standard definition.

In 1967, the *second* was redefined more precisely using a device known as an atomic clock, which uses specific frequencies associated with atomic transitions to determine one second with a precision of one part in 10^{12}, or equivalent less than one second every 30000 years.

Definition 2.5 (second). Thus, the SI unit of time (in 1967), the second, is defined from the characteristic frequency of the cesium atom, known as the reference clock, as the period of the radiation vibration from the cesium-133 atom multiplied by a factor of 9192631770.

Table 2.3 shows some approximate time intervals.

2.1.4 British engineering system of units

In addition to SI, the *British engineering system* of units is still used in the United States. According to this system, the length, mass, and time use as their units the *foot* (ft), *slug*, and *second*, respectively.

Table 2.3: Approximate values of some time intervals.

Age of the Universe	5×10^{17}
Age of the Earth	1.3×10^{17}
Age of a college student in average	6.3×10^{8}
One year	3.16×10^{7}
Rotation time of the Earth about its axis (or one day)	8.64×10^{4}
Time between normal heartbeats	8.0×10^{-1}
Period of audible sound waves	$\sim 10^{-3}$
Period of typical radio waves	$\sim 10^{-6}$
Period of atom vibration in solids	$\sim 10^{-13}$
Period of visible light waves	$\sim 10^{-15}$
Duration of a nuclear collision	$\sim 10^{-22}$
Time for light to cross a proton	$\sim 10^{-24}$

However, the SI units are the most often used in science and industry. On the other hand, British engineering units have some limited use in classical mechanics.

Besides the basic SI units of the meter, kilogram, and second, the subunits are also used, such as *millimeters* and *nanoseconds*. Here, the prefixes *milli-* and *nano-* characterize powers of ten. Table 2.4 presents the most frequently used prefixes or the various powers of ten and their abbreviations.

For example, 10^{-3} m is equivalent to 1 millimeter (mm), and 10^{3} m corresponds to 1 kilometer (km), and 1 kg is 10^{3} grams (g).

Table 2.4: Prefixes for SI units.

10^{-24}	yocto	y
10^{-21}	zepto	z
10^{-18}	atto	a
10^{-15}	femto	f
10^{-12}	pico	p
10^{-9}	nano	n
10^{-6}	micro	μ
10^{-3}	milli	m
10^{-2}	centi	c
10^{-1}	deci	d
10^{1}	deka	da
10^{3}	kilo	k
10^{6}	mega	M
10^{9}	giga	G
10^{12}	tera	T
10^{15}	peta	P
10^{18}	exa	E
10^{21}	zetta	Z
10^{24}	yotta	Y

2.2 Density

Definition 2.6 (Density). The density, often denoted with the Greek letter ρ, is a property of the substance that is defined as the ratio of mass (m) with the volume (V):

$$\rho = \frac{m}{V}. \tag{2.2}$$

For example, the aluminum density is $2.70\,\text{g/cm}^3$, and of the lead is $11.3\,\text{g/cm}^3$. Therefore, a volume of aluminum of $V = 10.0\,\text{cm}^3$ has a mass of $m = 27.0\,\text{g}$. The same volume of lead has a mass of $113\,\text{g}$.

The densities for some substances are given in Table 2.5.

Table 2.5: Densities of various substances.

Substance	Density ρ ($10^3\,\text{kg/m}^3$)
gold	19.3
uranium	18.7
lead	11.3
copper	8.92
iron	7.86
aluminum	2.70
magnesium	1.75
water	1.00
air	0.0012

This difference in densities between aluminum and lead is because of their differences in atomic mass. The atomic mass is defined as the average mass of an atom in a sample of this element containing all its isotopes.

Definition 2.7 (Atomic mass unit). The atomic mass unit is the so-called atomic mass unit (u), where

$$1\,\text{u} = 1.6605402 \times 10^{27}\,\text{kg}. \tag{2.3}$$

For lead, the atomic mass is $207\,\text{u}$, and for aluminum, it is $27.0\,\text{u}$. Furthermore, the ratio of atomic masses for the lead and aluminum is

$$\frac{207\,\text{u}}{27.0\,\text{u}} = 7.67. \tag{2.4}$$

One can see that it is not equal to the ratio of densities:

$$\frac{11.0\,\text{g/cm}^3}{2.70\,\text{g/cm}^3} = 4.19. \tag{2.5}$$

That is because of the difference in interatomic distances and arrangements of atoms in these two materials' crystal structure.

The carbon-12 isotope (^{12}C) has six protons and six neutrons. Other carbon isotopes have six protons but with different numbers of neutrons. Practically, the mass of an atom equals the mass of its nucleus. The atomic mass of ^{12}C is defined to be precisely 12 u, where the neutron and proton have an equal mass of about 1 u. Therefore, the mass of a nucleus is measured in terms of ^{12}C nucleus as the basis.

Definition 2.8 (Avogadro's number). By definition, in every mole of a substance, there is the same number of particles. For example, 1 mol of aluminum and 1 mol of lead include the same number of atoms. That number, based on the experimental results, denotes Avogadro's number, N_A

$$N_A = 6.022137 \times 10^{23} \text{ particles/mol.} \tag{2.6}$$

Avogadro's number defines that 1 mol of carbon-12 atoms has a mass of exactly 12 g.

Definition 2.9 (Mole). Based on the new definitions of the basic units, in May 2019, in a system, one mole (mol) equals the amount of substance that contains N_A specified elementary entities.

The mass in 1 mol of an element is its atomic mass in grams. For example, the mass of 1 mol of the element iron (with an atomic mass of 55.85 u) is 55.85 g, the molar mass of iron is 55.85 g/mol. Also, 1 mol of the element lead with atomic mass 207 u has a mass of 207 g, and hence its molar mass is 207 g/mol.

The mass per atom of an element, containing 6.02×10^{23} particles in 1 mol, is

$$m_{\text{atom}} = \frac{\text{molar mass}}{N_A}. \tag{2.7}$$

The mass of an iron atom can be determined as

$$m_{F_e} = \frac{55.85 \text{ g/mol}}{6.02 \times 10^{23} \text{ atoms/mol}} = 9.28 \times 10^{-23} \text{ g/atom.} \tag{2.8}$$

2.3 Dimensional analysis

In Physics, the word dimension has a special meaning as it denotes the physical nature. It is not as important if a given distance is measured in the unit feet of length or the unit meters of length since it will still be a distance. In physics nature, the dimension of the distance is *length*. Often the symbols used to specify the length, mass, and time are L, M, and T, respectively.

We often use brackets [$\cdots$] for the dimension of any physical quantity.

For example, the symbol v is often used for speed, and hence the dimensions notation of the speed can be written as

$$[v] = \frac{L}{T}.$$ (2.9)

In a second example, the dimensions of the area, A, are written as

$$[A] = L^2.$$ (2.10)

The dimensions of the area, volume, speed, and acceleration are shown in Table 2.6.

Table 2.6: Dimensions and common units of area, volume, speed, and acceleration.

SI	m^2	m^3	m/s	m/s^2
British engineering	ft^2	ft^3	ft/s	ft/s^2

A robust procedure for solving physics problems is dimensional analysis, where the dimensions are manipulated as algebraic quantities. Besides, the quantities that are added or subtracted must have the same dimensions. Similarly, in an equation, the terms on both sides must have the same dimensions. As a result of these simple rules, one can use dimensional analysis to determine whether an expression has the correct form by comparing the dimensions on both sides of the equation.

Example 2.1. Consider the derivation of a formula for the distance x traveled by car in time t if the car starts from rest and moves with constant acceleration a. In the next chapter, we will derive that

$$x = \frac{1}{2}at^2$$ (2.11)

assuming that the initial velocity is zero. Using the dimensional analysis, one can easier verify the validity of the expression. For instance, the quantity x on the left side has the dimension of length, L; the acceleration has L/T^2, and time t of T. The quantity on the right-hand side in eq. (2.11) must also have the dimension of the length, for eq. (2.11) to be dimensionally correct. We can do a dimensional check-up by substituting the dimensions for each quantity into the equation:

$$L = \frac{L}{T^2}T^2 = L.$$ (2.12)

The units are simplified as they were algebraic quantities, as shown, leaving the units on both sides of the relationship the same.

2.4 Conversion of units

Often it is the case to convert units from one system to another as shown here:

$$1\,\text{mi} = 1609\,\text{m} = 1.609\,\text{km} \tag{2.13}$$
$$1\,\text{ft} = 0.3048\,\text{m} = 30.48\,\text{cm} \tag{2.14}$$
$$1\,\text{m} = 39.37\,\text{in.} = 3.281\,\text{ft} \tag{2.15}$$
$$1\,\text{in.} = 0.0254\,m = 2.54\,\text{cm.} \tag{2.16}$$

We usually treat the units as algebraic quantities which can cancel each other.

Example 2.2. For example, suppose we would like to convert 22.0 in. into centimeters. Since 1 in. = 2.54 cm, then we can write

$$22.0\,\text{in.} = (22.0\,\text{in.}) \times (2.54\,\text{cm/in.}) = 55.88\,\text{cm} \approx 55.9\,\text{cm.} \tag{2.17}$$

2.5 Significant figures

Physical quantities are measured within the limits of the experimental uncertainty, depending on various factors, such as the apparatus's quality, the experimenter's skill, and the number of measurements performed.

Suppose that we will measure the area of a disk table using a meter stick as a measuring instrument. Let us assume that the accuracy we can measure with this stick is ±0.1 cm. If the length of the table is 5.5 cm, we can say that its range lies somewhere between 5.4 cm and 5.6 cm. It is noted that the measured value has two significant figures. For the same reasons, if the table width is 6.4 cm, the actual value lies in the interval of 6.3 cm and 6.5 cm. Hence, we could write the measured values as

$$5.5 \pm 0.1 \tag{2.18}$$
$$6.4 \pm 0.1. \tag{2.19}$$

Now, we can determine the area of the table as

$$(5.5\,\text{cm})(6.4\,\text{cm}) = 35.2\,\text{cm}^2. \tag{2.20}$$

which contains three significant figures, which is greater than the number of significant figures for each measured length.

A general rule in determining the number of significant figures
When multiplying several quantities, the number of significant figures in the final answer is the same as the number of significant figures in the quantities with the lowest number of significant figures being multiplied. For the division, the same applies.

Using the rule of the multiplication in the example above, we see that the result for the table area can have only two significant figures since we use only two significant figures for the measurements of the lengths. Thus, all we can say that the area is $35\,\mathrm{cm}^2$, realizing that the value can range between:

$$(5.4\,\mathrm{cm})(6.3\,\mathrm{cm}) = 34\,\mathrm{cm}^2 \qquad\qquad (2.21)$$

$$(5.6\,\mathrm{cm})(6.5\,\mathrm{cm}) = 36\,\mathrm{cm}^2. \qquad\qquad (2.22)$$

Therefore, we can finally write that the surface area of the table is

$$A = (35 \pm 1.0)\,\mathrm{cm}^2. \qquad\qquad (2.23)$$

Significant figures of zeros
Zeros may or may not be significant figures.

For instance, those used to position the decimal point in such numbers as 0.0011 and 0.000341 are not significant. Hence, there are two and three significant figures in these two values, respectively. However, the zeros coming after other digits may be ambiguous. For example, if the mass of an object is given as 2100 g, to avoid any ambiguity in the value because it is not clear whether the last two zeros are being used to locate the decimal point or represent significant figures in the measurement, we can use a scientific notation to present the number of significant figures. In this case, we would express the mass as 2.1×10^3 g if there are two significant figures in the measured value, 2.10×10^3 g if there are three significant figures, and 2.100×10^3 g if there are four.

The same rule applies when the number is less than one so that 2.3×10^{-4} has two significant figures (and so could be written 0.00023) and 2.30×10^{-4} has three significant figures (also written 0.000230).

Significant figures of addition and subtraction
When numbers are added or subtracted, the number of decimal places in the result should be equal to the smallest number of decimal places of any term in the sum.

For example, if we wish to compute 123 + 5.35, the answer given to the correct number of significant figures is 128 and not 128.35. If we calculate the sum 1.0001 + 0.0003 = 1.0004, the result has five significant figures, even though one of the terms in the sum, 0.0003, has only one significant figure.

Similarly, for the subtraction 1.002 − 0.998 = 0.004, the result of the subtraction will have three digits after the decimal point. It has only one significant figure, even though one term has four significant figures, and the other has three.

2.6 Exercises

Exercise 2.1. Calculate the area of a fenced yard 100 m long and 50.0 m wide. Express your answer in terms of square km.

Exercise 2.2. Calculate the volume of a raindrop with a radius of 2.0 mm. Express your answer in terms of mm^3. How many raindrops would fill a tank with a volume of $1 m^3$?

Exercise 2.3. The copies of the kg show variation in weight gain as much as 70 µ grams over a period of approximately 100 years. If the mass of the gunk build up on the kilogram replica as well as the original is due to hydrocarbon buildup and we assume it to be CH with a mass of $m_{CH} = 2.2 \times 10^{-26}$ kg, calculate how many CH molecules have made their home in the International Prototype Kilogram (IPK).

Exercise 2.4. When neutrons and protons bind to form nuclei, they release energy. The amount of energy released is equal to the energy that binds the nucleus. A deuteron is a heavy hydrogen wherein a neutron and a proton are bound together. If the binding energy (BE) of the deuteron is 0.002362 amu, find the mass of the deuteron. (amu is the Atomic Mass Unit, $1 amu = 1.66053886 \times 10^{-27}$ kg.)

Exercise 2.5. The lower limit of the proton lifetime is measured to be 1.01×10^{34} years. Calculate this number in ns. If the lifetime of a subatomic particle called the positive pion is 26 ns, what is the ratio of the lifetime of the proton to that of the pion?

Exercise 2.6. The size of an Amoeba is 150 µm. Express this length in nm and Å.

Exercise 2.7. The size of the H-atom is 1 Å. Assume the H-atom is spherical. Calculate its volume in m^3.

Exercise 2.8. A solid cube of aluminum (density 2.7 g/cm^3) has a volume of 0.20 cm^3. How many aluminum atoms are contained in the cube?

Exercise 2.9. Prove that the expression $v = at$ is dimensionally correct, where v represents speed, a acceleration, and t a time interval.

Exercise 2.10. The mass of a concrete cube, with a side length of 5.35 cm, is 856 g. Determine the density ρ of the cube in basic SI units.

Exercise 2.11. Calculate how many steps a person would walk from New York to Los Angeles.

Exercise 2.12. A rectangular has a length of (21.3 ± 0.2) cm and width of (9.80 ± 0.1) cm. Determine the area of the plate and the uncertainty in the calculated area.

Exercise 2.13. Approximate the following arithmetic:
(a) $100 + 1.51$
(b) $12/3.5$
(c) $99 + 13.65$.

Exercise 2.14. The mass of an object is 1234 g. Convert the mass in kilogram, and approximate it to three significant figures.

Exercise 2.15. The speed of the boat is 2321 cm/s. Find the speed of the boat in SI units, and approximate it to three significant figures.

3 One-dimensional motion

We will describe the moving objects as particles regardless of their size. A point-like mass particle is called a particle having infinitesimal size. For example, the motion of the Earth around the Sun can, in general, be described by considering the Earth as a point-like particle because the radius of the Earth's orbit around the Sun is much larger than the dimensions of the Earth and Sun.

3.1 Displacement

Definition 3.1 (Displacement). The displacement is defined as the change in the particle position,

$$\Delta x = x_f - x_i. \tag{3.1}$$

Here, x_f is the final position and x_i is the initial position.

Both these quantities are vectors, and thus Δx is a vector. That is, it has both *direction* and *magnitude*. There are only two possible directions in one dimension, specified either with a minus sign or a plus sign. From the definition one can see that Δx is positive if x_f is greater than x_i and negative if $x_f < x_i$.

A mistake easily made is to not recognize the difference between the *displacement* and the *distance* traveled.

Definition 3.2 (Distance). The distance is a scalar quantity that represents the length of the path traveled by the particle.

Example 3.1. For example, an object travels from A to B and turns back from B to A (see Fig. 3.1). What are the distance traveled by the object and its displacement?

Answer: The distance traveled by the object is

$$d = |AB| + |BA| = 2|AB| = 2|BA| \tag{3.2}$$

since, $|AB| = |BA|$. However, the displacement of the object is zero, because the final and initial position are identical:

$$\Delta x = x_f - x_i = x_A - x_A = 0 \tag{3.3}$$

where x_A is the position at A.

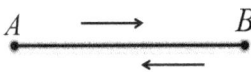

Figure 3.1: A round trip motion of an object from A to B.

https://doi.org/10.1515/9783110755824-003

3.2 The average velocity and speed

Definition 3.3 (Average velocity). The average velocity, $\bar{v}_x$, of an object (or a particle) is defined as the ratio of its displacement Δx with the time interval Δt during which that displacement took place:

$$\bar{v}_x = \frac{\Delta x}{\Delta t} \equiv \frac{x_f - x_i}{t_f - t_i} \tag{3.4}$$

where the subscript x indicates the motion along the x-axis. x_f and x_i are, respectively, the final and the initial positions, and t_f and t_i are the final and initial time, respectively.

One can see from the formula that the average velocity has the dimensions of the length divided by time; that is, in SI units, the average velocity has the units of meters per second (m/s).

It can be noticed that:
- The bar sign indicates the average value.
- Δt is always positive; therefore, the sign of $\bar{v}_x$ is determined always from the sign of Δx.
- If the coordinate decreases in time (i. e., $x_f < x_i$), then Δx is negative and hence $\bar{v}_x$ is negative. This case corresponds to motion along the negative direction of the x-axis. If the coordinate of the particle position increases in time (i. e., $x_f > x_i$), then Δx is positive and $\bar{v}_x$ is positive. This corresponds to motion in the positive x direction. This is illustrated in Fig. 3.2.

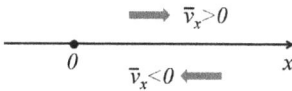

Figure 3.2: The motion along the x-axis. The arrows indicate the directions of the motion.

3.2.1 Graphical interpretation of the velocity

The average velocity can also be interpreted graphically, as illustrated in Fig. 3.3 in a position versus time plot. Let us consider a particle moving in one dimension from point P with coordinates (x_i, t_i) to the point Q with coordinates (x_f, t_f). We can draw the trajectory, as shown in Fig. 3.3 (blue line). The slope of the straight line connecting the points P and Q is

$$\tan\theta = \frac{\Delta x}{\Delta t}. \tag{3.5}$$

Therefore, by comparing it with eq. (3.4), we conclude that the average velocity can be interpreted as the straight-line slope connecting the points P and Q, representing, respectively, the initial and the final points of the trajectory.

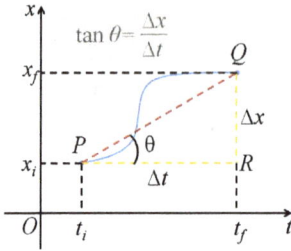

Figure 3.3: The motion of a particle in the (x, t) plan.

Often the terms *speed* and *velocity* are used interchangeably. However, in physics, there is a clear distinction between these two quantities.

To do this, we have to determine another quantity called *speed*.

Definition 3.4 (Average speed). By definition, the average speed of a particle is a scalar quantity given as the total distance traveled (ℓ) divided by the total time (T) it takes to travel that distance:

$$|\bar{v}_x| = \frac{\ell}{T}. \tag{3.6}$$

The SI unit of the average speed is the same as the unit of the average velocity: meters per second (m/s). Unlike the average velocity, the average speed has no direction, and hence it has no algebraic sign. Also, the particle's average speed does not tell us anything about the full motion details.

Example 3.2. Consider a runner who runs a distance d, let say between points A and B for time t. First from A to B, then he runs back from B to A for the same time. What are the displacement, average velocity, and average speed?

Answer: The displacement is $\Delta x = x_f - x_i = 0$, because $x_f = x_i$. Hence, the average velocity is $\bar{v}_x = 0$. The total distance traveled by the runner is

$$\ell = d + d = 2d \tag{3.7}$$

and the total time: $T = t + t = 2t$. Hence, the average speed is

$$|\bar{v}_x| = \frac{\ell}{T} = \frac{2d}{2t} = \frac{d}{t} \tag{3.8}$$

which tells us how fast the runner was running.

Example 3.3. Suppose it takes 3.00 h to travel 120 km in a car. What is the average speed?

Answer: The average speed for this trip is

$$|\bar{v}_x| = \frac{120 \text{ km}}{3.00 \text{ h}} = 40.0 \text{ km/h}. \tag{3.9}$$

However, one can travel at various speeds during the trip, and the average speed of 40.0 km/h may, in general, result from an infinite number of possible speed values.

3.3 Instantaneous velocity and speed

Usually, we are only interested in the velocity of an object at some instant in time, not over a finite time interval.

Example 3.4. For example, during a trip, you might want to know the average speed during the trip. Still, you would be especially interested in knowing your velocity at any instant, for example, when you see the police car parked alongside the road in front of you. In other words, you would like to know the velocity just as precisely as you can specify your position by not knowing what is happening at a specific clock reading, i. e., at some specific instant. That is the same as saying how fast a particle moves if we "freeze time" and talk only about an individual instant.

Definition 3.5 (Instantaneous velocity). The instantaneous velocity v_x is equal to the limiting value of the ratio $\Delta x/\Delta t$ as Δt approaches zero:

$$v_x = \lim_{\Delta t \to 0} \frac{\Delta x}{\Delta t}. \tag{3.10}$$

Note that the displacement Δx is also approaching zero as Δt approaches zero. As Δx and Δt become smaller and smaller, the ratio $\Delta x/\Delta t$ converges to a value equal to the slope of the line tangent to the x versus t curve.

Example 3.5. In the following table (see Table 3.1), we give some data about the position of a runner at different times. Find the instantaneous velocity of the runner at time $t = 1.00$ s.

Table 3.1: The position versus time for a runner.

t (s)	x (m)
1.00	1.00
1.01	1.02
1.10	1.21
1.20	1.44
1.50	2.25
2.00	4.00
3.00	9.00

Answer: First, we can determine the average velocity for the total running time:

$$\bar{v}_x = \frac{\Delta x}{\Delta t} = \frac{9.00\,\text{m} - 1.00\,\text{m}}{3.00\,\text{s} - 1.00\,\text{s}} = 4.00\,\text{m/s}. \tag{3.11}$$

Using the definition of the instantaneous velocity, we can find a good approximation by taking the shortest interval. As one can see from Table 3.1, the closest time next to $t = 1.00$ s is the time 1.01 s, hence, the instantaneous velocity at $t = 1.00$ s is

$$v_x(t = 1.00\,\text{s}) = \frac{\Delta x}{\Delta t} = \frac{1.02\,\text{m} - 1.00\,\text{m}}{1.01\,\text{s} - 1.00\,\text{s}} = 2.00\,\text{m/s}. \tag{3.12}$$

That example indicates that the average and instantaneous velocities can be quite different.

In calculus notation, the above limit is also called the *derivative* of x with respect to t, hence

$$v_x = \lim_{\Delta t \to 0} \frac{\Delta x}{\Delta t} = \frac{dx}{dt}. \tag{3.13}$$

As a conclusion, we can say that:

– The instantaneous velocity gives more information than the average velocity. From now, we will refer to the instantaneous velocity as, *velocity*.
– The velocity can take all possible values, e. g., it can be positive, negative, or zero.

Definition 3.6 (Instantaneous speed). The *magnitude* of the velocity of an object defines its instantaneous speed.

Like the average speed, instantaneous speed has no direction associated with it and does not carry any algebraic sign.

Example 3.6. If one particle has a velocity of +30 m/s along a given direction and another particle has a velocity of −30 m/s along the same line, but opposite direction, what is the speed of each particle?

Answer: Each has a speed of 30 m/s, because

$$|30| \frac{\text{m}}{\text{s}} = |{-30}| \frac{\text{m}}{\text{s}} = 30 \frac{\text{m}}{\text{s}}. \tag{3.14}$$

3.3.1 Graphical interpretation of the instantaneous velocity

We can also interpret the instantaneous velocity graphically, as illustrated in Fig. 3.4. As we explained above, the average velocity is the line's slope connecting to different positions of a particle. To calculate the instantaneous velocity, we will need to take the limit of $\Delta t \to 0$ or $B_2 \to B_1 \to B_0 \to B$. When $B_2 \to B$, the line connecting B_2 and B approaches the tangent with the curve at the point B, and hence $\Delta x \to 0$. Therefore, the slope of the tangent at the position B, $\tan\theta$, is the instantaneous velocity at the time t the particle is at the position B:

$$\tan\theta = \lim_{\Delta t \to 0} \frac{\Delta x}{\Delta t}. \tag{3.15}$$

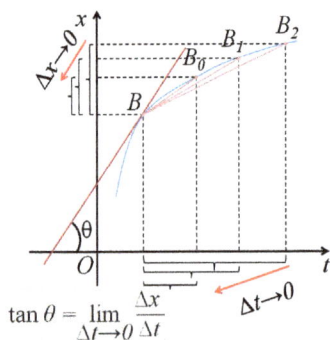

$$\tan\theta = \lim_{\Delta t \to 0} \frac{\Delta x}{\Delta t}$$

Figure 3.4: Graphical interpretation of the instantaneous velocity.

Note that, if the trajectory is a straight line, we will get $v_x = \bar{v}_x$, which is valid for all the path points.

3.4 Average acceleration

There are examples in which the velocity of a particle changes while the particle is moving. It is easy to quantify the changes in the velocity as a function of time similarly to quantifying the changes in position as a function of time,

$$\Delta v_x(t) = v_{xf}(t) - v_{xi}(t). \tag{3.16}$$

We can say that the particle accelerates when the velocity of a particle changes with time.

For example, the velocity increases when stepping on the gas and decreases when applying the car's brakes. Suppose the car is moving along the x-axis with velocity v_{xi} at time t_i and a velocity v_{xf} at time t_f, as illustrated in Fig. 3.5.

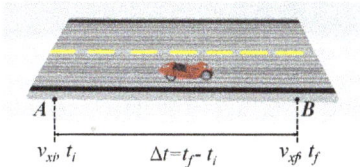

Figure 3.5: Illustration of motion of a racing car moving with velocity that changes along the x-axis.

Definition 3.7 (Average acceleration). By definition, the average acceleration of the particle is given as the change in the velocity $\Delta v_x = v_{xf} - v_{xi}$ divided by the time interval $\Delta t = t_f - t_i$ during which that change occurred:

$$\bar{a}_x = \frac{\Delta v_x}{\Delta t} = \frac{v_{xf} - v_{xi}}{t_f - t_i}. \tag{3.17}$$

Similar to the velocity, when the motion is analyzed in one-dimension, the direction of acceleration is determined by its sign, either positive and negative signs. Since the velocity dimensions are (m/s) and dimensions of time are (s), the acceleration has the dimensions of m/s² in SI units.

Example 3.7. Suppose that an object has an acceleration of $2\,\text{m/s}^2$. Describe the motion of that object.

Answer: It would be best if you imagined it as an object having a velocity along a straight line and increasing by $2\,\text{m/s}$ every $1\,\text{s}$ time interval. That is, if the object starts the motion from rest ($v_{xi} = 0$), then its velocity increases to $+2\,\text{m/s}$ after $1\,\text{s}$ and $+4\,\text{m/s}$ after $2\,\text{s}$ and so on.

3.4.1 Graphical interpretation of the average acceleration

The average acceleration can be interpreted graphically, similar to the average velocity. For example, if the object starts the motion from the point P at time t_i with velocity v_{xi}, and at time t_f it reaches the point Q with a velocity v_{xf}, then the average acceleration is the slope of the line connecting the points P and Q in a velocity versus time graph, as shown in Fig. 3.6.

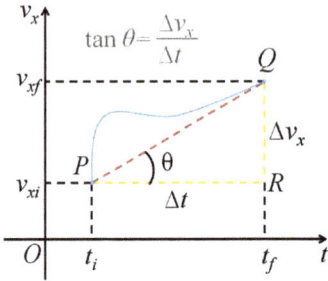

Figure 3.6: Graphical interpretation of the average acceleration.

3.5 Instantaneous acceleration

The average acceleration might have different values over different time intervals. Therefore, similar to instantaneous velocity, it is useful to define the instantaneous acceleration as the limit when Δt approaches zero of the average acceleration.

Definition 3.8 (Instantaneous acceleration). The instantaneous acceleration is determined as limit of the average acceleration as Δt approaches zero:

$$a_x = \lim_{\Delta t \to 0} \frac{\Delta v_x}{\Delta t}. \tag{3.18}$$

Hence, the instantaneous acceleration equals the *derivative* of the velocity with respect to the time:

$$a_x = \frac{dv_x}{dt}. \tag{3.19}$$

Note that as $\Delta t \to 0$ so does the change in the velocity, that is, $\Delta v_x \to 0$.

3.5.1 Graphical interpretation of instantaneous acceleration

Graphically, the expression given by eq. (3.19) can be interpreted as the slope of the tangent at an instant of time of the velocity–time graph (see Fig. 3.7):

$$\tan\theta = \lim_{\Delta t \to 0} \frac{\Delta v_x}{\Delta t}. \tag{3.20}$$

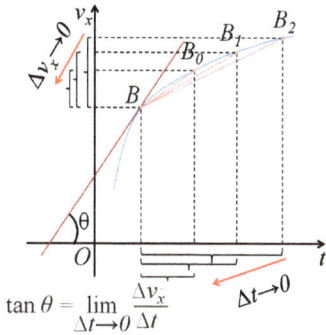

$$\tan\theta = \lim_{\Delta t \to 0} \frac{\Delta v_x}{\Delta t} \qquad \Delta t \to 0$$

Figure 3.7: Graphical interpretation of the instantaneous acceleration.

Note that:
- Acceleration is the change in particle velocity per unit of time.
- If $\Delta v_x > 0$ (that is, if the velocity increases with time), then $a_x > 0$; if $\Delta v_x < 0$ (that is, if the velocity decreases with time), then $a_x < 0$.
- The velocity increases in time, if the velocity and acceleration are in the same direction.
- Similar to the instantaneous velocity, from now on with acceleration we will understand simply the instantaneous acceleration.
- If the $a_x > 0$, then the acceleration is in the positive x direction; if $a_x < 0$, then the acceleration is in the negative x direction.

Since, $v_x = dx/dt$, then the acceleration can also be written as

$$a_x = \frac{dv_x}{dt} = \frac{d}{dt}\left(\frac{dx}{dt}\right) = \frac{d^2x}{dt^2}. \tag{3.21}$$

That is, the acceleration equals the *second derivative* of x with respect to time for one-dimensional case.

3.6 One-dimensional motion with constant acceleration

A common and straightforward one-dimensional motion is this with *acceleration constant*. In this case, constant acceleration means that the velocity increases or decreases with the same amount during the entire motion; for example, the movement of an object falling from a certain distance very close to the Earth's surface, if we ignore the air resistance. We derive the so-called kinematic equations of motion for the one-dimensional movement with constant acceleration in the following.

3.6.1 Kinematic equations of motion

We start with the definition of the average acceleration, given by eq. (3.17), and replacing $\bar{a}_x$ with a_x, and choosing $t_i = 0$, $x_i = 0$, and $t_f = t$. If a_x is constant, then

$$a_x = \frac{v_{xf} - v_{xi}}{t} \qquad (3.22)$$

or

$$v_{xf} = v_{xi} + a_x t. \qquad (3.23)$$

Equation (3.23) is an essential expression, which determines the velocity of the object at any time t if we know its initial velocity and its acceleration (which is assumed to be constant).

The velocity versus time graph for the case of constant acceleration motion is shown in Fig. 3.8. It can be seen that the graph is a straight line, the constant slope of which is the acceleration a_x, which is consistent with the fact that $a_x = dv_x/dt$ is a constant:

$$\tan \theta = a_x. \qquad (3.24)$$

From the graph (see Fig. 3.8), the slope is positive, indicating a positive acceleration, $a_x > 0$.

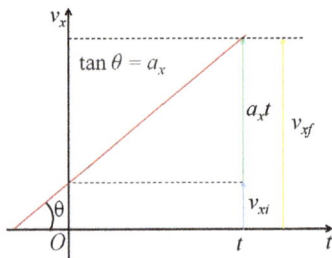

Figure 3.8: The motion with constant acceleration along the x-axis represented in (v_x, t) plan.

When the acceleration is constant, then the graph of the acceleration versus time is a straight line with a slope equal to zero, as shown in Fig. 3.9:

$$\tan \theta = 0. \qquad (3.25)$$

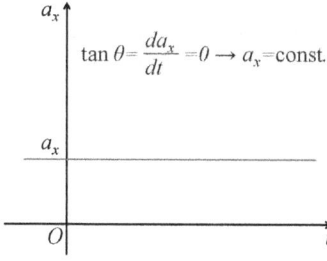

Figure 3.9: The motion with constant acceleration along the x-axis represented in (a_x, t) plan.

Since the velocity at constant acceleration varies linearly with time, we can express the average velocity at any time interval as the arithmetic mean of the initial velocity v_{xi} and the final velocity v_{xf}:

$$\bar{v}_x = \frac{1}{2}(v_{xf} + v_{xi}) \tag{3.26}$$

which is true if a_x is constant.

In addition, using eq. (3.4) for $t_i = 0$ and $t_f = t$, we obtain

$$\bar{v}_x = \frac{x_f - x_i}{t}. \tag{3.27}$$

From this,

$$x_f - x_i = \bar{v}_x t = (v_{xf} + v_{xi})t/2 \tag{3.28}$$

or

$$x_f - x_i = v_{xi}t + \frac{1}{2}a_x t^2. \tag{3.29}$$

Note that eq. (3.29) determines the displacement of an object, $\Delta x = x_f - x_i$ as a function of the initial velocity v_{xi}, time t and acceleration a_x. It represents a parabola in a displacement versus time plot. Besides, eq. (3.29) can be used to determine the position of the object as a function of its initial position x_i, the initial velocity v_{xi}, time t and the acceleration a_x as

$$x_f = x_i + v_{xi}t + \frac{1}{2}a_x t^2. \tag{3.30}$$

A graph of the position versus the time for the motion at constant (positive) acceleration is shown in Fig. 3.10. The curve represents a parabola (the blue line in Fig. 3.10). The slope of the line that is tangent to the curve at $t = t_i = 0$ equals the initial velocity $\tan \theta_i = v_{xi}$, and the slope of the line that is tangent to the curve at any later time t equals the velocity at that time $\tan \theta_f = v_{xf}$. Note that x_f is sum of two vector quantities, namely x_i and $v_{xi}t + a_x t^2/2$, as indicated in Fig. 3.10.

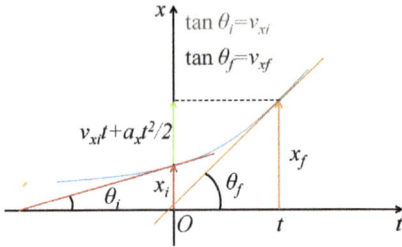

Figure 3.10: A graph of the position versus the time for the motion at constant (positive) acceleration.

Equation (3.23) can be re-arranged to obtain time t as

$$t = (v_{xf} - v_{xi})/a_x. \tag{3.31}$$

By plugging this equation into eq. (3.29), we get

$$x_f - x_i = v_{xi}\frac{v_{xf} - v_{xi}}{a_x} + \frac{1}{2}a_x\frac{(v_{xf} - v_{xi})^2}{a_x^2} \tag{3.32}$$

or

$$v_{xf}^2 - v_{xi}^2 = 2a_x(x_f - x_i). \tag{3.33}$$

Equation (3.23), eq. (3.29) (or eq. (3.30)) and eq. (3.33) form the set of kinematic equations of motion with constant acceleration in one dimension used to solve many problems.

For the motions with zero acceleration, $a_x = 0$, we have

$$v_{xf} = v_{xi} = v_x \tag{3.34}$$
$$x_f - x_i = v_x t. \tag{3.35}$$

That means that when $a_x = 0$, the velocity is constant and the displacement increases linearly with time.

3.7 Freely falling objects

We say that an object freely falling when it moves only upon the action of the gravity force (see also Fig. 3.11).

If we ignore the air resistance, all the objects falling freely upon the action of the gravity force move with constant acceleration which has the direction towards the center of the Earth, or normal to the Earth's surface and its magnitude is

$$|a_y| \equiv g = 9.8 \, \text{m/s}^2. \tag{3.36}$$

In eq. (3.36), g is the magnitude of the so-called gravitational acceleration.

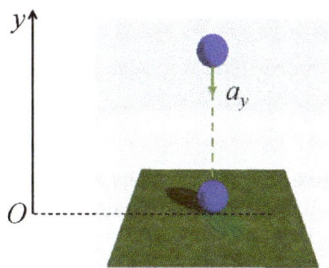

Figure 3.11: Illustration of the free fall.

Then the kinematic equations of motion (eq. (3.23), eq. (3.29) and eq. (3.33)) in this case become

$$v_{yf} = v_{yi} + a_y t \qquad (3.37)$$

$$y_f - y_i = v_{yi}t + a_y \frac{t^2}{2}, \qquad (3.38)$$

$$v_{yf}^2 - v_{yi}^2 = 2a_y(y_f - y_i). \qquad (3.39)$$

Note that the subscript has changed to y to indicate that the motion is along the y-axis.

If we choose the reference frame with positive y-axis direction upwards, as shown in Fig. 3.11, then

$$a_y = -g. \qquad (3.40)$$

The minus sign (see eq. (3.40)) indicates that the direction of the acceleration is along the negative y-axis. Then eq. (3.37), eq. (3.38), and eq. (3.39) are re-written for convenience (in particular, for use in solving problems) as follows:

$$v_{yf} = v_{yi} - gt \qquad (3.41)$$

$$y_f - y_i = v_{yi}t - g\frac{t^2}{2}, \qquad (3.42)$$

$$v_{yf}^2 - v_{yi}^2 = -2g(y_f - y_i). \qquad (3.43)$$

Here, g is given by eq. (3.36).

Note that, since the acceleration of the freely falling objects g is the same for every object, this indicates that the heavier objects do not fall faster than the lighter ones if we ignore the air resistance.

3.8 Exercises

Exercise 3.1. Determine the displacement, average velocity, and average speed of the car in Fig. 3.12 between positions A and F. The units of the displacement are given in

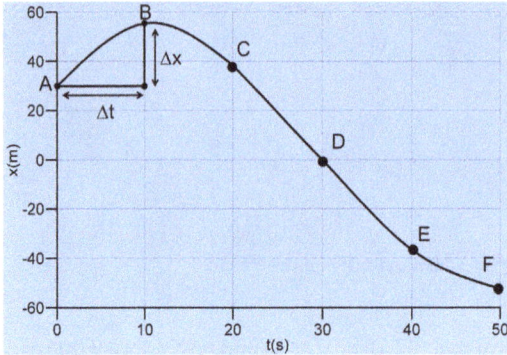

Figure 3.12: The movement of a car along a straight line indicated as the x axis. We are interested only in the translational motion of the car, and hence it is treated as a particle, and hence the graph presents a position–time plot of that particle.

meters, and the numerical results must be in the same order of magnitude presented in the graph.

Exercise 3.2. A particle moves along the x-axis with its x coordinate varying with time according to the expression $x = -4t + 2t^2$, where x is in meter, and t is in second. The position versus time graph for this motion is given in Fig. 3.13. Note that particle moves in the negative x direction for the first second of the motion, is at rest at the moment $t = 1\,$s, and moves in the positive x direction for $t > 1\,$s.

(a) Find the displacement of the particle in the following time intervals: Between $t = 0$ and $t = 1\,$s and between $t = 1\,$s and $t = 3\,$s.

(b) Calculate the average velocity during these two time intervals.

(c) Find the instantaneous velocity of the particle at $t = 2.5\,$s.

Use only one significant figure in all calculations.

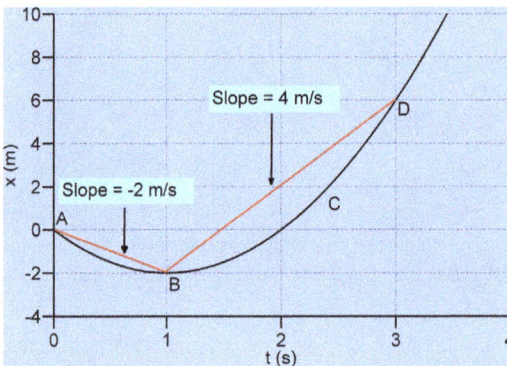

Figure 3.13: Position–time graph for a particle having an x coordinate that varies in time according to the expression $x = -4t + 2t^2$.

Exercise 3.3. Suppose that the velocity of some object moving along the x-axis varies in time according to the expression $v_x = (40 - 5t^2)\,$m/s, where t is in seconds.

(a) Find the average acceleration in the time interval $t = 0$ to $t = 2.0$ s.
(b) Determine the acceleration at $t = 2.0$ s.

Exercise 3.4. Consider driving in an entrance ramp to an interstate highway. Let us assume a final velocity of 100 km/h so you can merge with the traffic and the initial velocity is about one-third of the final velocity.
(a) Estimate the average acceleration assuming it takes around 10.0 s to merge to the traffic.
(b) How far did you go during the first half of the time interval during which you accelerated?

[Use only three significant figures in calculations.]

Exercise 3.5. Suppose a jet lands on at a velocity 140 mi/h.
(a) What is its acceleration if it stops in 2.00 s?
(b) What is the displacement of the plane while it is stopping?

[Use only three significant figures in calculations.]

Exercise 3.6. Suppose a ball is thrown vertically up at 25 m/s. Calculate its velocity after 1.0 s. [Use only two significant figures in calculations.]

Exercise 3.7. Consider a ball is thrown from the top of a building with an initial velocity of 20.0 m/s in the direction upward. Suppose the building is 50.0 m high. The ball passes the edge of the roof on its way down, as shown in Fig. 3.14. Let us assume that $t_A = 0$ is when the ball leaves the thrower's hand at position A.
(a) Determine the time at which the ball reaches its maximum height.
(b) Determine the maximum height.
(c) Determine the time at which the ball returns to the height from where it was thrown.
(d) Determine the velocity of the ball at this instant.
(e) Determine the velocity and position of the ball at $t = 5.00$ s.

[Use only three significant figures in calculations.]

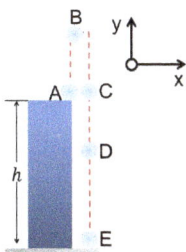

Figure 3.14: Position and velocity versus time for a free falling ball thrown initially upward with velocity $v_{yi} = 20.0$ m/s.

Exercise 3.8. Can average and instantaneous velocity ever be equal for a specific type of motion? Explain.

Exercise 3.9. Assume that the average velocity is nonzero for some time interval. Does that mean that the instantaneous velocity can never be zero during this interval? Explain.

Exercise 3.10. Prove that, if the average velocity equals zero for some time interval Δt, and if $v_x(t)$ is a continuous function, the instantaneous velocity must go to zero at some time in this interval. Explain.

Exercise 3.11. Can we assume that velocity and acceleration may also have an opposite sign? If so, sketch a velocity–time graph to prove your point.

Exercise 3.12. Suppose the velocity of a particle is nonzero. Can its acceleration be zero? Explain.

Exercise 3.13. If the velocity of a particle is zero, can its acceleration be nonzero? Explain.

Exercise 3.14. Can an object having constant acceleration ever stop and stay stopped?

Exercise 3.15. Consider a stone is thrown straight up from the top of a building. Does the stone's displacement depend on the location of the origin of the coordinate system? Does the stone's velocity depend on the origin? (Assume that the coordinate system is stationary with respect to the building.) Explain.

Exercise 3.16. A student throws a ball upward from the top of a building, which has a height h, with an initial speed v_{yi} and then throws a second ball downward with the same initial speed. Compare the final speeds of the balls when they reach the ground.

Exercise 3.17. Can the magnitude of the instantaneous velocity of an object ever be greater than the magnitude of its average velocity? Can it ever be less?

Exercise 3.18. Assume that the average velocity of an object is zero in some time interval. Determine the displacement of the object in that interval.

Exercise 3.19. A car's position was observed at various times; the results are summarized in Tab. 3.2. Calculate the average velocity of that car (a) during the first second, (b) the last 3 s, and (c) the entire period of observation.

Table 3.2: The results of the observation.

x (m)	0	2.3	9.2	20.7	36.8	57.5
t (s)	0	1.0	2.0	3.0	4.0	5.0

Exercise 3.20. A motorist drives north for 35 min at 85 km/h and then stops for 15 min. Then he continues north, traveling 130 km in 2.00 h. (a) What is his total displacement? (b) What is his average velocity?

Exercise 3.21. The displacement versus time graph of a particle moving along the x-axis is shown in Fig. 3.15. What is the average velocity in the following time intervals (a) 0 to 2 s, (b) 0 to 4 s, (c) 2 s to 4 s, (d) 4 s to 7 s, (e) 0 to 8 s? [Use only one significant figure for the results.]

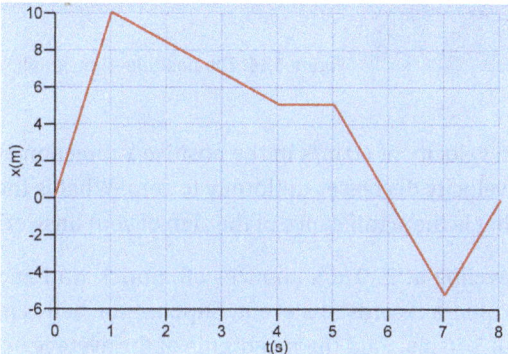

Figure 3.15: A diagram of trajectory.

Exercise 3.22. A particle moves along the x-axis with a position varying as a function of time according to $x = 10t^2$, where x is in meters, and t is in seconds. (a) Determine the average velocity for the time interval from 2.0 s to 3.0 s. (b) Determine the average velocity for the time interval from 2.0 s to 2.1 s. [Use only two significant figures in calculations.]

Exercise 3.23. A person walks initially with constant speed of v_1 along a straight line from point A to point B, and then turns back along the same line from B to A at another constant speed of v_2. (a) What is the average speed over the entire trip? (b) What is the average velocity over the entire trip?

Exercise 3.24. At $t = 1.00$ s, a particle moving with constant velocity is located at $x = -3.00$ m, and at $t = 6.00$ s the particle is located at $x = 5.00$ m. (a) From this information, plot the position as a function of time. (b) Determine the velocity of the particle from the slope of this graph.

Exercise 3.25. A position–time graph for a particle moving along the x-axis is shown in Fig. 3.16. (a) Find the average velocity from time $t = 2.0$ s to $t = 4.0$ s. (b) Find the instantaneous velocity at instant of time $t = 2.0$ s by measuring the slope of the tangent line, as shown in the graph. (c) At what value of t is the velocity zero?

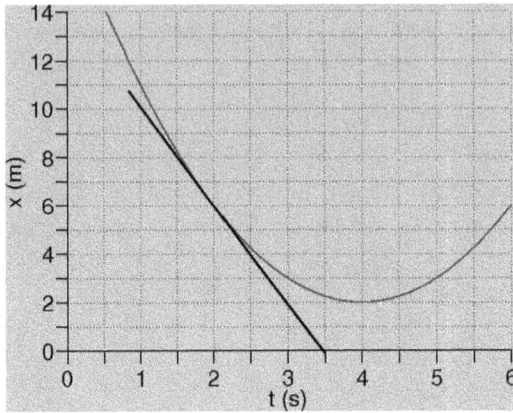

Figure 3.16: The position–time graph.

Exercise 3.26. A particle moves with velocity of 60 m/s in the positive x direction at $t = 0$. During an interval of 15 s, the velocity decreases uniformly to zero. What is the acceleration during that interval? What is the significance of the sign of your answer?

Exercise 3.27. A 50.0 g superball traveling at 25.0 m/s bounces off a brick wall and rebounds at 22.0 m/s. A high-speed camera records this event. Suppose the ball is in contact with the wall for some time of 3.50 ms. Find the magnitude of the average acceleration of the ball during this time interval. (Note: $1\,\text{ms} = 10^{-3}\,\text{s}$).

Exercise 3.28. Consider a particle that starts the motion from rest, and it accelerates, as shown in Fig. 3.17. (a) Determine the particle's speed at $t = 10\,\text{s}$ and at $t = 20\,\text{s}$. (b) Determine the distance traveled during the first 20 s.

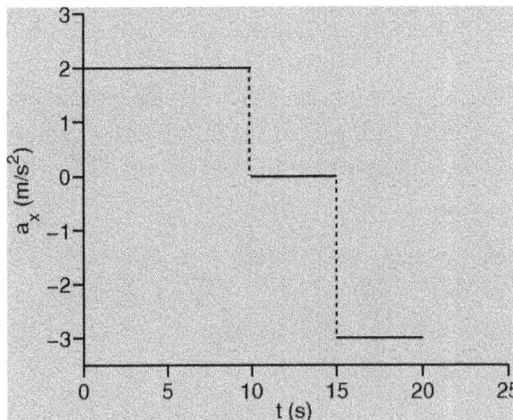

Figure 3.17: The acceleration–time graph.

Exercise 3.29. A graph of the velocity–time for an object moving along the x-axis is shown in Fig. 3.18. (a) Plot a graph of the acceleration as a function time. (b) Determine the average acceleration of the object during the time intervals between $t = 5.00$ s and $t = 15.0$ s; $t = 0$ and $t = 20.0$ s.

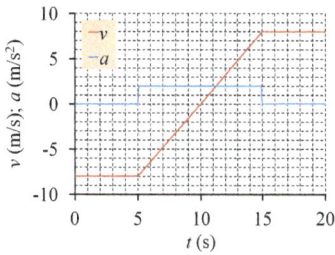

Figure 3.18: The velocity–time and acceleration–time graphs.

Exercise 3.30. Assume a particle moves along the x-axis with a position versus the time equation given as $x = 2.00 + 3.00t - t^2$, where x is in meters, and t is in seconds. At $t = 3.00$ s, find (a) its position, (b) its velocity, and (c) its acceleration.

Exercise 3.31. An object moves along the x-axis where its position is given by the equation $x = (3.00t^2 - 2.00t + 3.00)$ m. (a) Determine average velocity between $t = 2.00$ s and $t = 3.00$ s. (b) What are the instantaneous velocities at $t = 2.00$ s and $t = 3.00$ s? (c) Find the average acceleration between $t = 2.00$ s and $t = 3.00$ s, and (d) the instantaneous acceleration at $t = 2.00$ s and $t = 3.00$ s.

Exercise 3.32. Assume that a car will accelerate from rest to a speed of 42.0 m/s in 8.00 s with a constant acceleration. (a) Determine the acceleration of the car. (b) Find the distance traveled by the car during the first 8.00 s. (c) What is the speed of the car 10.0 s after it begins its motion, if it moves with the same acceleration?

Exercise 3.33. A truck travels 40.0 m in 8.50 s while smoothly slowing down to a final speed of 2.80 m/s. (a) Find its actual speed. (b) Find its acceleration.

Exercise 3.34. A body is moving with a constant acceleration having a velocity of 12.0 cm/s along the positive x direction when its x coordinate is 3.00 cm. If its x coordinate after 2.00 s is –5.00 cm, what is the magnitude of its acceleration?

Exercise 3.35. A particle moves along the x-axis. Its position versus time is given as $x = 2.00 + 3.00t - 4.00t^2$ with x in meters and t in seconds. Determine (a) its position at the instant it changes the direction and (b) its velocity when it returns to the same position at $t = 0$.

Exercise 3.36. The initial velocity of a body is 5.20 m/s. What is its velocity after 2.50 s (a) if its accelerates uniformly at 3.00 m/s^2 and (b) if it accelerates uniformly at -3.00 m/s^2?

Exercise 3.37. A car starts the motion from the rest and accelerates at $10.0 \, \text{m/s}^2$ for the entire distance of 400 m. (a) How long did it take for the car to travel this distance? (b) What is the speed of the car at the end of its run?

Exercise 3.38. Suppose a car is approaching a hill at speed 30.0 m/s where its engine, just at the bottom of the hill, stops due to a failure. The car moves with a constant acceleration of $-2.00 \, \text{m/s}^2$ while coasting up the hill. (a) Write equations for the position and velocity along the slope as a function of time. Assume that $x = 0$ at the bottom of the hill, where $v_i = 30.0$ m/s. (b) Determine the maximum distance the car moves up the hill.

Exercise 3.39. A golf ball is released from rest from the top of a very tall building. Calculate the position and the ball's velocity after 1.00 s, 2.00 s, and 3.00 s.

Exercise 3.40. A student throws a set of keys straight up to his roommate, who is catching them from a window 4.00 m above. Those are caught after 1.50 s by the roommate's outstretched hand. (a) What is the initial velocity that the keys are thrown? (b) What is the velocity of the keys just before they are caught?

Exercise 3.41. A ball is thrown directly downward from 30.0 m above the ground with an initial speed of 8.00 m/s. How many seconds later does the ball strike the ground?

Exercise 3.42. A ball dropped from the rest at a height h above the ground. Another ball is thrown vertically upward from the Earth surface at the same time the first ball was released. Determine the speed of the second ball if the two balls are to meet at a height of $h/2$ above the ground.

4 Two- and three-dimensional motion

In this chapter, we will discuss the kinematics of a particle moving in two and three dimensions. Utilizing two- and three-dimensional motion, we will be able to examine a variety of movements, starting with the motion of satellites in orbit to the flow of electrons in a uniform electric field. We will begin studying in more detail the vector nature of displacement, velocity, and acceleration. Similar to one-dimensional motion, we will also derive the kinematic equations for three-dimensional motion from these three quantities' fundamental definitions. Then the projectile motion and uniform circular motion will be described in detail as particular cases of the movements in two dimensions.

4.1 The displacement, velocity, and acceleration vectors

When we discussed the one-dimensional motion (see Chapter 3), we mentioned that the movement of an object along a straight line is thoroughly described in terms of its position as a function of time, $x(t)$. For the two-dimensional motion, we will extend this idea to the movement in the xy plane.

As a start, we describe a particle's position by the position vector $\mathbf{r}$ pointing from the origin of some coordinate system to the particle located in the xy plane, as shown in Fig. 4.1. At time t_i the particle is at point P, and at some later time t_f it is at the position Q. The path from P to Q generally is not a straight line. As the particle moves from P to Q in the time interval $\Delta t = t_f - t_i$, its position vector changes from $\mathbf{r}_i$ to $\mathbf{r}_f$.

Definition 4.1 (Displacement vector). The displacement is a vector, and the displacement of the particle is the difference between its final position and its initial position. We now formally define the displacement vector for the particle as the difference between its final position vector and its initial position vector:

$$\Delta \mathbf{r} = \mathbf{r}_f - \mathbf{r}_i. \tag{4.1}$$

The direction of $\Delta\mathbf{r}$ is indicated in Fig. 4.1 from P to Q. Note that the magnitude of $\Delta\mathbf{r}$ is smaller than the distance traveled by the particle along the curved path.

Often, an object's motion is quantified by the ratio of displacement with the time interval during which that displacement took place. In contrast to one-dimensional kinematics, where we used the signs plus or minus to indicate the motion direction, in two-dimensional (or three-dimensional) kinematics, we will use vectors. All the other quantities have the same meanings.

Definition 4.2 (Average velocity). By definition, the average velocity of a particle during the time interval Δt is the displacement of the particle divided by that time interval:

$$\bar{\mathbf{v}} = \frac{\Delta \mathbf{r}}{\Delta t}. \tag{4.2}$$

https://doi.org/10.1515/9783110755824-004

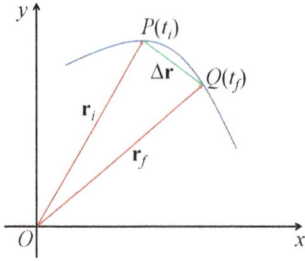

Figure 4.1: Two-dimensional motion of a particle moving in the *xy* plane.

We know that multiplying or dividing a vector quantity by a positive scalar number changes only the vector's magnitude, not its direction. Therefore, since the displacement is a vector quantity and the time interval is a positive scalar quantity (because $t_f > t_i$). The average velocity is a vector directed along $\Delta \mathbf{r}$.

From the definition, the average velocity is proportional to displacement, which, in turn, depends only on the initial and final position vectors but not on the path is taken. Therefore, the average velocity between any two points is independent of the path that is taken, but it depends only on the initial and final positions. Like one-dimensional motion, if a particle starts its motion at the point P and returns to the same location, P, taking any path, then the displacement is zero. Hence, its average velocity is zero.

Definition 4.3 (Magnitude of average velocity). The magnitude of the average velocity vector

$$\bar{v} = |\bar{\mathbf{v}}| \tag{4.3}$$

is called the *average speed*, which is a scalar quantity.

Consider again the motion of a particle between two points, $P(t)$ and $Q(t_k)$ (for $k = 1, 2, 3, \ldots$), in the xy plane, as shown in Fig. 4.2. Taking the limit when the time interval $\Delta t = t_k - t$ over which we observe the motion becomes infinitesimally small ($\Delta t \to 0$), the direction of the displacement, $\Delta \mathbf{r} = \mathbf{r}_k - \mathbf{r}_i$, approaches the line tangent with the path at P and its magnitude becomes infinitesimally small, $|\Delta \mathbf{r}| \to 0$.

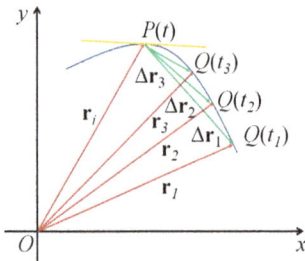

Figure 4.2: A particle moving in the *xy* plane as a case of a two-dimensional motion.

Definition 4.4 (Instantaneous velocity). The instantaneous velocity **v** is defined as the limit of the average velocity $\Delta\mathbf{r}/\Delta t$ as Δt approaches zero:

$$\mathbf{v} = \lim_{\Delta t \to 0} \frac{\Delta\mathbf{r}}{\Delta t} = \frac{d\mathbf{r}}{dt}. \tag{4.4}$$

The definition of the instantaneous velocity shows that this quantity equals the *derivative* of the position vector with respect to time (see Fig. 4.3). As such, the direction of the particle's instantaneous velocity vector at any point in the path is tangent with it at that point and along the direction of motion.

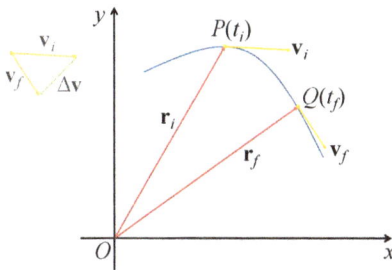

Figure 4.3: Instantaneous velocity of the two-dimensional motion of a particle moving in the *xy* plane.

Definition 4.5 (Magnitude of velocity). The magnitude of the instantaneous velocity vector

$$v = |\mathbf{v}| \tag{4.5}$$

is called the *instantaneous speed* (or, simply *speed*), which, as you should remember, is a scalar quantity.

When a particle moves from one point to another along a path, its instantaneous velocity vector changes from $\mathbf{v}_i$ at time t_i to $\mathbf{v}_f$ at time t_f, as shown in Fig. 4.3. Note that, in general, since the trajectory curve is not linear, the velocities $\mathbf{v}_i$ and $\mathbf{v}_f$ are not the same (even if the speeds are the same). Therefore, in general, $\Delta\mathbf{v} = \mathbf{v}_f - \mathbf{v}_i \neq 0$. If we suppose that we know the velocity at these points, we can determine the particle's average acceleration as follows.

Definition 4.6 (Average acceleration). By definition, the average acceleration of a particle as it moves between any two positions is given as the change in the instantaneous velocity vector $\Delta\mathbf{v}$ divided by the interval of time Δt during which that change occurred:

$$\bar{\mathbf{a}} \equiv \frac{\mathbf{v}_f - \mathbf{v}_i}{t_f - t_i} = \frac{\Delta\mathbf{v}}{\Delta t}. \tag{4.6}$$

Velocity vector change

It can be seen that the average acceleration is the ratio of a vector quantity $\Delta \mathbf{v}$ and a positive scalar quantity Δt, hence the average acceleration $\mathbf{a}$ is a vector quantity directed along $\Delta \mathbf{v}$. As indicated in Fig. 4.3, the direction of $\Delta \mathbf{v}$ is found by adding the vector $-\mathbf{v}_i$ (the negative of $\mathbf{v}_i$) to the vector $\mathbf{v}_f$, because by definition

$$\Delta \mathbf{v} = \mathbf{v}_f - \mathbf{v}_i. \tag{4.7}$$

If the average acceleration of a particle changes during different time intervals, it is useful to define its instantaneous acceleration $\mathbf{a}$.

Definition 4.7 (Instantaneous acceleration). The instantaneous acceleration is defined as the limiting value of the ration $\Delta \mathbf{v}/\Delta t$ as Δt approaches 0:

$$\mathbf{a} \equiv \lim_{\Delta t \to 0} \frac{\Delta \mathbf{v}}{\Delta t} = \frac{d\mathbf{v}}{dt}. \tag{4.8}$$

That is, the instantaneous acceleration is the *derivative* of the velocity vector with respect to time.

Note that there exist various changes when a particle accelerates.
- First, the change with time on the magnitude of the velocity vector; that is, the speed may cause the particle to accelerate, similar to the straight-line (one-dimensional) motion.
- Second, the change with time on the direction of the velocity vector even if its magnitude remains constant, as in curved-path (two-dimensional) motion, can also cause the particle to accelerate.
- Finally, if both the magnitude and the direction of the velocity vector change with time simultaneously, the particle accelerates.

4.2 Three-dimensional motion with constant acceleration

We focus on three-dimensional motion with constant acceleration in both magnitude and direction to derive the kinematic equations.

For the three-dimensional motion, the position vector of a particle can be written as

$$\mathbf{r} = x\mathbf{i} + y\mathbf{j} + z\mathbf{k} \tag{4.9}$$

where x, y, z, and $\mathbf{r}$ are functions of time as the particle moves while $\mathbf{i}$, $\mathbf{j}$ and $\mathbf{k}$ are constant and unitary vectors, that is,

$$|\mathbf{i}| = 1, \quad |\mathbf{j}| = 1, \quad |\mathbf{k}| = 1. \tag{4.10}$$

The velocity of a particle can be obtained from its position vector as

$$\mathbf{v} = \frac{d\mathbf{r}}{dt} = \frac{d}{dt}(x\mathbf{i} + y\mathbf{j} + z\mathbf{k}) \tag{4.11}$$

$$= \frac{dx}{dt}\mathbf{i} + \frac{dy}{dt}\mathbf{j} + \frac{dz}{dt}\mathbf{k}$$
$$= v_x\mathbf{i} + v_y\mathbf{j} + v_z\mathbf{k}$$

where

$$v_x = \frac{dx}{dt} \tag{4.12}$$

$$v_y = \frac{dy}{dt} \tag{4.13}$$

$$v_z = \frac{dz}{dt} \tag{4.14}$$

are the components of the velocity vector $\mathbf{v}$ along the x-, y- and z-axis, respectively.

Now, since $\mathbf{a}$ is assumed constant, its components a_x, a_y and a_z, respectively, along the x-, y- and z-axis are constants, too. Therefore, applying the equations of kinematics to the x, y and z components of the velocity vector, we obtain

$$v_{xf} = v_{xi} + a_x t \tag{4.15}$$
$$v_{yf} = v_{yi} + a_y t$$
$$v_{zf} = v_{zi} + a_z t.$$

Substituting these expressions into eq. (4.11) to determine the final velocity at any time t, we obtain

$$\mathbf{v_f} = v_{xf}\mathbf{i} + v_{yf}\mathbf{j} + v_{zf}\mathbf{k} \tag{4.16}$$
$$= (v_{xi} + a_x t)\mathbf{i} + (v_{yi} + a_y t)\mathbf{j} + (v_{zi} + a_z t)\mathbf{k}$$
$$= (v_{xi}\mathbf{i} + v_{yi}\mathbf{j} + v_{zi}\mathbf{k}) + (a_x\mathbf{i} + a_y\mathbf{j} + a_z\mathbf{k})t$$
$$= \mathbf{v_i} + \mathbf{a}t.$$

Therefore, we derived the change on the velocity vector as a function of time.

Velocity vector as a function of time

$$\mathbf{v_f} = \mathbf{v_i} + \mathbf{a}t. \tag{4.17}$$

Equation (4.17) indicates that the velocity of a particle at the time t equals the vector sum of its initial velocity $\mathbf{v_i}$ and the additional velocity at acquired in the time t as a result of constant acceleration.

Similarly, we know that the x, y and z coordinates of a particle moving with constant acceleration are

$$x_f = x_i + v_{xi}t + \frac{1}{2}a_x t^2 \tag{4.18}$$

$$y_f = y_i + v_{yi}t + \frac{1}{2}a_yt^2 \tag{4.19}$$

$$z_f = z_i + v_{zi}t + \frac{1}{2}a_zt^2. \tag{4.20}$$

Substituting these expressions into the equation $\mathbf{r}_f = x_f\mathbf{i} + y_f\mathbf{j} + z_f\mathbf{k}$, we get

$$\mathbf{r}_f = \left(x_i + v_{xi}t + \frac{1}{2}a_xt^2\right)\mathbf{i} \tag{4.21}$$

$$+ \left(y_i + v_{yi}t + \frac{1}{2}a_yt^2\right)\mathbf{j}$$

$$+ \left(z_i + v_{zi}t + \frac{1}{2}a_zt^2\right)\mathbf{k}$$

$$= (x_i\mathbf{i} + y_i\mathbf{j} + z_i\mathbf{k}) + (v_{xi}\mathbf{i} + v_{yi}\mathbf{j} + v_{zi}\mathbf{k}) + (a_x\mathbf{i} + a_y\mathbf{j} + a_z\mathbf{k})\frac{t^2}{2}$$

or we can state the following.

Position vector as a function of time

$$\mathbf{r}_f = \mathbf{r}_i + \mathbf{v}_it + \frac{1}{2}\mathbf{a}t^2. \tag{4.22}$$

This equation, eq. (4.22), indicates that the displacement vector $\Delta\mathbf{r} = \mathbf{r}_f - \mathbf{r}_i$ can be written as follows.

Displacement vector as a function of time

$$\mathbf{r}_f - \mathbf{r}_i = \mathbf{v}_it + \frac{1}{2}\mathbf{a}t^2 \tag{4.23}$$

where the first term is the displacement arising from the initial velocity of the particle and the second term is the displacement resulting from the uniform acceleration of the particle. According to eq. (4.23) and the rules of analytic geometry, motion with constant acceleration takes place on a constant plane (i. e., is plane motion).

Graphical representations of eq. (4.17) and eq. (4.22) are shown in Fig. 4.4 in xy plane for the case of the two-dimensional motion. To simplify the drawings of the figure, we assume that $\mathbf{r}_i = 0$, and hence the particle is placed at the origin initially at $t_i = 0$. Note from Fig. 4.4(a) that $\mathbf{r}_f$ is generally not along the direction of either $\mathbf{v}_i$ or $\mathbf{a}$ because the relationship between these quantities is a vector expression. For the same reason, from Fig. 4.4(b) we see that $\mathbf{v}_f$ is generally not along the direction of $\mathbf{v}_i$ or $\mathbf{a}$. Finally, note that $\mathbf{v}_f$ and $\mathbf{r}_f$ are generally not in the same direction.

Since eq. (4.17) and eq. (4.22) are vector expressions, we can write their components, respectively, as follows:

$$\mathbf{v}_f = \mathbf{v}_i + \mathbf{a}t \quad \leftrightarrow \quad \begin{cases} v_{xf} = v_{xi} + a_xt \\ v_{yf} = v_{yi} + a_yt \\ v_{zf} = v_{zi} + a_zt \end{cases} \tag{4.24}$$

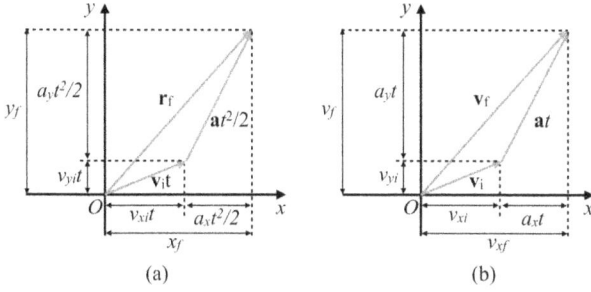

Figure 4.4: Vector representations and components of (a) the displacement and (b) the velocity of a particle moving with a uniform acceleration. To simplify the drawing, we have set $r_i = 0$.

and

$$\mathbf{r}_f = \mathbf{r}_i + \mathbf{v}_i t + \frac{1}{2}\mathbf{a}t^2 \quad \leftrightarrow \quad \begin{cases} x_f = x_i + v_{xi}t + \frac{1}{2}a_x t^2 \\ y_f = y_i + v_{yi}t + \frac{1}{2}a_y t^2 \\ z_f = z_i + v_{zi}t + \frac{1}{2}a_z t^2. \end{cases} \tag{4.25}$$

Similarly to one-dimensional motion, we can find time t from eq. (4.17):

$$t = \frac{\mathbf{a} \cdot (\mathbf{v}_f - \mathbf{v}_i)}{a^2}. \tag{4.26}$$

Substituting eq. (4.26) into eq. (4.23), we obtain

$$v_f^2 - v_i^2 = 2\mathbf{a} \cdot (\mathbf{r}_f - \mathbf{r}_i). \tag{4.27}$$

In eq. (4.27), v_f and v_i represent the final and the initial speed, respectively, of the moving body.

Knowing that $v^2 = v_x^2 + v_y^2 + v_z^2$ and $\mathbf{a} \cdot (\mathbf{r}_f - \mathbf{r}_i) = a_x(x_f - x_i) + a_y(y_f - y_i) + a_z(z_f - z_i)$, eq. (4.27) is equivalent to the following three equations, representing the motion in each direction:

$$\begin{aligned} v_{xf}^2 - v_{xi}^2 &= 2a_x(x_f - x_i) \\ v_{yf}^2 - v_{yi}^2 &= 2a_y(y_f - y_i) \\ v_{zf}^2 - v_{zi}^2 &= 2a_z(z_f - z_i). \end{aligned} \tag{4.28}$$

Note that eq. (4.24), eq. (4.25), and eq. (4.28) indicate that the three-dimensional motion of an arbitrary body can be described by three independent motions representing, respectively, the movement along the x, y and z axes.

4.3 Projectile motion

We often have observed a baseball in motion or any other object thrown into the air, such as an arrow shown in Fig. 4.5. By definition, this kind of movement is called projectile motion.

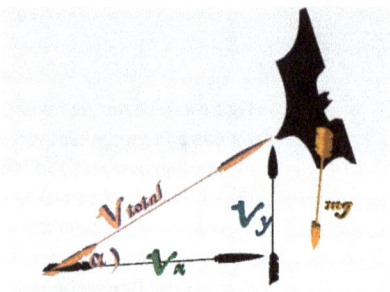

Figure 4.5: An example of the projectile motion. v_{total} represents the total velocity at any point of the projectile motion; v_x and v_y are, respectively, the x and y components. α is the angle formed by the direction of v_{total} and x-direction, which varies from point to point along the trajectory.

Assumptions of projectile motion

In this motion, the ball follows a curved path, and its motion is simple to analyze if we make two assumptions:
1. The free-fall acceleration **g** is constant over the range of motion and is directed downward.
2. The effect of air resistance is negligible.

Taking into account these assumptions, we will find in the following that the projectile motion's path of a projectile (or the trajectory) is always a parabola. To show this, we consider the projectile motion in a reference frame such that the y-axis is vertical, and its positive direction is upward. Since the air resistance is neglected, we know that $a_y = -g$ (as in one-dimensional free fall) and that $a_x = 0$. Also, the projectile is initially ($t = 0$) positioned at the origin ($x_i = y_i = 0$) with velocity $\mathbf{v}_i$, as shown in Fig. 4.6. The vector $\mathbf{v}_i$ makes an angle θ_i with the horizontal, where θ_i is the initial angle (at the origin) between the direction of $\mathbf{v}_i$ vector and the x-direction.

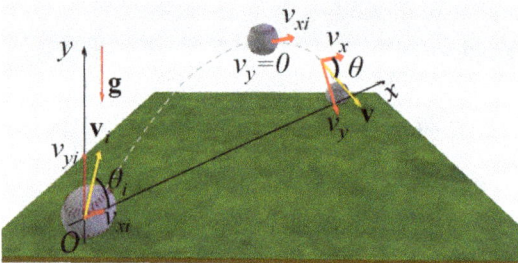

Figure 4.6: Projectile motion of the baseball.

From the definitions of the cosine and sinus functions, we obtain the x and y components of the initial velocity:

$$v_{xi} = v_i \cos \theta_i \tag{4.29}$$
$$v_{yi} = v_i \sin \theta_i.$$

Substituting the x component from eq. (4.29) into eq. (4.25) and taking $x_i = 0$, $a_x = 0$, we get

$$x_f = v_{xi}t = (v_i \cos \theta_i)t. \tag{4.30}$$

Solving eq. (4.30) for t, we obtain

$$t = \frac{x_f}{v_i \cos \theta_i}. \tag{4.31}$$

Similarly, substituting y component from eq. (4.29) into eq. (4.25), and considering that $y_i = 0$ and $a_y = -g$, we get

$$y_f = v_{yi}t + \frac{1}{2}a_y t^2 = (v_i \sin \theta_i)t - \frac{1}{2}gt^2. \tag{4.32}$$

Replacing the time t from eq. (4.31) into eq. (4.32), we get

$$y = (\tan \theta_i)x - \left(\frac{g}{2v_i^2 \cos^2 \theta_i}\right)x^2. \tag{4.33}$$

Equation (4.33) is valid for initial angles in the range $0 < \theta_i < \pi/2$. Furthermore, eq. (4.33) is valid for any point (x, y) along the path of the projectile, and thus the subscripts of x and y are omitted. It can be seen that eq. (4.33) has the form

$$y = ax - bx^2. \tag{4.34}$$

That indicates that $y = f(x)$ is a parabola, passing through the origin of the system. Thus, the trajectory of a projectile is a parabola, and it is completely defined if we know both the initial velocity, $\mathbf{v}_i$, and the launching angle, θ_i.

The position vector of the projectile versus time follows from eq. (4.22), where $\mathbf{r}_i = 0$ and $\mathbf{a} = \mathbf{g}$:

$$\mathbf{r} = \mathbf{v}_i t + \frac{1}{2}\mathbf{g}t^2. \tag{4.35}$$

This expression is shown graphically in Fig. 4.7.

Note that the motion of a particle can be considered the superposition of two terms. The term $\mathbf{v}_i t$, which is the displacement if no acceleration was present, and the term $\mathbf{g}t^2/2$, representing the acceleration because of gravity. That is, if there were no gravitational acceleration ($g = 0$), the particle would continue to move in a straight line in the same direction as $\mathbf{v}_i$. Therefore, the vertical distance of $\mathbf{g}t^2/2$ through which the particle "falls" off the straight-line path equals the distance that a freely falling body would fall during the same time interval. Therefore, projectile motion is the superposition of two independent motions:

1. constant-velocity motion in the horizontal direction;
2. free-fall motion in the vertical direction.

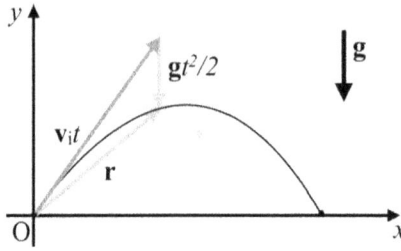

Figure 4.7: The position vector **r** of a projectile whose initial velocity at the origin is $\mathbf{v}_i$. The vector $\mathbf{v}_i t$ would be the displacement of the projectile if gravity were absent, and the vector $\mathbf{g}t^2/2$ is its vertical displacement due to its downward gravitational acceleration.

Except for t, which is the time of flight, both the horizontal and vertical components of the projectile's motion are completely independent of each other.

Moreover, at the highest point of the parabola (see also Fig. 4.6), $v_y = 0$; therefore, using the expression of the velocity for the y component from eq. (4.24), we obtain

$$t_{1/2} = \frac{v_i \sin \theta_i}{g} \tag{4.36}$$

where $t_{1/2}$ is half of the time of the full trajectory. Therefore, the total time of the entire trajectory of a projectile motion is (because of the symmetry of parabola):

$$t = 2t_{1/2} = \frac{2v_i \sin \theta_i}{g}. \tag{4.37}$$

Combining eq. (4.30) and eq. (4.37), we obtain the horizontal displacement of the object during the entire projectile motion:

$$x = \frac{v_i^2 \sin(2\theta_i)}{g}. \tag{4.38}$$

In addition, the maximum vertical displacement of the object can be obtained by substituting $t_{1/2}$ from eq. (4.36) into eq. (4.32) as

$$h = y_f(t_{1/2}) = (v_i \sin \theta_i)\left(\frac{v_i \sin \theta_i}{g}\right) - \frac{1}{2}g\left(\frac{v_i \sin \theta_i}{g}\right)^2 \tag{4.39}$$

$$= \frac{v_i^2 \sin^2 \theta_i}{g} - \frac{v_i^2 \sin^2 \theta_i}{2g}$$

$$= \frac{v_i^2 \sin^2 \theta_i}{2g}.$$

Equation (4.38) and eq. (4.39) indicate the following:
-   For $\theta_i = 90°$, which corresponds to the case when the object is thrown initially vertically up, we obtain

$$x = 0 \tag{4.40}$$

$$h = \frac{v_i^2}{2g}.$$

This corresponds to the case of maximum displacement vertically and minimum horizontal displacement of the object. Note that this does not represent a projectile motion by definition. Indeed, it represents a free-fall motion as described in Chapter 3 for one-dimensional motion. That is, initially the object will reach the maximum height h, then it freely falls down until hits the ground with a speed (since the air resistance is ignored):

$$v = \sqrt{2hg} = v_i. \tag{4.41}$$

– For $\theta_i = 45°$, we obtain

$$x = \frac{v_i^2}{g} \tag{4.42}$$

$$h = \frac{v_i^2 (\sqrt{2}/2)^2}{2g} = \frac{v_i^2}{4g}.$$

This corresponds to the maximum displacement along the horizontal direction.
– For $\theta_i = 0°$, we do not obtain a projectile motion, but a straight-line motion with constant velocity v_i, if we ignore friction with horizontal surface.

4.4 Uniform circular motion

Another example of planar motion is circular motion, which represents the motion of a particle along a circular path.

Uniform circular motion

The motion of a particle along a circular path is defined by the velocity and the acceleration of that particle. The velocity vector of the particle is tangent to the circular trajectory at every instant. The velocity is a vector quantity, and hence, it has a magnitude and direction. By definition, the circular motion is uniform if the velocity vector's magnitude is constant, but its direction changes as the particle moves along the circular path.

Centripetal acceleration

Since the direction of the velocity changes, this causes the particle to accelerate along the circle. By definition, that acceleration is called centripetal acceleration, which stands for "acceleration seeking a center". Therefore, the acceleration vector always has a direction towards the center of the circle. Fig. 4.8 presents the velocity and acceleration vectors as the particle moves along a circular path from point P to the point Q with an angle $\Delta\theta$ in a circle with radius r.

To derive an expression for the centripetal acceleration as a function of the velocity and the radius r, we will refer to the displacement in Fig. 4.8 from point P to Q along

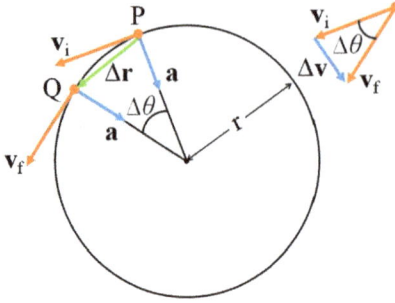

Figure 4.8: The circular motion of a particle. The velocity and centripetal acceleration of the particle moving along the circle with radius r from point P to Q with an angle θ.

the circle. We denote by Δs the arc along the circle from point P to Q traveled during the time interval Δt, then

$$\Delta s = r\Delta\theta \tag{4.43}$$

where r is constant and $\Delta\theta$ is the time dependent variable. We divide both sides of eq. (4.43) by Δt:

$$\frac{\Delta s}{\Delta t} = r\frac{\Delta\theta}{\Delta t}. \tag{4.44}$$

Since the magnitude of the velocity vector $\mathbf{v}$ is constant, $\Delta s = v\Delta t$, and hence we obtain

$$v\Delta t = r\Delta\theta. \tag{4.45}$$

In addition, see also Fig. 4.8, the change on the velocity vector is given as

$$\Delta\mathbf{v} = \mathbf{v}_f - \mathbf{v}_i. \tag{4.46}$$

It can be seen that, if we take the limit when Δt approaches zero (or, equivalently, when $\Delta\theta$ approaches zero), the vector $\Delta\mathbf{v}$ has the direction toward the center of the center of the circle. Therefore, if we denote by $\hat{\mathbf{r}}$ the inward unit vector along the radial direction, then

$$\Delta\mathbf{v} = |\mathbf{v}_f - \mathbf{v}_i|\hat{\mathbf{r}}. \tag{4.47}$$

Dividing both sides of eq. (4.47) by Δt and taking the limit when Δt approaches zero, we obtain the centripetal acceleration vector as

$$\mathbf{a}_r = \lim_{\Delta t\to 0}\frac{\Delta\mathbf{v}}{\Delta t} = \lim_{\Delta t\to 0}\frac{\Delta v}{\Delta t}\hat{\mathbf{r}}. \tag{4.48}$$

The magnitude of $\Delta\mathbf{v}$, for small angle $\Delta\theta$, is given as

$$\Delta v = v\Delta\theta \tag{4.49}$$

since the magnitude of the velocity vector is constant, v. Combining eq. (4.45) and eq. (4.49), we get

$$v\Delta t = r\frac{\Delta v}{v}. \tag{4.50}$$

Rearranging the expression in eq. (4.50), we obtain

$$\frac{\Delta v}{\Delta t} = \frac{v^2}{r}. \tag{4.51}$$

Substituting eq. (4.51) into eq. (4.48), we obtain the centripetal acceleration vector as

$$\mathbf{a}_r = \frac{v^2}{r}\hat{\mathbf{r}}. \tag{4.52}$$

It can be seen from eq. (4.52) that $\mathbf{a}_r$ has the direction toward the center of the circle, and hence sometime is called radial acceleration. The magnitude of the centripetal acceleration is given as

$$a_r = \frac{v^2}{r}. \tag{4.53}$$

Frequency and period

The circular motion of a particle is repetitive by nature; that is, the particle may pass through the same point in the circle during its movement many times. By definition, the frequency is the number of times per unit of time that the particle visits the same state. The frequency is denoted by the letter f and has the units of 1/s or Hz.

By definition, the period, denoted by T, is the time it takes a particle to complete one full cycle along the circle.

To find a relationship between the period T and the frequency f, suppose Δt is the interval of time that we observe the particle moving along the circle. We denote with N the number of times that the particle visited a particular point on the circular path, then, by definition, the frequency is

$$f = \frac{N}{\Delta t} \tag{4.54}$$

where $\Delta t = NT$. Replacing that in eq. (4.54), we obtain the relationship between the frequency and the period as

$$f = \frac{1}{T}. \tag{4.55}$$

It can be seen that the unit of period is second.

4.5 Non-uniform circular motion

Consider a particle moving along a curved path, as shown in Fig. 4.9, where both the direction and magnitude of the velocity vector change. The velocity vector is tangent to the curve at any instant of time; however, the acceleration vector **a** changes along the path from some point P to another point Q. The acceleration vector can be composed into two components, namely the tangential component $\mathbf{a}_t$ and radial component $\mathbf{a}_r$:

$$\mathbf{a} = \mathbf{a}_t + \mathbf{a}_r. \tag{4.56}$$

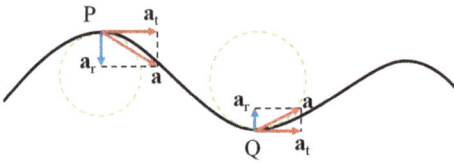

Figure 4.9: Illustration of the trajectory of a particle moving along a curved path.

Tangential acceleration
The tangential acceleration arises from the change in the magnitude of the velocity vector along the curve, and its magnitude is defined as

$$a_t = \frac{dv}{dt}. \tag{4.57}$$

The tangential acceleration has the same direction as the velocity vector; that is, $\mathbf{a}_t$ is tangent to the curve. Denoting by $\hat{\mathbf{t}}$ a unit vector along the tangential direction to the curve, then

$$\mathbf{a}_t = a_t \hat{\mathbf{t}} = \frac{dv}{dt} \hat{\mathbf{t}}. \tag{4.58}$$

Radial acceleration
On the other hand, we derived that the radial acceleration arises from the change in the velocity vector direction. It has the direction toward the center of the curvature of the path at the instant of time. Its magnitude is v^2/r, where r is the radius of the curvature. Therefore,

$$\mathbf{a}_r = \frac{v^2}{r} \hat{\mathbf{r}}. \tag{4.59}$$

Then the total acceleration can be written as

$$\mathbf{a} = \mathbf{a}_t + \mathbf{a}_r = \frac{dv}{dt} \hat{\mathbf{t}} + \frac{v^2}{r} \hat{\mathbf{r}}. \tag{4.60}$$

Since the vectors $\hat{\mathbf{t}}$ and $\hat{\mathbf{r}}$ are perpendicular, the magnitude of the total acceleration is given as

$$a = \sqrt{a_t^2 + a_r^2} = \sqrt{\left(\frac{dv}{dt}\right)^2 + \left(\frac{v^2}{r}\right)^2}. \tag{4.61}$$

Non-uniform circular motion

A special case of the curved path non-uniform motion is the non-uniform circular motion, as presented in Fig. 4.10. In this motion, it is convenient to express the acceleration of the particle moving along a circle with radius r in terms of unit vectors $\hat{\mathbf{t}}$, tangent to the circle, and $\hat{\mathbf{r}}$ a unit vector along the radius vector inward from the circle to the center of the circle, as depicted in Fig. 4.10. Then the acceleration $\mathbf{a}$ of the particle is given by eq. (4.60). Note that both unit vectors, $\hat{\mathbf{t}}$ and $\hat{\mathbf{r}}$, move as the particle moves along the circular path.

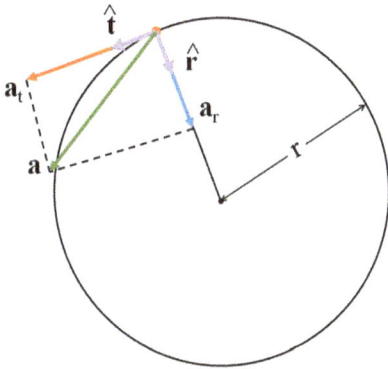

Figure 4.10: Illustration of the non-uniform circular motion.

4.6 Relative motion

We will also discuss the measurement of the displacement, velocity, and acceleration relative to observers in different frames of reference related to each other. For example, consider two observers in reference frames S_0 and S, respectively, where S is moving relative to S_0 with constant velocity $\mathbf{v}$ along the positive x-axis, as shown in Fig. 4.11. Suppose the origin initially and the axes of both reference frames coincide. After a time t, the reference frame S is displaced to the right, along the positive x-axis by the vector $\mathbf{v}t$, as depicted from Fig. 4.11. We denote with $\mathbf{r}$ the position of the particle P at time t measured by the observer in frame S, and with $\mathbf{r}_0$ its position measured by the observer in the frame S_0.

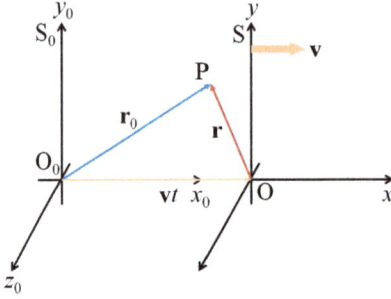

Figure 4.11: The first observer is located at the fixed reference frame S_0 and the second observer at the moving reference frame S with constant velocity $\mathbf{v}$ along the positive x-axis. The particle is located at point P. Initially, both reference frames coincide with the origin and axes of both reference frames.

Galilean transformations

Then, using the rule of the vector addition,

$$\mathbf{r}_0 = \mathbf{r} + \mathbf{v}t \tag{4.62}$$

or

$$\mathbf{r} = \mathbf{r}_0 - \mathbf{v}t. \tag{4.63}$$

Taking the derivative of both sides of eq. (4.63) with respect to time t, we obtain relationship between the velocities measured by the observers in two references frames as

$$\mathbf{u} = \mathbf{u}_0 - \mathbf{v} \tag{4.64}$$

where

$$\mathbf{u} = \frac{d\mathbf{r}}{dt} \tag{4.65}$$

$$\mathbf{u}_0 = \frac{d\mathbf{r}_0}{dt} \tag{4.66}$$

give the velocity of the particle P measured by the observer in frame S and frame S_0, respectively. Equations given by eq. (4.63) and eq. (4.64) are known as Galilean transformation equations.

Taking the derivative of eq. (4.64) with respect to time, again, we get

$$\mathbf{a} = \mathbf{a}_0 \tag{4.67}$$

where

$$\mathbf{a} = \frac{d\mathbf{u}}{dt} \tag{4.68}$$

$$\mathbf{a}_0 = \frac{d\mathbf{u}_0}{dt} \tag{4.69}$$

since $d\mathbf{v}/dt = 0$. Therefore, the observers moving relative to each other with constant velocity will measure the same acceleration. For example, the acceleration of a particle measured by an observer in the reference frame of Earth is the same as the acceleration measured by an observer moving with constant velocity relative the reference frame of Earth.

It is important to note that, if the relative velocity between the reference frames is not constant, then $d\mathbf{v}/dt \neq 0$, and hence

$$\mathbf{a}_{rel} = \frac{d\mathbf{v}}{dt}. \tag{4.70}$$

Therefore, the transformation of the acceleration between the two reference frames is given as

$$\mathbf{a} = \mathbf{a}_0 - \mathbf{a}_{rel}. \tag{4.71}$$

For example, consider a ball dropping from the ceiling of a moving train that is accelerating at a rate $\mathbf{a}_{rel}$ (see also Fig. 4.12). The acceleration of the ball relative to the train (that is, measured by the observer in S) is $\mathbf{a} = \mathbf{g}$ and relative to the Earth is $\mathbf{a}_0$ (that is, measured by the observer in S_0):

$$\mathbf{g} = \mathbf{a}_0 - \mathbf{a}_{rel}. \tag{4.72}$$

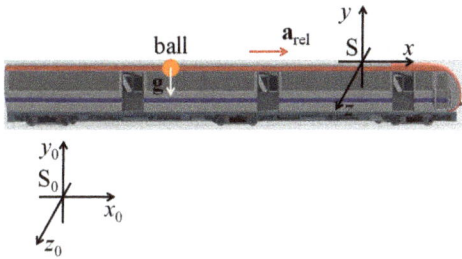

Figure 4.12: Consider a ball dropping from the ceiling of a moving train that is accelerating at a rate $\mathbf{a}_{rel}$. The acceleration of the ball relative to the train is $\mathbf{a} = \mathbf{g}$ and relative to the Earth is $\mathbf{a}_0$.

4.7 Exercises

Exercise 4.1. Consider a motorist is driving south at 20.0 m/s for 3.00 min. Then the motorist turns and drives west at 25.0 m/s for 2.00 min. Finally, the motorist travels northwest at 30.0 m/s for 1.00 min. For this 6.00 min trip, find (a) the total vector displacement, (b) the average speed, and (c) the average velocity.

Exercise 4.2. Suppose a particle with the position vector given as $\mathbf{r} = x\mathbf{i} + y\mathbf{j}$, where $x = at + b$ and $y = ct^2 + d$. Assume that $a = 1.00$ m/s, $b = 1.00$ m, $c = 0.125$ m/s^2, and $d = 1.00$ m. (a) Calculate average velocity during time interval from $t = 2.00$ s to $t = 4.00$ s. (b) Determine its velocity and its speed at $t = 2.00$ s.

Exercise 4.3. The x and y coordinates versus time for a golf ball hitting a tree at the edge of a cliff are given by the following expressions:

$$x = (18.0 \, \text{m/s})t \tag{4.73}$$

and

$$y = (4.00\,\text{m/s})t - (4.90\,\text{m/s}^2)t^2. \tag{4.74}$$

(a) Write a vector expression for the position of ball versus time, using the unit vectors $\mathbf{i}$ and $\mathbf{j}$. Calculating the derivatives of y for time t, give the expressions for (b) the velocity vector as a function of time and (c) the acceleration vector as a function of time. Using the notation of the unit vector, derive the expressions for (d) the position, (e) the velocity, and (f) the acceleration of the ball, all at $t = 3.00\,\text{s}$.

Exercise 4.4. An object is moving in the xy plane with coordinates varying with time according to the equations

$$x = -(5.00\,\text{m}) \sin \omega t \tag{4.75}$$

and

$$y = (4.00\,\text{m}) - (5.00\,\text{m}) \cos \omega t \tag{4.76}$$

where t is in seconds and ω has units of s^{-1}. (a) Determine the components of velocity and acceleration at $t = 0$. (b) Write expressions for the position vector, velocity vector, and the acceleration vector at any time $t > 0$. (c) Using a xy graph, describe the path of the object.

Exercise 4.5. Consider a particle is moving in the xy plane with constant acceleration, which at $t = 0$ has an initial velocity of $\mathbf{v}_i = (3.00\mathbf{i} - 2.00\mathbf{j})\,\text{m/s}$. At some later time, $t = 3.00\,\text{s}$, the particle has a velocity of $\mathbf{v} = (9.00\mathbf{i} + 7.00\mathbf{j})\,\text{m/s}$. (a) Calculate the acceleration of the particle and (b) its coordinates at any time t.

Exercise 4.6. Assume a fish is swimming in a horizontal plane with a velocity of $\mathbf{v}_i = (4.00\mathbf{i} + 1.00\mathbf{j})\,\text{m/s}$ at the position from a certain rock of $\mathbf{r}_i = (10.0\mathbf{i} - 4.00\mathbf{j})\,\text{m}$. After the fish swims with constant acceleration for $20.0\,\text{s}$, its velocity becomes $\mathbf{v} = (20.0\mathbf{i} - 5.00\mathbf{j})\,\text{m/s}$. Determine (a) the components of the acceleration, (b) the direction of the acceleration with respect to the unit vector $\mathbf{i}$, and (c) the position of the fish at $t = 25.0\,\text{s}$ if it maintains its original acceleration. In what direction is it moving?

Exercise 4.7. The acceleration of a particle initially at the origin is $\mathbf{a} = 3.00\mathbf{j}\,\text{m/s}^2$ and the initial velocity is $\mathbf{v}_i = 5.00\mathbf{i}\,\text{m/s}$. (a) What are the vector position and velocity at any time t? (b) What are the coordinates and speed of the particle at $t = 2.00\,\text{s}$?

Exercise 4.8. A customer slides a mug off the counter, which strikes the floor at distance d from the base of the counter. If the height of the counter is h, (a) with what velocity did the mug leave the counter and (b) what was the direction of the mug's velocity just before it hit the floor?

Exercise 4.9. As a strategy, in a snowball fight, the snowball is thrown at a high angle over level ground. As the opponent is watching the first one, a second snowball is thrown at a low angle such that it arrives at the opponent before or at the same time as the first one. Assume both snowballs are thrown with speed of 25.0 m/s. Furthermore, assume that the first ball is thrown at an angle of 70.0° with respect to the horizontal. (a) What is the angle that the second snowball is thrown if it is to land at the same point as the first? (b) What time after the first snowball is the second snowball thrown if it is to land at the same time as the first?

Exercise 4.10. A tennis payer standing 12.6 m from the net hits the ball at 3.00° above the horizontal. To cross the net, the ball must rise at least 0.330 m. If the ball just crosses the net at the apex of its trajectory, what is the velocity when the ball left the racket?

Exercise 4.11. An artillery shell is fired with some initial velocity of 300 m/s at 55.0° above the horizontal, which explodes on a mountain side 42.0 s after firing. What are the components x and y of the shell's position where it explodes, relative to its firing point?

Exercise 4.12. An astronaut on another planet finds that can jump a maximum vertical distance of 15.0 m if the initial speed is 3.00 m/s. Determine the free-fall acceleration on that planet.

Exercise 4.13. A projectile is fired such that its horizontal range is equal to three times its maximum height. What is the angle of projection?

Exercise 4.14. Consider a ball tossed from an upper-story window of a building. The ball has an initial velocity of 8.00 m/s and an angle of 20.0° below the horizontal line. After 3.00 s, it strikes the ground. (a) How far horizontally does the ball fly until it strikes the ground? (b) What is the height from which the ball was thrown? (c) How long does it take the ball to reach the point that is 10.0 m below the level of launching?

Exercise 4.15. Suppose a cannon has a muzzle speed of about 1000 m/s, which used to start an avalanche on a mountain slope. Assume that the target is 2000 m from the cannon along the horizontal line and 800 m above the cannon. Find the angle, above the horizontal, that the cannon fires.

Exercise 4.16. A projectile motion starts at the origin of a xy coordinate system with speed v_i at an initial angle θ_i above the horizontal. Note that at the apex of its trajectory, the projectile is moving horizontally, so that the slope of its path is zero. Use the expression for the trajectory to find the x coordinate that corresponds to the maximum height. Use this x coordinate and the symmetry of the trajectory to determine the horizontal range of the projectile.

Exercise 4.17. A placekicker kicks a football from a point 36.0 m from the goal, and he hopes the ball will cross the crossbar, which is 3.05 m high. The ball is kicked with a speed of 20.0 m/s at an angle of 53.0° to the horizontal. (a) By how much does the ball cross over or fall short of crossing the crossbar? (b) Does the ball reach the crossbar while still rising or while falling?

Exercise 4.18. Assume a firefighter, who is 50.0 m away from a burning building, points a stream of water from a fire hose with an angle of 30.0° above the horizontal direction. If we assume that the speed of the stream is 40.0 m/s, at what height will the water strike the building?

Exercise 4.19. Assume a soccer player kicks a ball horizontally off a cliff 40.0 m high into a pool of water. Suppose the player hears the sound of the splash after 3.00 s. Determine the initial speed given to the ball. Assume the speed of sound in air is 343 m/s.

Exercise 4.20. Consider the circular orbit of the Moon about the Earth, with a mean radius of 3.84×10^8 m. It takes 27.3 days for the Moon to complete one revolution about the Earth. (a) What is the mean orbital speed of the Moon? (b) What is its centripetal acceleration?

Exercise 4.21. An athlete rotates a 1.00 kg disc along a circular path of radius 1.06 m with a maximum speed of 20.0 m/s. What is the magnitude of the maximum radial acceleration of the disc?

Exercise 4.22. A tire of radius 0.500 m rotates at a constant speed of 200 rev/min. What are the speed and acceleration of a small stone on the outer edge of the tire?

Exercise 4.23. During a liftoff, Space Shuttle astronauts typically feel accelerations up to 1.4 g, where $g = 9.8$ m/s^2. In the training, an astronaut is in a device that has a given centripetal acceleration. Specifically, the astronaut is fastened securely at the end of a mechanical arm that then turns at constant speed in a horizontal circle. Determine the rotation rate, in revolutions per seconds, required to give an astronaut a centripetal acceleration of 1.4 g while the astronaut moves in a circle of radius 10.0 m.

Exercise 4.24. Assume that someone could revolve a sling of length 0.600 m at the rate of 8.00 rev/s. If the length increased to 0.900 m, he could revolve the sling only 6.00 times per second. (a) Compare the speeds of the stone at the end of the sling for the two rotation rates. (b) Find the centripetal acceleration of stone at 8.00 rev/s. (c) Find the centripetal acceleration at 6.00 rev/s.

Exercise 4.25. Suppose an astronaut is orbiting the Earth with a satellite, moving in a circular orbit of radius 600 km above the surface of the Earth, where the free-fall acceleration is 8.21 m/s^2. Considering the radius of the Earth is about 6400 km. What is the speed of the satellite and the time required to complete one orbit around the Earth?

Exercise 4.26. Suppose a train slowing down, as it rounds a sharp horizontal curve, from 90.0 km/h to 50.0 km/h in 15.0 s. The radius of the curve is 150 m. Calculate the acceleration at the moment the train speed reaches 50.0 km/h. Assume that the train slows down at a uniform rate during the interval of 15.0 s.

Exercise 4.27. The speed of an automobile increases at a rate of 0.600 m/s² travels along a circular road. The radius of the road is 20.0 m. When the instantaneous speed of the automobile is 4.00 m/s, find (a) the tangential acceleration component, (b) the radial acceleration component, and (c) the magnitude and direction of the total acceleration.

Exercise 4.28. In Fig. 4.13 is shown at an instant of time the total acceleration and velocity of a particle in a circular clockwise motion of radius 2.50 m. At this instant, (a) what is the radial acceleration? (b) What is the speed of the particle? (c) What is its tangential acceleration?

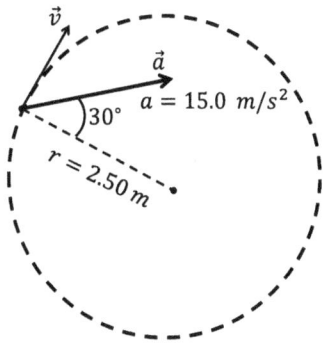

Figure 4.13: Direction of the total acceleration and velocity vectors.

Exercise 4.29. Consider a string of length of 0.600 m, where a ball is attached at the end of the string. Then a student swings the ball along a vertical circle. The speed of the ball is 4.30 m/s at its highest point and 6.50 m/s at its lowest point. What is the acceleration of the ball when the string is straight up, and the ball is at (a) its highest point and (b) its lowest point?

Exercise 4.30. A ball swings in a vertical circle at the end of a rope 1.50 m long. When the ball is 36.9° past the lowest point and on its way up, its total acceleration is $(-22.5\mathbf{i} + 20.2\mathbf{j})$ m/s². At that instant, (a) sketch a vector diagram showing the components of this acceleration, (b) determine the magnitude of its radial acceleration, and (c) determine the speed and velocity of the ball.

Exercise 4.31. Suppose the car 1 accelerates at the rate of $(3.00\mathbf{i} - 2.00\mathbf{j})$ m/s², while the car 2 accelerates at $(1.00\mathbf{i} + 3.00\mathbf{j})$ m/s². Both cars start the motion from rest at the origin of an xy coordinate system. After 5.00 s, (a) what is the speed the car 1 with respect to car 2, (b) how far apart are they, and (c) what is the acceleration of car 1 relative to car 2?

Exercise 4.32. A river has a uniform speed of 0.500 m/s. A student swims upstream for a distance of 1.00 km and swims back to the same starting point. Assume that the student swims at a uniform speed of 1.20 m/s in still water. How long does the trip take? Compare this with the time the trip would take if the water were still.

Exercise 4.33. How long does it take to the automobile traveling in the left lane at 60.0 km/h to pass a car traveling in the right lane at 40.0 km/h? The cars are initially 100 m apart.

Exercise 4.34. The pilot in an airplane finds out that the compass indicates a heading due west. The airplane's speed relative to the air is 150 km/h. If there is a wind of 30.0 km/h toward the north, find the velocity of the airplane relative to the ground.

Exercise 4.35. Suppose a child is carried downstream by the current of a river that has a steady speed of 2.50 km/h. The child is 0.600 km from shore and 0.800 km upstream of a boat landing when a rescue boat sets out. (a) If the boat proceeds at its maximum speed of 20.0 km/h relative to the water, what heading relative to the shore should the pilot take? (b) What angle does the boat velocity make with the shore? (c) How long does it take the boat to reach the child?

Exercise 4.36. Consider a bolt dropping from the ceiling of a moving train that is accelerating at a rate 2.50 m/s^2. What is the acceleration of the bolt relative to the train and to the Earth?

Exercise 4.37. Consider a student who is on a train traveling along a straight horizontal track with a constant speed equal to 10.0 m/s. The student throws a ball at an initial angle of 60.0° with the horizontal line of the track. The professor, standing on the ground nearby, observes the ball to rise vertically. How high does the professor see the ball rise?

5 The laws of motion

So far, we described motion in terms of displacement, velocity, and acceleration without considering the cause of that motion. For example, what causes one particle to stay at rest and another particle to accelerate? Here, we investigate what causes changes in motion. The two main factors that are usually considered are the forces acting on an object and its mass. We will discuss the three fundamental laws of motion, which are related to forces and masses. These laws were formulated more than three centuries ago by Isaac Newton. These laws allow understanding the mechanism of changing the state of motion and the degree of acceleration of different objects.

5.1 The concept of force

Our basic understanding of the concept of force comes from everyday experience. For example, consider the experiment of pushing or pulling a desk, we exert a force on it. Similarly, when we throw a ball, we exert a force on the ball to throw it. In these typical examples, the force is related to muscular activity and some changes in the velocity of the object. However, it is important to understand that forces do not always cause motion. For example, we can push the wall of the building, but we are not able to move it.

A body accelerates because of an external force
Newton was able to answer the question of what force causes the planets to orbit around the Sun, and related questions by stating that forces are what cause any change in the velocity of an object. Thus, no force is required for the motion with a constant velocity of an object to be maintained. For example, the Moon's velocity is not constant because it moves in a nearly circular orbit around the Earth because of the force exerted on the Moon by the Earth. Therefore, since the velocity of objects changes whenever a force is applied to them, then the forces make the objects accelerate. Here, we are concerned with the relationship between the force exerted on an object and the acceleration of that object.

Equilibrium
In the case when more than one force acts upon an object simultaneously, if the net force acting on it is different from zero, then the object accelerates. By definition, the net (or resultant) force acting on an object is given by the vector sum of all vectors of the forces acting on the object:

$$\mathbf{F}_R = \sum_i \mathbf{F}_i \qquad (5.1)$$

https://doi.org/10.1515/9783110755824-005

where $\mathbf{F}_R$ is the net force as the over all force, the resultant force, or the unbalanced force, and the sum runs overall forces $\mathbf{F}_i$ acting upon the object. If the resultant force, $\mathbf{F}_R$, exerted on an object is zero, then its acceleration is zero, and hence the velocity of that object remains unchanged. In other words, if $\mathbf{F}_R = 0$, then the object either remains at the state of rest or continues moving with the same velocity. If an object moves with constant velocity or it is at rest (that is, its velocity is zero), the object is said to be in *equilibrium*.

When a coiled spring is pulled, as in Fig. 5.1, the spring stretches and hence changes its state of rest. That is the case of contact forces.

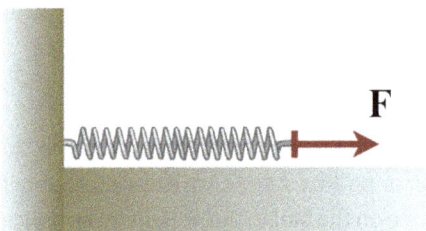

Figure 5.1: Example of applied contact force.

To overcome the conceptual problem of non-contact forces, Michael Faraday (1791–1867) introduced the concept of a *field*. Based on this concept, if any object is placed near another object, then the objects interact with each other through the gravitational field. In this case, we say that a gravitational field is created by each object at the position of the other object. All objects create a gravitational field in the space around themselves. Field forces are those types of forces, which do not have physical contact between two objects, but they are acting through empty space. An example of the field force is the gravitational attraction force between two objects, illustrated in Fig. 5.2, which keeps the objects bound on the surface of the Earth. The planets of our Solar System are bound to the Sun by the action of gravitational forces. Other examples of the field forces include the electrostatic forces between the charged particles.

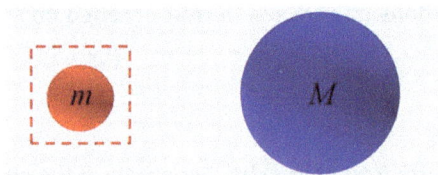

Figure 5.2: Example of applied non-contact force.

5.2 Measuring the strength of a force

The strength of the force can conveniently be measured using the deformation of a spring. For that, suppose we apply a horizontal force to a spring scale that has a fixed left end, as shown in Fig. 5.3(a).

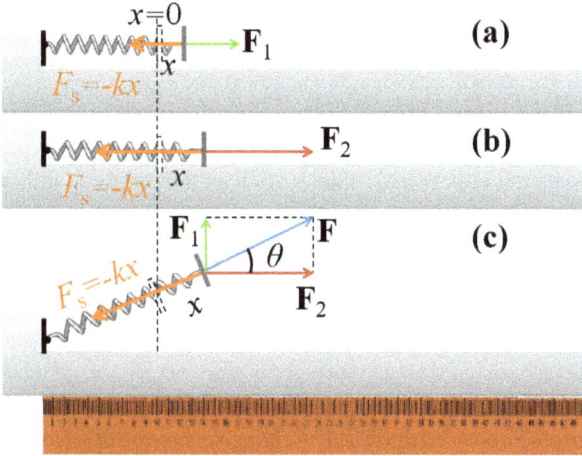

Figure 5.3: Testing the vector nature of the force using the spring, which is fixed on the left-hand side and on the right-hand side, it experiences an external force. Under the influence of the external force on some arbitrary direction, as indicated, the spring extends by x until an equilibrium is established between the external force and the spring force $F_s = -kx$, where k represents the stiffness of the spring. The mass of the spring is ignored, and hence gravitational force on the spring is omitted. External forces and their directions with respect to the horizontal direction are: (a) $\mathbf{F}_1$ parallel to horizontal line; (b) $\mathbf{F}_2$ ($F_2 = 2F_1$) parallel to horizontal line; (c) $\mathbf{F} = \mathbf{F}_1 + \mathbf{F}_2$ forming an angle θ with horizontal line.

If we apply a force on the spring, then it elongates. A pointer can then read on the scale the value of the force acting on the spring. Calibration of the spring can be performed by defining the unit force $\mathbf{F}_1$, which represents the force that produces a pointer reading of 1.00 cm. If we now apply a different horizontal force to the right $\mathbf{F}_2$ whose magnitude is 2 units, as seen in Fig. 5.3(b), the pointer will move to 2.00 cm. Fig. 5.3(c) shows that, if the forces are not linear (for example, they are perpendicular to each other), then their net effect is the sum of the two forces. Suppose that two forces act on the spring simultaneously with $\mathbf{F}_1$ upward and $\mathbf{F}_2$ horizontal, as illustrated in Fig. 5.3(c). In this case, the pointer reads $\sqrt{5.00\,\text{cm}^2} = 2.24$ cm. The single force $\mathbf{F}$ that would produce this same reading is the sum of the two vectors $\mathbf{F}_1$ and $\mathbf{F}_2$, as described in Fig. 5.3(c). That is,

$$|\mathbf{F}| = \sqrt{F_1^2 + F_2^2} = 2.24\,\text{units} \qquad (5.2)$$

and its direction relative to horizontal level is

$$\theta = \tan^{-1}(-0.500) = -26.6°$$ (5.3)

where the minus sign indicates that the angle is measure clockwise.

Because forces are vectors, the addition rule of vectors apply, and hence the net force acting on an object can be obtained:

$$\mathbf{F} = \mathbf{F}_1 + \mathbf{F}_2.$$ (5.4)

5.3 Newton's first law and inertia frame

To describe a force, we can use its net effect on an object, such as the force capability to change the object's shape or its position and accelerate or decelerate the object. These effects are described by the laws of the motion introduced by Sir Isaac Newton (1642–1727).

Newton's laws

Every body continues in its state of rest, or uniform motion in a straight line, unless compelled to change that state by forces impressed upon it. The change of motion of an object is proportional to the force impressed upon it and made in the direction of the straight line in which the force is impressed. To every action, there is always opposed an equal reaction; or, the mutual actions of two bodies upon each other are always equal and directed contrary.

In mechanics, Newton's laws are formulated in what is known as Newton's first law of motion, Newton's second law of motion, and Newton's third law of motion, respectively. In this chapter, we will examine each of these laws in detail, and then give some simple illustrations of their use.

It was first Aristotle, then after him Galileo, who recognized that the motion of an object is because of the forces exerted upon it, and because of an external cause its motion stops. That was then perfected by Descartes, who added that the motion must be in a straight line, which is formulated in terms of Newton's first law, which states that, if there are no external forces exerted on an object to disturb its motion, then the object moves with constant velocity.

Newton's first law of motion

If an object is at rest or moving along a straight line with uniform speed, then it will remain at rest, or it will continue moving along the straight line with constant speed unless an external net force is applied on it to change its existing state of motion.

Thus, if no forces are acting on the object, then its acceleration is zero. In other words, if no force acts to change the object's motion, then its velocity remains constant. Based on the first law, we can say that any isolated object is either at rest or moving with constant velocity along a straight line. Here, an isolated object is the one that does not interact with its environment.

Law of inertia
The tendency of an object to resist any attempt to change its velocity is called the inertia of the object.

An object that is moving can be observed from any number of *reference frames*. Sometimes Newton's first law is also called the *law of inertia*, which defines a particular set of reference frames called *inertial frames*.

Inertial frame
An inertial frame of reference is one relative to which an observer is either at rest or moving with constant velocity (i. e., the observer is not accelerating).

Since Newton's first law holds only with objects that are not accelerating, it holds only in inertial frames. Besides, a reference frame that moves with constant velocity relative to an inertial frame is also inertial.

For an observer in one inertial frame (say, one at rest relative to the object), if the object moves with constant velocity, the acceleration of the object and the net force acting on it are zero. Also, an observer in any other inertial frame claims for that object

$$\mathbf{a} = 0 \qquad\qquad (5.5)$$

and

$$\sum_i \mathbf{F}_i = 0. \qquad\qquad (5.6)$$

According to the first law, a body at rest and one moving with constant velocity are equivalent.

Example 5.1. For example, a passenger in a car moving along a straight road at a constant speed of, let say, 100 km/h can easily pour coffee into a cup. On the other hand, if the driver steps on the gas or brake pedal or turns the steering wheel, the coffee may be poured onto the floor—why?

Answer: Because the car accelerates and it is no longer an inertial frame, and hence the laws of motion do not work the same in both car's reference frame (which is accelerating frame) and passenger reference frame (an inertial frame), the coffee pours out of the cup.

5.4 Newton's second law of motion

According to Newton's first law, when no net force acts on an object, it stays at rest or moves with a constant velocity along a straight line. The second law, on the other hand, tells us what happens when this force is not zero. The concept of *motion* used by Newton is related to the concept of *momentum* used nowadays. It is necessary to take into account since the change in the *momentum* in time is the net force applied to the object as we will demonstrate this later on.

Example 5.2. Consider you are pushing a block of ice across a frictionless horizontal surface. When you exert some horizontal force, let say **F**, the block moves with some acceleration **a**. If you apply a force twice as big, **2F**, the acceleration doubles. If you increase the applied force to **3F**, the acceleration will also be **3a**, and so on. These examples indicate that the acceleration of an object is directly proportional to the resultant force acting on it.

The acceleration of an object should also depend on its mass. For example, consider the following experiment: If you apply a force **F** to a block of ice moving on a horizontal frictionless surface, then the block undergoes some acceleration, **a**. If you double the mass of the block, then the same applied force produces an acceleration of **a**/2. If the mass is tripled, then the same applied force delivers acceleration of **a**/3, and so on. This observation indicates that the magnitude of the acceleration of an object is inversely proportional to its mass.

These observations lead to Newton's second law.

Newton's second law

The acceleration of an object is directly proportional to the net force acting on it and inversely proportional to its mass.

We can relate mass and force using the following expression, which represents a mathematical statement of the second law of the Newton:

$$\sum_i \mathbf{F}_i = m\mathbf{a}.$$
(5.7)

Since this expression is a vectorial equation, we can project along the three coordinate axes as follows:

$$\sum_i F_{xi} = ma_x$$
(5.8)

$$\sum_i F_{yi} = ma_y$$
(5.9)

$$\sum_i F_{zi} = ma_z.$$
(5.10)

In the SI unit system, the force has the units of *newton* (N), which is defined as the force acting on a mass of 1 kg gaining an acceleration of 1 m/s². From the definition of Newton's second law, we see that the newton can be expressed in terms of the units of mass, length, and time as follows:

$$1\,\mathrm{N} \equiv 1\,\mathrm{kg} \cdot \frac{\mathrm{m}}{\mathrm{s}^2}.$$
(5.11)

In the engineering system or British system, the unit of force is *pound*, which is defined as the force acting on a mass of 1 slug[1] to produce an acceleration of $1\,\text{ft/s}^2$:

$$1\,\text{lb} \equiv 1\,\text{slug} \cdot \text{ft/s}^2. \tag{5.12}$$

An approximation is

$$1\,\text{lb} \approx \frac{1}{4}\,\text{N}. \tag{5.13}$$

5.5 Newton's third law

When we press against a corner of a textbook with the fingertip, the book pushes back and makes a small dent in our skin. If we press harder, then the book does the same, and the dent in our skin gets a little larger. In general, consider the experiment of two bodies in contact; each exerting a force on the other at the contact point. If the contact point is considered as a body of mass zero, based on the second law of Newton, the net force is zero, $\mathbf{F}_R = 0$. If we assume that only these two forces act at the body with mass zero, then their sum is zero, and hence, they have the same magnitude and oppositely directed. This experiment is what is known as Newton's third law.

Newton's third law
If two objects interact, the force $\mathbf{F}_{12}$ exerted by object 1 on object 2 is equal in magnitude to and opposite in direction to the force $\mathbf{F}_{21}$ exerted by object 2 on object 1:

$$\mathbf{F}_{12} = -\mathbf{F}_{21}. \tag{5.14}$$

This law, which is illustrated in Fig. 5.4, states that a force that affects the motion of an object must come from a second, external object. An equal magnitude but oppositely directed force exerted on the second object, too.

Therefore, the forces cannot exist isolated in nature. The force acting from object 1 on object 2 is called the *action force*, and the force acting from object 2 on object 1 is called the *reaction force*. Note that either force can be called the action or the reaction force, and always the action force equals in magnitude the reaction force, and they have the opposite direction. Thus, the action and reaction forces always act on different objects.

---

1 The *slug* is used as a unit of mass in the British engineering system; it is the counterpart of the SI unit the *kilogram*: 1 slug = 14.59383741601181 kg.

$\mathbf{F}_{12} = -\mathbf{F}_{21}$

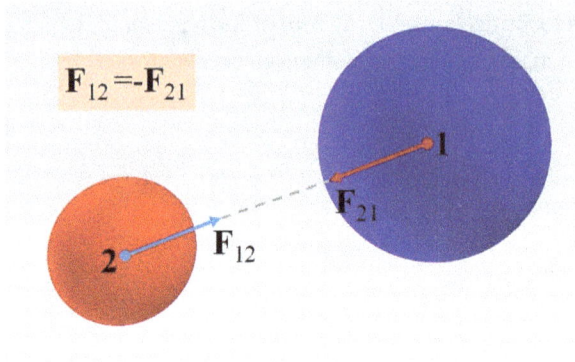

Figure 5.4: Illustration of Newton's third law.

5.6 Mass and weight

In fact, the terms *mass* and *weight* are often interchanged with one another. But, in physics, their meanings are quite distinct.

The mass measures the inertia of an object. As such, the mass is the resistance of an object to deviate from uniform straight-line motion under the influence of external forces. According to Newton's second law, see eq. (5.7), if on two objects of different masses are acting forces of the same magnitude, then the object with smaller mass gains more acceleration than that of the object with larger mass. In other words, the object with larger mass exhibits a more significant resistance to deviate from its previous state of uniform motion in a straight line. Besides, in classical mechanics, the mass of an object represents an intrinsic property of the object, and hence it does not change if the object is moved to a different place.

Consider an object with mass m as in Fig. 5.5. We know that all objects are attracted to the Earth due to the force of gravity exerted by the Earth on an object, $\mathbf{F}_g$. The direction of $\mathbf{F}_g$ is toward the center of the Earth, and it has a magnitude, often called the *weight* of the object. We saw that a freely falling object experiences an acceleration $\mathbf{g}$ acting toward the center of the Earth. Applying Newton's second law

$$\mathbf{F} = m\mathbf{a} \tag{5.15}$$

to a freely falling object of mass m, with $\mathbf{a} = \mathbf{g}$ and $\mathbf{F} = \mathbf{F}_g$, we obtain

$$\mathbf{F}_g = m\mathbf{g}. \tag{5.16}$$

The block exerts this force to the supporting floor below it, which is stopping it from accelerating downwards. That is, the block exerts a downward force $\mathbf{F}_W$ on the supporting floor immediately beneath it. This force denotes the *weight* of the block. According to Newton's third law, the ground below the block exerts an upward reaction force $\mathbf{F}_R$ on the block. Therefore, based on Newton's third law:

$$\mathbf{F}_R = -\mathbf{F}_W \tag{5.17}$$

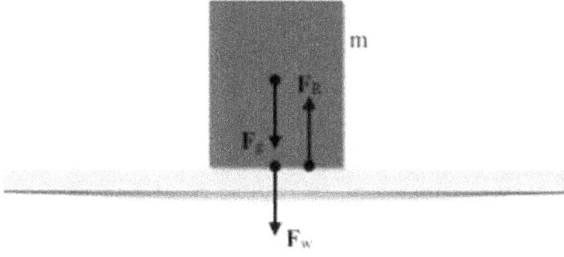

Figure 5.5: Weight.

and

$$|\mathbf{F}_R| = |\mathbf{F}_W|. \tag{5.18}$$

The resultant force acting on the block, which continues to remain at rest, is

$$\mathbf{F}_g + \mathbf{F}_R = 0. \tag{5.19}$$

Projecting along the vertical y-axis, which is oriented vertically up, we get

$$F_{gy} \equiv F_g = -mg \tag{5.20}$$
$$F_{Ry} = |\mathbf{F}_R|.$$

Substituting eq. (5.20) into eq. (5.19), we obtain

$$|\mathbf{F}_R| = mg. \tag{5.21}$$

Combining eq. (5.21), eq. (5.17) and eq. (5.18), we get

$$|\mathbf{F}_W| = mg \tag{5.22}$$
$$\mathbf{F}_W = m\mathbf{g}.$$

Equation (5.22) indicates that $\mathbf{F}_W = \mathbf{F}_g$, and that is why $|\mathbf{F}_g|$ is often called weight of the object.

Note that $\mathbf{F}_g$ is the gravitational force exerted on the block and directed toward the center of the Earth. Based on the third law of Newton, a reaction force is exerted at the center of the Earth. In other words, the Earth exerts an attractive force on the block, and the block exerts an attractive force on the Earth by an equal magnitude but opposite direction. However, since the Earth is much larger than the block, the force exerted by the block at the center of the Earth has no observable consequence.

So far, we have established that the weight F_W of a body is the magnitude of the downward force this body exerts on any supporting table such that $F_W = mg$ with m being the mass of the body and g being the local acceleration due to gravity.

Note that the weight as a force is measured in newtons. Furthermore, the weight of an object depends on its location, and therefore, it is not an intrinsic property of that object. For instance, a body weighing 10 N on the surface of the Earth will only have a weight of about 3.8 N on the surface of Mars, due to the weaker surface gravity of Mars relative to the Earth.

Consider a block of mass m at rest on the floor of an elevator, as shown in Fig. 5.6. We will assume that the elevator is accelerating upwards with acceleration **a**. Let us see how this acceleration affects the weight of the block. The block experiences a downward force $\mathbf{F}_g = m\mathbf{g}$ due to gravity. If F_W is the weight of the block, then it gives the magnitude of the downward force acting on the floor of the elevator by the block. Using Newton's third law, the floor of the elevator exerts an upward reaction force of magnitude F_R on the block such that eq. (5.17) and eq. (5.18) hold.

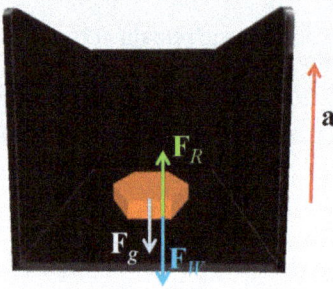

Figure 5.6: Mass m on the elevator.

Let us apply Newton's second law, eq. (5.7), for the motion of the block with mass m, and a upward acceleration of the block of **a**. Furthermore, there are two forces acting on the block: a downward force $\mathbf{F}_g$ due to gravity, and an upward reaction force $\mathbf{F}_R$. Hence,

$$\mathbf{F}_R + m\mathbf{g} = m\mathbf{a}. \tag{5.23}$$

The projection of eq. (5.23) along the vertical y-axis gives

$$F_R - mg = ma. \tag{5.24}$$

Equation (5.24) can be rearranged as

$$F_R = m(g + a). \tag{5.25}$$

Combining eq. (5.25) and eq. (5.18), we obtain

$$F_W = m(g + a). \tag{5.26}$$

Note that here a is an algebraic value, that is, it has a sign: $a < 0$ if **a** is downward and $a > 0$ if **a** is upward.

From eq. (5.26), the weight, F_W, will be greater than the actual weight (i. e., the weight $F_W = mg$ when $a = 0$) if $a > 0$. Hence, when the elevator has an upward acceleration, the weight F_W of the block increases. For example, when the elevator accelerates upward at $g = 9.80 \text{ m/s}^2$, then the weight of the block is doubled. Conversely, when the elevator accelerates downward (i. e., when a becomes negative), then the weight of the block is reduced. For example, when the elevator accelerates downward at $g/2$, then the weight of the block is halved, $F_W = mg/2$.

To measure these weight changes, we could have a mass scale placed on the elevator between the block and its floor.

When the downward acceleration of the elevator matches the acceleration due to gravity, i. e. $a = -g$, we get $F_W = 0$. In other words, the block becomes weightless.

Now, suppose that the accelerator has a downward acceleration that exceeds the acceleration due to gravity, i. e. $a < -g$. In this case, the block has a negative weight, $F_W < 0$, which means that the block flies off the floor of the elevator and slams into the ceiling.

5.7 Elastic forces

The elastic force is the force that arises from the deformation of a solid such as a spring or a rubber band. The elastic force is proportional to the deformation of the object through the following equation:

$$\mathbf{F} = -k\mathbf{x}. \tag{5.27}$$

This equation is known as Hooke's law. k is called the spring constant, which is a property of the spring, and it is a measure of the spring stiffness. The minus sign means that the direction of the pull (or push) of the spring is always opposite to that of the applied force.

5.8 Friction and dissipation forces

The frictional forces are *retarding forces*, which oppose the motion of an object. Mathematically this can be written as

$$f = \mu N \tag{5.28}$$

where μ is the friction coefficient, and it is a property of the material or the substance. It is a measure of the roughness of a given surface. For example, we expect that it is easier to move a piece of ice on a glass surface than to move the rock on a gravel road.

There are two types of friction, *static* friction and *kinetic* friction, which are described in the following.

5.8.1 Static friction

The static friction is the frictional force, which appears when two objects are on the verge of motion. The force is given by

$$f_s = \mu_s N \tag{5.29}$$

where μ_s is called the static friction coefficient, which also is a property of the material.

To measure μ_s consider the following simple experiment: Assume we have a block on a given surface, for example, a board. Then, we raise the end of the board slowly until the block is just about to slide (see Fig. 5.7). We measure the angle of the incline and take its tangent, and we will prove below that this is the static friction coefficient, μ_s.

The gravity force, $\mathbf{F}_g$ acting on the block is given as

$$\mathbf{F}_g = m\mathbf{g} \tag{5.30}$$

which has two components, $\mathbf{N}$ normal to the inclined plane, and $\mathbf{F}_p$ parallel to the inclined plane, as shown in Fig. 5.7. From the figure, we can write their magnitudes as

$$N = mg\cos\theta; \quad F_p = mg\sin\theta. \tag{5.31}$$

The other forces acting on the block are the board resistance force $\mathbf{F}_R$ (based on Newton's third law to balance the weight of the block) and the friction force $\mathbf{f}_s$. Then, the resultant force acting on the block is given by

$$\mathbf{F} = \mathbf{F}_g + \mathbf{F}_R + \mathbf{f}_s. \tag{5.32}$$

From Newton's second law, we have

$$\mathbf{F} = m\mathbf{a}. \tag{5.33}$$

Projection along the x-axis (see also Fig. 5.7) gives

$$-F_p + f_s = 0 \tag{5.34}$$

or

$$F_p = f_s = mg\sin\theta. \tag{5.35}$$

On the other hand, from the definition

$$f_s = \mu_s N = \mu_s mg\cos\theta. \tag{5.36}$$

Combining these last two equations, we get

$$\mu_s = \tan\theta. \tag{5.37}$$

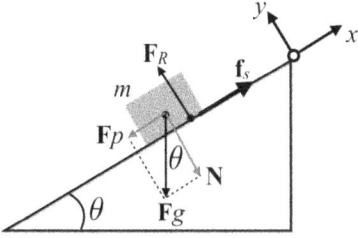

Figure 5.7: The forces acting on a block of mass m about to slide on a rough inclined plane.

5.8.2 Kinetic friction

So far, we have completely ignored any interaction between the object and the medium through which it moves, which can be either a liquid or a gas. Here, we will consider the effect of that medium in the motion of the objects. The medium exerts a resistance force $\mathbf{R}$ on the object moving through it. For example, the air exerts resistance to moving vehicles, which is called air drag, and the liquids exert the viscous forces on objects moving through them. The magnitude of $\mathbf{R}$ depends on factors such as the speed of the object, and the direction of $\mathbf{R}$ is always opposite the direction of motion of the object relative to the medium. The magnitude of $\mathbf{R}$ nearly always increases with increasing speed.

The magnitude of the resistance force can have various dependencies on speed:
1. The resistance force can be proportional to the speed of the moving object, which is valid for slowly falling objects through a liquid and for tiny objects, such as dust particles, moving through the air.
2. The resistance force can be proportional to the square of the speed of the moving object, which is valid for large objects, such as a skydiver moving through the air in free fall.

5.8.3 Case 1

We assume an object is moving through the liquid or gas. Furthermore, we assume that the resistance force acting on the object is proportional to the speed of the object. In that case, the magnitude of the resistance force can be written as follows:

$$R = \gamma v. \tag{5.38}$$

In eq. (5.38), v is the speed of the object, and γ is a constant, which depends on the properties of the medium, shape and dimensions of the object. For example, if the object is a sphere with radius r, then

$$\gamma \propto r. \tag{5.39}$$

Consider a small sphere of mass m released from rest in a liquid, as shown in Fig. 5.8.

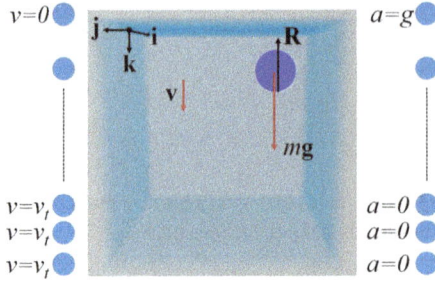

Figure 5.8: A small sphere falling through a liquid. **i, j,** and **k** are unit vectors along x-, y-, and z-axis, respectively.

If we assume that the resistance force, $\mathbf{R} = -\gamma v \mathbf{k}$, and the force of gravity, $\mathbf{F}_g = mg\mathbf{k}$, are the only forces acting on the sphere, applying Newton's second law, we get

$$\mathbf{R} + \mathbf{F}_g = m\mathbf{a}. \tag{5.40}$$

Projecting eq. (5.40) along the vertical axis and choosing the downward direction to be positive, we get

$$mg - \gamma v = ma = m\frac{dv}{dt} \tag{5.41}$$

where $\mathbf{a}$ has a downward direction. If we solve eq. (5.41) for dv/dt, we get

$$\frac{dv}{dt} = g - (\gamma/m)v. \tag{5.42}$$

Equation (5.42) represents a first order differential equation for the speed v, which is solved as follows. First, its homogenous form is solved, given as:

$$\frac{dv}{dt} = -\frac{\gamma}{m}v. \tag{5.43}$$

Integration is performed using the following steps:

$$\int \frac{dv}{v} = -\int \frac{\gamma}{m}dt, \tag{5.44}$$

$$\ln v(t) - \ln C_0 = -\frac{\gamma}{m}t,$$

$$v(t) = C_0 \exp\left(-\frac{\gamma}{m}t\right)$$

where C_0 is an integration constant. Next, we consider that the integration constant C_0 is a function of time t, and we write:

$$v(t) = C_0(t) \exp\left(-\frac{\gamma}{m}t\right). \tag{5.45}$$

Substituting eq. (5.45) into eq. (5.42), we obtain:

$$\frac{dC_0}{dt}\exp\left(-\frac{\gamma}{m}t\right) - C_0\frac{\gamma}{m}\exp\left(-\frac{\gamma}{m}t\right) = g - \frac{\gamma}{m}C_0\exp\left(-\frac{\gamma}{m}t\right), \tag{5.46}$$

$$\frac{dC_0}{dt} = g \exp\left(+\frac{\gamma}{m}t\right),$$

$$C_0(t) = \frac{mg}{\gamma}\exp\left(+\frac{\gamma}{m}t\right) + C$$

where C is another integration constant, which is obtained using the initial condition for the speed ($t = 0$). Substituting the last expression of eq. (5.46) into eq. (5.45), we get:

$$v(t) = \left[\frac{mg}{\gamma}\exp\left(+\frac{\gamma}{m}t\right) + C\right]\exp\left(-\frac{\gamma}{m}t\right). \tag{5.47}$$

Or,

$$v(t) = \frac{mg}{\gamma} + C\exp\left(-\frac{\gamma}{m}t\right). \tag{5.48}$$

For $v(0) = 0$, we find the integration constant is:

$$C = -\frac{mg}{\gamma}. \tag{5.49}$$

Substituting eq. (5.49) into eq. (5.48), we obtain the expression for the speed as:

$$v(t) = \frac{mg}{\gamma}(1 - e^{-\gamma t/m}). \tag{5.50}$$

Using eq. (5.50), the acceleration is $a = dv/dt$:

$$a = \frac{dv}{dt} = ge^{-\gamma t/m}. \tag{5.51}$$

As shown in Fig. 5.9, at $t = 0$, $v = 0$, so the resistance is $R = 0$, and the acceleration $dv/dt = g$. As t increases, R increases and the acceleration dv/dt decreases, and eventually the acceleration dv/dt becomes zero and resistance $R = mg$, which is equal to the weight of the sphere.

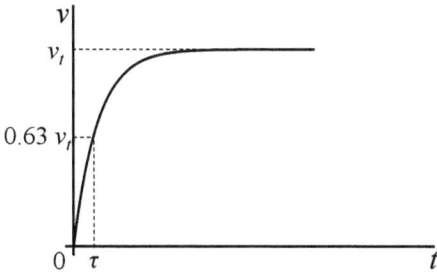

Figure 5.9: The speed as a function of time.

At this point, the speed of the particle becomes v_t, terminal speed, and after that, the sphere moves with a constant velocity of v_t.

As an example, we can consider the water droplets of the rain, which are created in the atmosphere, far from the surface of the Earth. As they approach the surface of Earth, their speeds increase, and when they are very close to Earth's surface, the speed of each water droplet reaches the plateau value, namely v_t. Therefore, the water droplets of the rain reach the surface of Earth with about a constant speed.

5.8.4 Case 2

In the second case, we consider objects moving at high speeds through the air, such as airplanes, skydivers, cars, and baseballs. In that case, the resistance force is approximately proportional to the square of the speed, and therefore, the magnitude of the resistance force can be expressed as

$$R = \frac{1}{2}D\rho A v^2. \tag{5.52}$$

In eq. (5.52), ρ denotes the density of air, A denotes the cross-sectional surface area of the falling object, which is measured in a plane perpendicular to the direction of its motion, and D denotes a dimensionless empirical quantity called the *drag coefficient*. The drag coefficient depends on the shape of objects; for instance, it has a value of about 0.5 for spherical objects and a value of about 2 for irregularly shaped objects.

In the following, we analyze the motion of an object in free fall subject to an upward air resistance force of magnitude R. The object has a mass m and it is released from the rest (i. e., the initial speed is $v_i = 0$), as in Fig. 5.10.

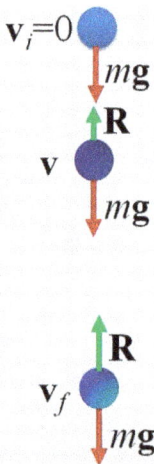

Figure 5.10: An object falling through air.

There are two external forces acting on the object: the downward force of gravity:

$$\mathbf{F}_g = m\mathbf{g} \tag{5.53}$$

and the upward resistance force $\mathbf{R}$ with magnitude given by eq. (5.52). In general, there is also an upward buoyant force, which in practice can be neglected. Hence, the resultant force is

$$\mathbf{F}_{\text{total}} = \mathbf{F}_g + \mathbf{R}. \tag{5.54}$$

Using Newton's second law,

$$\mathbf{F}_g + \mathbf{R} = m\mathbf{a}. \tag{5.55}$$

The projection of eq. (5.55) along the vertical axis with positive direction downward gives

$$mg - \frac{1}{2}D\rho Av^2 = ma \tag{5.56}$$

where the direction of the acceleration vector is downward. Solving eq. (5.56) for a, we get

$$a = g - \left(\frac{D\rho A}{2m}\right)v^2. \tag{5.57}$$

To calculate the terminal velocity, v_t, we consider that when v_t is reached, then $a = 0$, which is equivalent to (see also eq. (5.56)):

$$mg = \frac{1}{2}D\rho Av_t^2. \tag{5.58}$$

Solving eq. (5.58) for v_t, we get

$$v_t^2 = \frac{2mg}{D\rho A} \tag{5.59}$$

thus the speed v_t is

$$v_t = \sqrt{\frac{2mg}{D\rho A}}. \tag{5.60}$$

Equation (5.60) indicates that the terminal speed, v_t, as a function of the dimensions of the object can be determined. For example, for an object in the form of a sphere of radius r, we have $A = \pi r^2$ and $m = \frac{4}{3}\rho\pi r^3$. Therefore,

$$v_t \propto \sqrt{r}. \tag{5.61}$$

That result indicates, in contrast to what we have said for free-falling objects under the gravity force if we neglect the air resistance, that more massive objects reach the surface of Earth with greater speed.

5.9 Exercises

Exercise 5.1. Due to the force $\mathbf{F}$ applied to an object of mass m_1, the object gains an acceleration of $3.00\,\text{m/s}^2$. If we apply the same force to a second object of mass m_2, it gains an acceleration of $1.00\,\text{m/s}^2$. (a) Find the value of the ratio m_1/m_2. (b) If m_1 and m_2 are combined in one object with mass $m_1 + m_2$, find their acceleration under the action of the force $\mathbf{F}$.

Exercise 5.2. A force of $10.0\,\text{N}$ is applied on an object of mass $2.00\,\text{kg}$. Determine (a) the body's acceleration, (b) its weight in newton, and (c) its acceleration if the force is doubled.

Exercise 5.3. A body of mass $3.00\,\text{kg}$ undergoes an acceleration given by $\mathbf{a} = (2.00\mathbf{i} + 5.00\mathbf{j})\,\text{m/s}^2$. Find the resultant force $\sum \mathbf{F}$ and its magnitude.

Exercise 5.4. A train has a mass of 15000 tons. Assume the locomotive can pull with a force of $750000\,\text{N}$. Find the time it takes to increase the speed from 0 to $80.0\,\text{km/h}$.

Exercise 5.5. A bullet of $5.00\,\text{g}$ leaves a rifle with a speed of $320\,\text{m/s}$. The expanding gases behind it exert what force on the bullet while it is traveling down the barrel of the rifle, $0.820\,\text{m}$ long. Assume constant acceleration and negligible friction.

Exercise 5.6. A pitcher releases a baseball of weight $-F_g\mathbf{j}$ with a velocity $v\mathbf{i}$, after uniformly accelerating his arm for a time t. Assume the ball starts from the rest. Determine (a) the distance the ball accelerates before its release, and (b) the force the pitcher exerts on the ball.

Exercise 5.7. Assume that one pound is the weight of an object of mass $0.45359237\,\text{kg}$ at a place where the acceleration due to gravity is $32.174\,\text{ft/s}^2$. Express the pound as a quantity with a SI unit.

Exercise 5.8. An object with mass $4.00\,\text{kg}$ has a velocity of $3.00\mathbf{i}\,\text{m/s}$ at one instant. After 8 seconds, its velocity increases to $(8.00\mathbf{i} + 10.0\mathbf{j})\,\text{m/s}$. If the object is subject to a constant total force, find (a) the components of the force and (b) its magnitude.

Exercise 5.9. Consider a nitrogen molecule in air with an average speed about $6.70 \times 10^2\,\text{m/s}$, and mass of $4.68 \times 10^{-26}\,\text{kg}$. Assume it takes $3.00 \times 10^{-13}\,\text{s}$ for the molecule to hit a wall and rebound with the same speed but moving in the opposite direction. Find (a) the average acceleration of the molecule during this time interval, and (b) the average force the molecule exerts on the wall.

Exercise 5.10. An electron of mass $9.11 \times 10^{-31}\,\text{kg}$ has an initial speed of $3.00 \times 10^5\,\text{m/s}$ along a straight line. Assume that its speed increases to $7.00 \times 10^5\,\text{m/s}$ in a distance of $5.00\,\text{cm}$ moving with constant acceleration. (a) Determine the force acting on the electron and (b) compare this force with the weight of the electron, which we neglected.

Exercise 5.11. Suppose a person weighs 120 lb. Determine (a) the weight in newton and (b) the mass in kilogram.

Exercise 5.12. If someone weighs 900 N on the Earth, what would the weight be on Jupiter? The acceleration due to gravity in Jupiter is 25.9 m/s².

Exercise 5.13. Two forces F_1 and F_2 act on a body with mass 5.00 kg. If F_1 = 20.0 N and F_2 = 15.0 N, find the accelerations in (a) and (b) of Fig. 5.11.

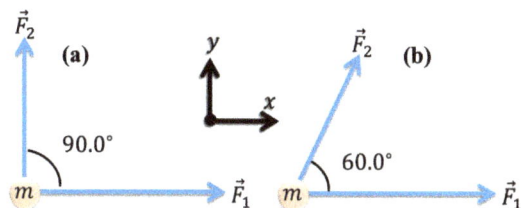

Figure 5.11: (a) Case 1 and (b) Case 2 graphs.

Exercise 5.14. Besides its weight, on an object with mass of 2.80 kg is applied a constant force. The object is initially at rest, and in 1.20 s experiences a displacement of $(4.20 \text{ m})\mathbf{i} - (3.30 \text{ m})\mathbf{j}$, where the direction of $\mathbf{j}$ is the upward vertical direction. Determine the other force.

Exercise 5.15. Three forces of 10.0 N north, 20.0 N east, and 15.0 N south are simultaneously applied to an object of mass 4.00 kg as it rests on an air table. Find the object's acceleration.

Exercise 5.16. Assume a boat moving through water with two horizontal forces acting on it. The first force is 2000 N forward push caused by the motor; the other is a constant 1800 N resistance force caused by the water. (a) What is the acceleration of the 1000 kg boat? (b) If the motion starts from rest, how far will it move in 10.0 s? (c) What will then be its speed?

Exercise 5.17. Three forces, given by F_1 = $(-2.00\mathbf{i} + 2.00\mathbf{j})$ N, F_2 = $(5.00\mathbf{i} - 3.00\mathbf{j})$ N, and F_3 = $(-45.00\mathbf{i})$ N, are applied on an object producing an acceleration of magnitude 3.75 m/s². Determine (a) the direction of the acceleration, (b) the mass of the object, (c) its speed after 10.0 s if the object is initially at rest, and (d) the velocity components of the object after 10.0 s.

Exercise 5.18. A body of mass 3.00 kg is moving in a plane, with its x and y coordinates given by $x = 5t^2 - 1$ and $y = 3t^3 + 2$, where x and y are in meter and t is in second. Find the magnitude of the net force acting on this mass at t = 2.00 s.

Exercise 5.19. The distance between the two telephone poles is 50.0 m. When a 1.00 kg bird lands on the telephone wire midway between the poles, the wire sags about

0.200 m. (a) Draw a free-body diagram of the bird. (b) Find the tension the bird produces in the wire. Ignore the weight of the wire.

Exercise 5.20. A bag of cement of weight F_g hangs from three wires as shown in Fig. 5.12. Two of the wires make angles θ_1 and θ_2 with the horizontal. If the system is in equilibrium, show that the tension in the left-hand wire is

$$T_1 = F_g \cos\theta_2 / \sin(\theta_1 + \theta_2). \tag{5.62}$$

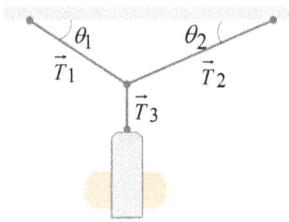

Figure 5.12: A graphical representation of the forces acting upon system.

Exercise 5.21. An object is resting on a spring with $k = 1.0 \times 10^3$ N/m. If the spring is compressed by $x = 2.0$ cm, calculate the mass of the object.

Exercise 5.22. Consider a small sphere of mass 2.00 g, which is released from rest in a large vessel filled with oil. Assume the sphere experiences a resistance force proportional to its speed, and it reaches the terminal speed of 5.00 cm/s. Determine the time constant τ and the time it takes for the sphere to reach 90 % of its terminal speed.

6 Circular motion and other applications of Newton's laws

In Chapter 5, we introduced Newton's laws. In this chapter, we will discuss some applications of Newton's laws, including circular motion.

6.1 Newton's second law applied to uniform circular motion

Centripetal acceleration

As discussed in Chapter 5, a particle moving with a constant speed v in a circular path of radius r experiences an acceleration $\mathbf{a}_r$ with magnitude

$$a_r = \frac{v^2}{r}. \tag{6.1}$$

This acceleration is called the *centripetal acceleration* because $\mathbf{a}_r$ is directed toward the center of the circle, and it is always perpendicular to $\mathbf{v}$ (see Chapter 5).

To illustrate the problem, consider a sphere of mass m that is tied to a string of length r that is being rotated at a constant speed in a horizontal circular path, as shown in Fig. 6.1. A low-friction table supports its weight.

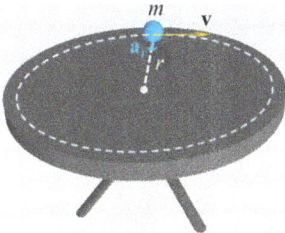

Figure 6.1: A sphere moving in a circular path.

Next, we can explain that the sphere moves in a circle because of its inertia; the sphere will tend to move in a straight line; however, the string prevents motion along a straight line by exerting on the sphere a force that makes it move in a circular orbit. This force is directed along the string toward the center of the circle, as shown in Fig. 6.2. In general, that force can be any of the known forces causing an object to follow a circular path. If we apply Newton's second law, we write

$$\sum_i \mathbf{F}_i = m\mathbf{a}. \tag{6.2}$$

Projecting eq. (6.2) along the radial direction, we find that the value of the net force causing the centripetal acceleration can be evaluated:

$$\sum_i F_{ri} = ma_r = m\frac{v^2}{r}. \tag{6.3}$$

https://doi.org/10.1515/9783110755824-006

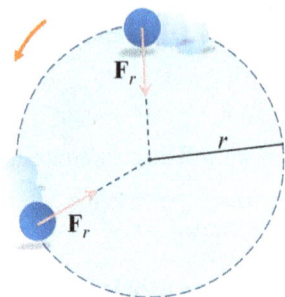

Figure 6.2: Forces acting on a sphere moving in a circular path.

The force causing the centripetal acceleration of a particle moving in a circular path acts toward the center and causes a change in the direction of the velocity vector. If that force vanishes, the object would no longer move in its circular path; instead, it would move along a straight line path tangent to the circle, as illustrated in Fig. 6.3. If, at some instant, the string breaks at some point in the circle, the object moves along the straight line that is tangent to that point.

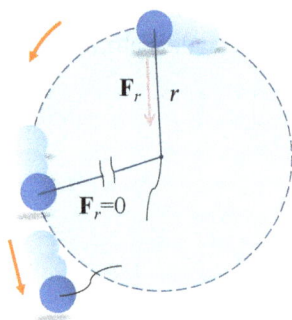

Figure 6.3: A sphere moving in a circular path.

6.2 Non-uniform circular motion

In Chapter 5, we mentioned that if a particle moves with varying speed in a circular path, there is, in addition to the centripetal (radial) component of acceleration, a tangential component having magnitude dv/dt. Therefore, the force acting on the particle must also have a tangential and a radial component. The total acceleration is

$$\mathbf{a} = \mathbf{a}_r + \mathbf{a}_t. \tag{6.4}$$

The total force exerted on the particle is

$$\mathbf{F} = \mathbf{F}_r + \mathbf{F}_t. \tag{6.5}$$

This is illustrated in Fig. 6.4.

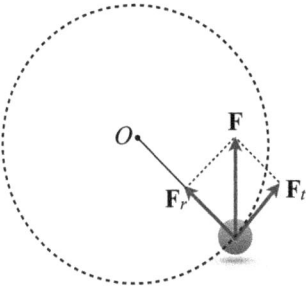

Figure 6.4: Nonuniform circular motion.

The vector $\mathbf{F}_r$ is directed toward the center of the circle and is responsible for the centripetal acceleration. The vector $\mathbf{F}_t$, which is tangent to the circle, is responsible for gaining a tangential acceleration. That tangential acceleration gives the change in the speed of the particle with time.

6.3 Motion in accelerated frames

Newton's laws of motion are valid only when observations are made in an inertial frame of reference. Here, we analyze how an observer in a non-inertial frame of reference (one that is accelerating) applies Newton's second law.

Consider a car traveling along a highway at high speed and approaching a curved exit ramp, as shown in Fig. 6.5. When the car turns left onto the ramp, a person sitting in the passenger seat slides to the right and hits the door. At that point, the force exerted on the person by the door keeps the person from being ejected from the car. The question we may want to ask is: What causes this person to move toward the door? The answer is that a force, often called the "centrifugal" force, acts upon the person. The passenger invents this fictitious force to explain what is going on in the accelerated frame of reference, as shown in Fig. 6.5.

To explain the phenomenon correctly: Before the car enters the ramp, the vehicle and passenger are moving in a straight-line path. As the car comes to the ramp and travels a curved path, the passenger tends to move along the original straight-line path. This is in agreement with Newton's first law: The natural tendency of a body is to continue moving in a straight line. However, if a sufficiently large force (toward the center of curvature) acts on the passenger, as shown in Fig. 6.5, the person will move in a curved path along with the car. The origin of this force is the force of friction between the person and the seat of the car. If that frictional force is not large enough, the person will slide to the right as the vehicle turns to the left under the passenger. Eventually, the person encounters the door, which provides a force large enough to make the passenger follow the curved path taken by the car. The passenger slides toward the door not because of some mysterious outward force but because the force

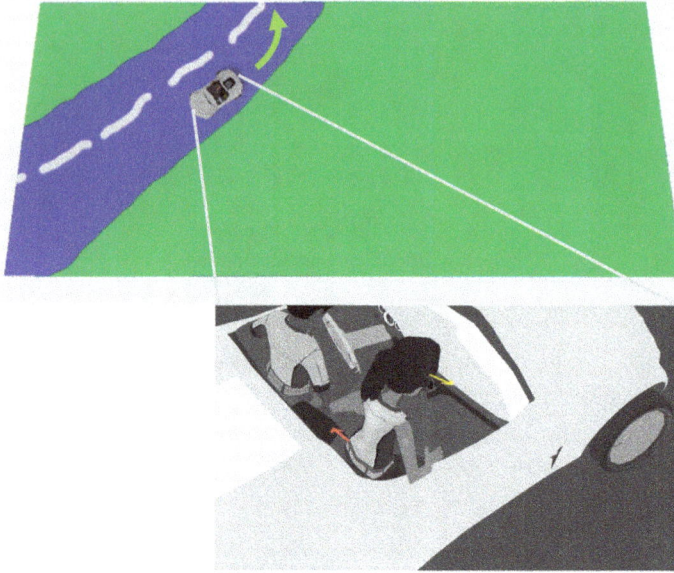

Figure 6.5: Motion of a car entering a ramp by rotating to the left, indicated by the green curved arrow. The red arrow indicates the direction of the friction force between the sit and the passenger, and the yellow arrow indicates the direction of the centrifugal (fictitious) force.

of friction is not sufficiently high to allow the person to travel along the circular path followed by the car.

In general, for a particle moving with an acceleration $\mathbf{a}$, relative to an observer at the origin of an inertial frame, we can use Newton's second law of motion,

$$\sum \mathbf{F} = m\mathbf{a}. \tag{6.6}$$

If another observer is in an accelerated frame and tries to apply Newton's second law to the motion of the particle, the fictitious forces have to be introduced to make Newton's second law work. These forces, "invented" by the observer in the accelerating frame, appear to be real. However, we emphasize that these fictitious forces do not exist when the motion is observed in an inertial frame.

For example, in the frame related to the particle, which is accelerated with $\mathbf{a}$, the second law of Newton given by eq. (6.6) can be written as

$$\sum \mathbf{F} - m\mathbf{a} = 0 \tag{6.7}$$

or

$$\sum \mathbf{F} + \mathbf{F}_{\text{fictitious}} = 0. \tag{6.8}$$

In eq. (6.8), $\mathbf{F}_{\text{fictitious}}$ is

$$\mathbf{F}_{\text{fictitious}} = -m\mathbf{a}. \tag{6.9}$$

6.4 Exercises

Exercise 6.1. Assume a ball of mass 0.500 kg is attached to the end of a cord 1.50 m long. The ball is whirled in a horizontal circle, as shown in Fig. 6.6. Suppose the cord can withstand a maximum tension of 50.0 N. Find the maximum speed the ball can attain before the cord breaks. Assume that the string remains horizontal during the motion.

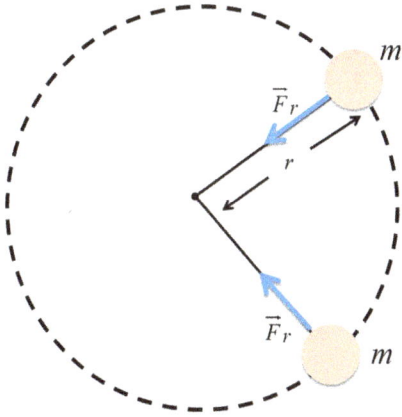

Figure 6.6: A graphical representation of the motion of a mass *m* in a circular horizontal plane.

Exercise 6.2. A sphere of mass 0.500 kg is attached to the end of a cord 1.50 m long. The sphere is moving in a horizontal circle, as shown in Fig. 6.6.

(a) Suppose the cord can withstand a maximum tension of 50.0 N. Determine the maximum speed the ball can attain before the cord breaks. Assume that the string remains horizontal during the motion.

(b) What is the tension in the cord if the speed of the sphere is 5.00 m/s?

Exercise 6.3. The conical pendulum: Assume a small object of mass m is suspended from a string of length L. The object moves with constant speed v in a horizontal circle of radius r, as shown in Fig. 6.7. The system is also known as the conical pendulum. Determine an expression for v.

Exercise 6.4. Assume a sphere of mass m is attached to the end of a cord of length R. The sphere rotates in a vertical circle about a fixed point of O, as illustrated in Fig. 6.8. What is the tension in the cord at any instant? Assume that the speed of the sphere is v and the cord makes an angle θ with the vertical.

Exercise 6.5. The orbital velocity of Mars is 24.1 km/s. If the average distance of Mars from the Sun is 228000000 km and its mass is 0.11 that of the Earth, find the centripetal force experienced by Mars.

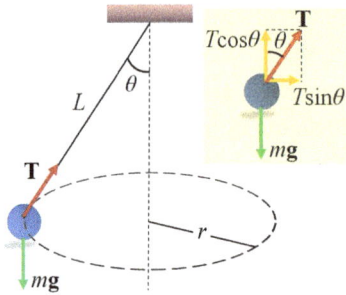

Figure 6.7: The conical pendulum and the free body diagram of forces.

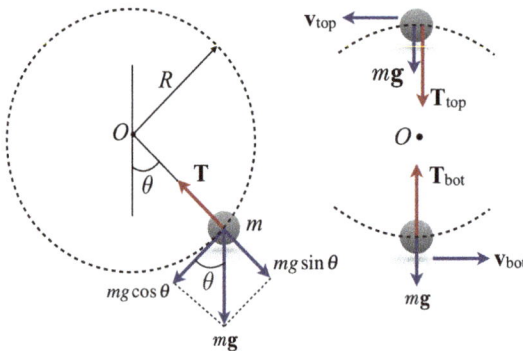

Figure 6.8: (a) All forces applied on the sphere of mass m attached to the end of the cord of length R and rotating in a vertical circle centered at O. (b) Forces acting on the sphere at the top and bottom of the circle. The tension is a maximum at the bottom and a minimum at the top.

Exercise 6.6. The force causing centripetal acceleration is also called a *centripetal force*. We can mention a variety of forces in nature—friction, gravity, normal forces, and tension. Should we add centripetal force to this list?

Exercise 6.7. A mass m is attached to one end of a string of length 1.0 m. It swings in a vertical circle under the gravitational force, as shown in Fig. 6.9. When the string makes the angle $\phi = 30°$ with the vertical direction, the mass has a speed of 2.0 m/s. Find the magnitudes of the radial and tangential components, and the total acceleration at this instant.

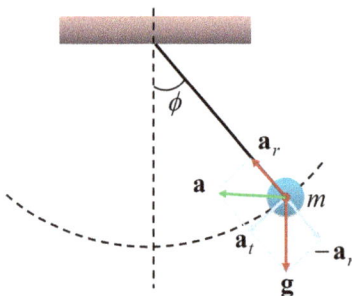

Figure 6.9: The oscillatory motion of the mass m under the gravitational force with acceleration **g**.

Exercise 6.8. The orbit of the Moon about the Earth is approximately circular, with a mean radius of 3.8×10^5 km. The Moon completes one full rotation around the Earth in 27.3 days. What are the mean orbital speed and centripetal acceleration of the Moon.

Exercise 6.9. An athlete rotates a disc of mass 1.00 kg along a circular path of radius 1.00 m. The maximum speed of the disc is 20.0 m/s. Find the magnitude of the maximum radial acceleration of the disc.

Exercise 6.10. Consider a tire with a radius of 0.500 m that rotates at a constant rate of 200 rev/min. What are the speed and acceleration of a small stone lodged in the tread of the tire?

7 Work and kinetic energy

We learned everything about Newtonian physics, such as kinematics in which the motion itself is investigated and dynamics in which the reasons and effects of the movement are investigated. Next, we are going to learn what is derived from what we have seen so far. However, the results obtained here are more refined and possibly will need a deeper effort for conceptual understanding.

7.1 Work done by a constant force

Physical quantities, such as velocity, acceleration, and force, have an almost similar meaning in physics to their meaning in real life. Now, however, we encounter a term whose purpose in physics is distinctly different from its ordinary meaning. That new term is *work*.

Let us examine the situation in Fig. 7.1, where an object under the influence of a constant force **F** undergoes a displacement **d** along a straight line. We suppose that the force **F** makes an angle θ with **d**.

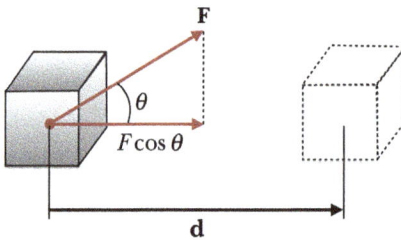

Figure 7.1: An object under the force **F**.

Definition 7.1 (Work done by a constant force). The work W done on some object by an external constant force exerted on the object equals the product of the projection of the force in the direction of the displacement and the magnitude of the displacement:

$$W = Fd\cos\theta. \tag{7.1}$$

Note from eq. (7.1) that the work done by an external force acting on an object that is moving vanish when the force applied is perpendicular to the object's displacement. In other words, if $\theta = 90°$, then $W = 0$ because

$$\cos 90° = 0. \tag{7.2}$$

For example, in Fig. 7.2, the work done by the normal force acting on the object and the work done by the gravity force on the object are both zeroes because both forces are

https://doi.org/10.1515/9783110755824-007

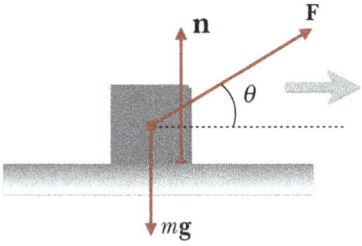

Figure 7.2: An object under the force of gravity.

perpendicular to the displacement and hence have zero components in the direction of **d**.

Besides, the sign of the work depends on the direction of **F** relative to **d**, namely the work done by the applied force is positive when the vector associated with the component $F\cos\theta$ is along the direction of the displacement vector. For example, if an object is lifted, the work done by the applied force is positive, $W > 0$, because the direction of that force is upward, and thus in the same direction as the displacement.

If the vector associated with the component $F\cos\theta$ is in the opposite direction with the displacement vector, then $W < 0$. For example, when an object is lifted, the work done by the gravitational force exerted on the object is negative, $W < 0$, because the gravity force is downward and the displacement vector is upward (that is, $\theta = 180°$, and hence $\cos\theta = -1$). The factor $\cos\theta$ in the definition of W determines its sign. Note that work is equivalent to the energy transfer. Furthermore, if energy is transferred to the system, then W is positive, and if energy is transferred from the system, then W is negative.

If an applied force **F** acts along the direction of the displacement **d**, then $\theta = 0$ and $\cos 0° = 1$. In this case, from eq. (7.1), we get

$$W = Fd. \tag{7.3}$$

Work represents a scalar algebraic quantity with its units defined as a force multiplied by the length; therefore, in the SI unit, the work is defined as the newton·meter (N·m). This unit has a name of its own: the joule (J):

$$1\,J = 1\,N \cdot m. \tag{7.4}$$

In general, an object may have either a constant or a varying velocity moving under the influence of more than one force. Therefore, because work is a scalar algebraic quantity, the total work on the object, which undergoes a displacement, equals the algebraic sum of the amounts of work done by each of the forces exerted on the object:

$$W_{total} = \sum_i W_i = \left(\sum_i F_i\right) d\cos\theta. \tag{7.5}$$

The relationship of the force and displacement vectors in eq. (7.1) implies the use of a more convenient mathematical formula, namely the scalar product. That allows us to indicate how **F** and **d** are related from the perspective of how close to being parallel they are. We write

$$W = \mathbf{F} \cdot \mathbf{d}. \tag{7.6}$$

In general, a particle is be moving under the influence of several forces, the total work done as the particle undergoes some displacement is

$$W_{\text{total}} = \sum_i W_i = \left(\sum_i \mathbf{F}_i\right) \cdot \mathbf{d}. \tag{7.7}$$

7.2 The work done by non-constant forces

Above, we discussed the case of the constant force. However, forces are not always constant. For example, when a particle is displaced along the x-axis under the influence of a non-constant force from $x = x_i$ to $x = x_f$. In such a case, we cannot use $W = Fd \cos\theta$ to estimate the work done by the force, F, since this formula is valid only for F being constant in both its magnitude and its direction.

However, if we consider an infinitesimally small displacement of the particle Δx, as shown in Fig. 7.3, such that the x component of the force F_x remains approximately constant over all this interval, the work done by that force is

$$\Delta W = F_x \Delta x. \tag{7.8}$$

Note that ΔW represents just the area of the shaded rectangle in Fig. 7.3. If we consider that F_x as a function of x is partitioned into a large countable number of such small intervals, then the total net work done for a displacement from x_i to x_f equals, approximately, the following sum of terms:

$$W \approx \sum_{x_i}^{x_f} F_x \Delta x. \tag{7.9}$$

If the displacement Δx approaches 0, then the sum goes to the integral

$$\lim_{\Delta x \to 0} \sum_{x_i}^{x_f} F_x \Delta x = \int_{x_i}^{x_f} F_x dx. \tag{7.10}$$

Therefore, the work done by the force F_x as the particle moves from x_i to x_f is

$$W = \int_{x_i}^{x_f} F_x dx. \tag{7.11}$$

Equation (7.11) indicates that the total work done by the non-constant force $F_x(x)$ equals the area under the curve from x_i to x_f.

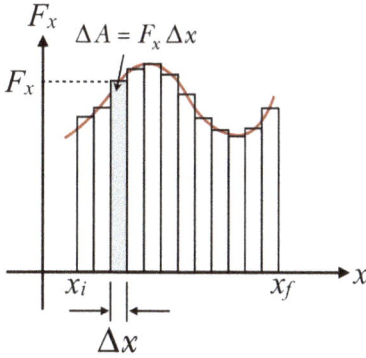

Figure 7.3: The work done by the force component F_x for the small displacement Δx is $F_x\Delta x$, which is equal to the area of the shaded rectangle. The net work done for a displacement from x_i to x_f can be approximated by the sum of the areas of all the rectangles.

7.3 Work done by a spring

An often seen physical system for which the force varies with the position is shown in Fig. 7.4. This system is composed of a block of mass m on a horizontal, frictionless surface, which is attached to a spring (see Fig. 7.4). We assume that the spring is either stretched or compressed by small displacement from its equilibrium position. Then the spring exerts on the mass a force with a magnitude given as

$$F_s = -kx. \tag{7.12}$$

In eq. (7.12), x is the displacement of the block from its equilibrium position (i. e., $x = 0$) and k denotes the so-called the force constant of the spring such that $k > 0$. That is, the force necessary to pull or compress a spring is proportional to the amount of the displacement of x from the equilibrium position (unstretched spring). That is also known as Hooke's law. Note that this law is only valid in the limiting case of small displacements from the equilibrium position.

The value of k measures of the stiffness of the spring; large values of k indicate a stiff spring, and small values of k indicate a soft spring. The minus sign in eq. (7.12) suggests the direction of the applied force on the block, which is always the opposite of the displacement of x.

Suppose that the block moves to the left by a distance x_{max} from equilibrium position and then it is released (i. e., $v = 0$). The work, W_s, done by the force F_s on the block as it moves from initial position $x_i = -x_{max}$ to $x_f = 0$ is

$$W_s = \int_{x_i}^{x_f} F_s dx = \int_{-x_{max}}^{0} (-kx)dx = \frac{1}{2}kx_{max}^2 \tag{7.13}$$

where this definition is used

$$\int x^m dx = \frac{x^{m+1}}{m+1} \tag{7.14}$$

with $m = 1$.

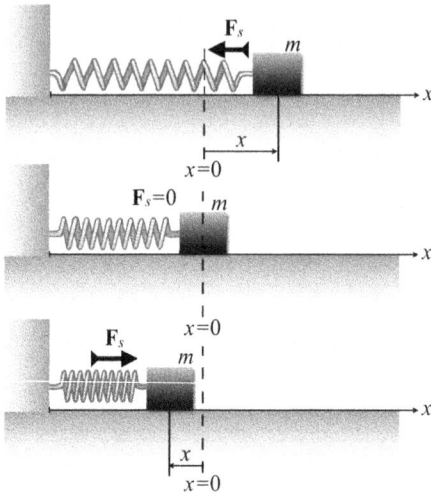

Figure 7.4: The force exerted by a spring on a block varies with the block's displacement x from the equilibrium position $x = 0$.

The force does the work F_s, which is positive in this case since the force F_s and the displacement x are in the same direction (both are to the right).

Now, we can consider the work done by the spring force, F_s, as the block displaces from its initial position $x_i = 0$ to $x_f = x_{max}$:

$$W_s = \int_{x_i}^{x_f} F_s dx = \int_0^{x_{max}} (-kx)dx = -\frac{1}{2}kx_{max}^2. \tag{7.15}$$

Note that this work is now negative because for this displacement, the spring force is in the opposite direction with displacement (i.e., the force is to the left and the displacement to the right).

The net work done by the spring force as the block moves from $x_i = -x_{max}$ to $x_f = x_{max}$ is

$$W_s = \int_{x_i}^{x_f} F_s dx = \int_{-x_{max}}^{x_{max}} (-kx)dx = 0. \tag{7.16}$$

If the block undergoes any arbitrary displacement from some initial position $x = x_i$ to a final position $x = x_f$, the work done by the spring force is

$$W_s = \int_{x_i}^{x_f} F_s dx = \int_{x_i}^{x_f} (-kx)dx = \frac{1}{2}kx_i^2 - \frac{1}{2}kx_f^2. \tag{7.17}$$

Next, we will consider the work done on the spring by some external agent as the spring stretches very slowly from its initial position $x_i = 0$ to the final position $x_f =$

x_{max}, as shown in Fig. 7.5. The work done by the external force $\mathbf{F}_p = -\mathbf{F}_s = -(-kx) = kx$ for any value of the displacement x equals

$$W_p = \int_{x_i}^{x_f} F_p dx = \int_0^{x_{max}} kx dx = \frac{1}{2}kx_{max}^2. \tag{7.18}$$

It can be seen that the external force F_p does the work W_p, which is equal to minus the work is done by the spring force, F_s, for the same displacement:

$$W_p = -W_s. \tag{7.19}$$

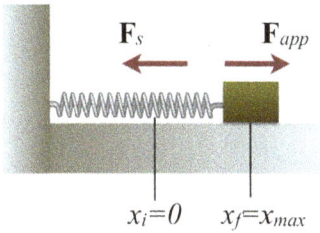

$x_i=0$ $x_f=x_{max}$

Figure 7.5: A block being pulled from $x_i = 0$ to $x_f = x_{max}$ on a frictionless surface by a force $\mathbf{F}_p$.

7.4 Work–energy relation

Consider a particle of mass m moving to the right under the action of a constant net force $\sum_i \mathbf{F}_i$ (see Fig. 7.6). Since the force is constant, using Newton's second law for a particle moving with a constant acceleration $\mathbf{a}$ and displaced at a distance d, we have

$$\sum_i \mathbf{F}_i = m\mathbf{a}. \tag{7.20}$$

Projecting eq. (7.20) along the direction of the displacement (see also Fig. 7.6), we obtain

$$\sum_i F_i = ma. \tag{7.21}$$

Then the net work done by the resultant force is

$$\sum_i W_i = \left(\sum_i F_i\right) d \cos\theta = \left(\sum_i F_i\right) d = (ma)d \tag{7.22}$$

since $\theta = 0°$.

Furthermore, we know that

$$v_f^2 - v_i^2 = 2ad. \tag{7.23}$$

From eq. (7.23), we get

$$ad = \frac{v_f^2 - v_i^2}{2}.$$ (7.24)

Substituting eq. (7.24) into the expression for the work (eq. (7.22)), we get

$$\sum_i W_i = \frac{1}{2}mv_f^2 - \frac{1}{2}mv_i^2$$ (7.25)

where

$$K = \frac{1}{2}mv^2$$ (7.26)

represents the energy associated with motion of a particle and it is called *kinetic energy*.

The work–kinetic energy theorem
Therefore, we obtain

$$\sum_i W_i = K_f - K_i = \Delta K.$$ (7.27)

This result is known as the *work–kinetic energy theorem*.

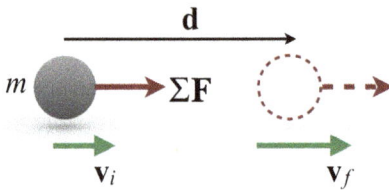

Figure 7.6: A particle undergoing a displacement.

7.5 Power

The *power* is the rate of the work done on an object. From a mechanical systems point of view, the efficiency of the engines, for example, is not characterized by the quantity of the work they can do, but rather by the rate at which they can perform the work.

Definition 7.2 (Power). By definition, the quantity representing the rate at which work is done is power. Thus, the average power of a system doing an amount of work, ΔW, over some time, Δt, is given by

$$\bar{P} = \frac{\Delta W}{\Delta t}.$$ (7.28)

Note that eq. (7.28) determines the average power during an interval of time Δt, and hence it is not the instantaneous power. Besides, the work, ΔW, increases with displacement Δx, even if the applied force is constant; therefore, the work done by an applied force increases with the displacement Δx, too. That is, the power does not remain constant. The instantaneous power can be found after some mathematics as we now show.

The instantaneous power is calculated for the time intervals of the work done that are infinitesimally small; then we should differentiate the work with respect to time:

$$P = \lim_{\Delta t \to 0} \frac{\Delta W}{\Delta t} = \frac{dW}{dt} \tag{7.29}$$

where

$$dW = \mathbf{F} \cdot d\mathbf{s}. \tag{7.30}$$

Thus

$$P = \frac{dW}{dt} = \mathbf{F} \cdot \frac{d\mathbf{s}}{dt} = \mathbf{F} \cdot \mathbf{v} \tag{7.31}$$

where we have used the relation

$$\frac{d\mathbf{s}}{dt} = \mathbf{v}. \tag{7.32}$$

In the SI, the unit of power is joules per second (J/s), also called the watt (W) after the inventor James Watt:

$$1\,W = 1\,J/s = 1\,kg \cdot m^2/s^3. \tag{7.33}$$

7.6 Exercises

Exercise 7.1. A person cleaning a floor pulls a vacuum cleaner with force having a magnitude $F = 50.0$ N. The force is along the direction forming an angle of $30.0°$ with the horizontal direction (see Fig. 7.7). Find the work done by $\mathbf{F}$ on the vacuum cleaner as it displaces 3.00 m to the right.

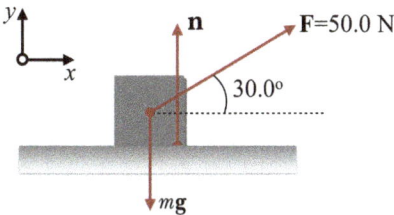

Figure 7.7: Free-body diagram of the forces acting on the vacuum cleaner being pulled at an angle of $30.0°$ with the horizontal.

Exercise 7.2. Assume a shopper in a supermarket who pushes a cart with a force of 35.0 N directed at an angle of 25.0° with the horizontal line. Find the work done by the shopper as cart moves down an aisle about 50.0 m.

Exercise 7.3. Assume a force acting on a particle varies with x (see Fig. 7.8). What is the work done by the force as the particle moves from $x = 0$ to $x = 6.0$ m?

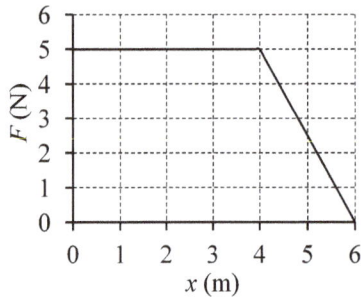

Figure 7.8: Force (in newton) versus x (in meters) plot.

Exercise 7.4. A block of mass 6.0 kg is initially at rest. Then it is pulled to the right along a horizontal, frictionless surface by a constant horizontal force of 12 N (see Fig. 7.9). Determine the speed of the block if it has moved 3.0 m.

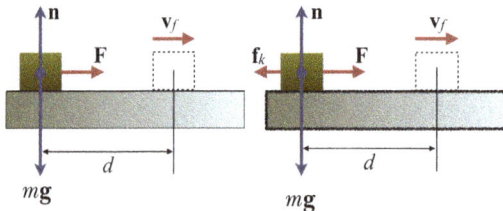

Figure 7.9: The motion of the block.

Exercise 7.5. Consider again a block of mass 6.0 kg initially at rest. The block is pulled to the right along a horizontal surface by a constant horizontal force of 12 N. Determine the speed of the block after it is displaced to the right by 3.0 m, assuming that the surface is not frictionless, but instead has a coefficient of kinetic friction of 0.15.

Exercise 7.6. Consider a block of mass 1.6 kg attached to the end of a horizontal spring with force constant of 1.0×10^3 N/m, as shown in Fig. 7.10. The spring is first compressed 2.0 cm, and then it is released from rest. Find the speed of the block at the equilibrium position $x = 0$. Assume that the surface is frictionless.

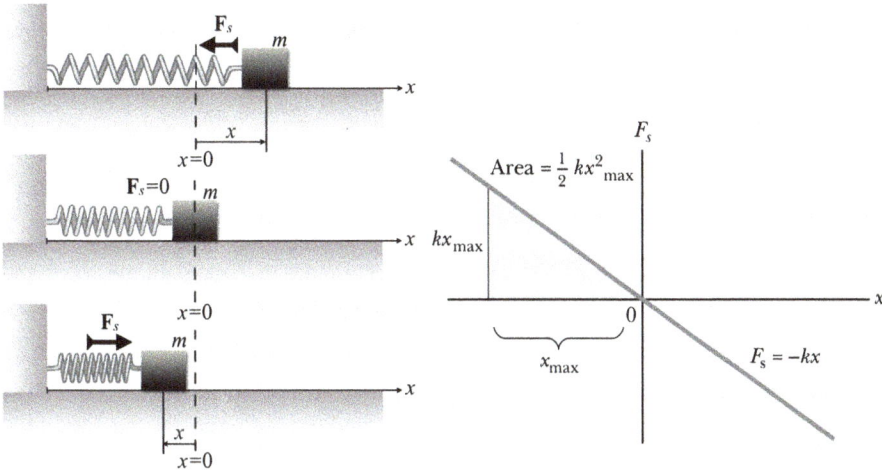

Figure 7.10: The motion of the block attached to a spring.

Exercise 7.7. A car of mass m is accelerating uphill (see also Fig. 7.11). The magnitude of the total resistance force is

$$f_t = (218 + 0.70v^2)\,\text{N}$$

with v the speed in meters per second. What is the power the engine must deliver to the wheels as a function of speed?

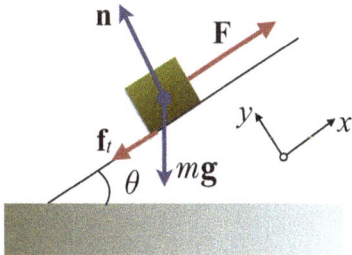

Figure 7.11: The motion of the car uphill.

Exercise 7.8. Consider an elevator car that has a mass of 1000 kg carrying passengers having a combined mass of 800 kg. Besides, there is a constant frictional force of 4000 N, which retards its motion upward. Find the minimum power delivered by the motor to lift the elevator car at a constant speed of 3.00 m/s.

Exercise 7.9. Find the potential energy of a block of mass 100 kg moved over 10.0 m along a 30° inclined plane.

Exercise 7.10. Find the elastic potential energy stored in a spring with a spring constant $k = 800\,\text{N/m}$, which is stretched by 10.0 cm.

Exercise 7.11. A spring gun has a spring constant $k = 20.0 \times 10^3\,\text{N/m}$. Suppose the spring is compressed by 10.0 cm. If a steel ball with a $m = 20.0\,\text{g}$ is then dropped into the barrel, calculate the velocity of the steel ball as it exits the barrel. Neglect all frictions.

Exercise 7.12. Suppose a person is holding a block of mass m at the height h above the ground by applying the external force $\mathbf{F}$, as shown in Fig. 7.12. Determine (a) the work done by the applied force $\mathbf{F}$ and (b) the work done by the gravity force $m\mathbf{g}$. Neglect all frictions.

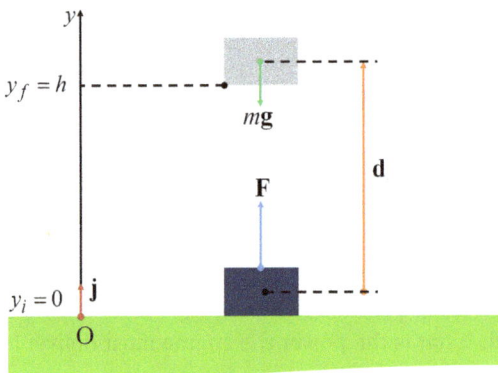

Figure 7.12: A person holding a block at the height h above the ground by applying a force **F**.

Exercise 7.13. Suppose a person is pulling a block of mass m along an inclined plane for a distance d along the plane and it is holding it at the height h above the ground by applying the external force $\mathbf{F}$, as shown in Fig. 7.13. Determine (a) the work done by the gravity force $m\mathbf{g}$ and (b) the work done by the applied force $\mathbf{F}$. Neglect all frictions.

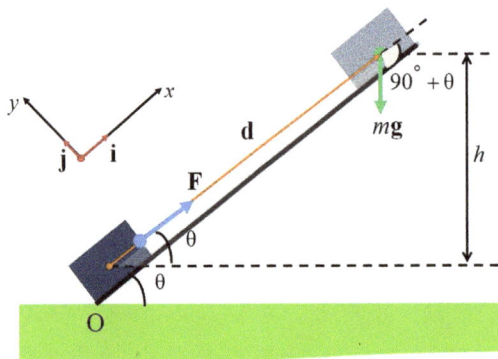

Figure 7.13: A person holding a block at the height h above the ground by applying a force **F** after the block is pulled along the inclined plane for a distance d.

8 Potential energy and conservation of energy

In this chapter, we will discuss the potential energy and the total energy or mechanical energy of the system. Also, we will introduce the conservation law of total energy.

If the quantity of an entity does not change with time, it means that the quantity is "conserved". The amount of that entity remains constant, even if its change between the two points is different. The best way to explain that quantity is energy. If the energy is conserved in a system, then it is known that the total amount of this energy remains constant, even if its form changes.

From Chapter 7, energy is the capacity to do work. The work is a kind of energy. The work is related to the force that is applied to a body to change its position. So, there is a relationship between energy and the force.

8.1 Potential energy

Every object that has some kinetic energy can do work on other objects, such as a hammer moving to drive a nail into the wall. Another form of energy is the so-called *potential energy U*, which, on the other hand, is the energy associated with a system of objects.

Definition of a system
Before we describe specific forms of potential energy, we must first define a system, which consists of two or more objects that exert forces on one another. If the arrangement of the system changes, then the potential energy of the system changes. Let us consider two dimensionless objects that exert forces on each other. The work done by the force acting on one of the objects causes a transformation of energy between the object's kinetic energy and other forms of the system's energy.

8.1.1 Gravitational potential energy

When the object, for example, a ball, falls toward the Earth, the Earth exerts a gravitational force, namely $\mathbf{F}_g = m\mathbf{g}$, on the ball. The direction of the force $\mathbf{F}_g$ is along the direction of the displacement of the object. Due to the work of the gravitational force on the object, the kinetic energy of the object increases.

Definition 8.1 (Gravitational potential energy). By definition, the gravitational potential energy is called the product of the magnitude of the gravitational force (i. e., mg) acting on the object with the height y of the object from the surface of the Earth. Usually, the gravitational potential energy is denoted by U_g, and mathematically it is given

https://doi.org/10.1515/9783110755824-008

by

$$U_g = mgy. \tag{8.1}$$

The system for the gravitational potential energy is the object and Earth. This potential energy transforms into the kinetic energy of the system by the gravitational force.

To relate the work done on any object by the gravitational force with the gravitational potential energy of the object–Earth system, we will consider a brick of mass m, initially at the height y_i above the surface (see also Fig. 8.1). If we neglect air resistance, then the gravitational force is the only force doing work on the brick as it falls. The gravitational force exerted on the brick is given as

$$\mathbf{F}_g = -mg\mathbf{j} \tag{8.2}$$

where $\mathbf{j}$ is a unit vector along the positive y-axis (that is, it is upward), and the minus sign indicates that the gravitational force is downward. Therefore, the work W_g done by it as the brick undergoes a downward displacement $\mathbf{d} = -(y_i - y_f)\mathbf{j}$ is

$$W_g = (m\mathbf{g}) \cdot \mathbf{d} \tag{8.3}$$
$$= (-mg\mathbf{j}) \cdot [-(y_i - y_f)\mathbf{j}]$$
$$= mgy_i - mgy_f$$

where we have used

$$\mathbf{j} \cdot \mathbf{j} = 1. \tag{8.4}$$

Using eq. (8.1), we obtain

$$W_g = U_{g,i} - U_{g,f} = -(U_{g,f} - U_{g,i}) = -\Delta U_g. \tag{8.5}$$

That indicates that the work done on any object by the gravitational force equals minus the change in the system's gravitational potential energy, or decrease in gravitational potential energy of the system.

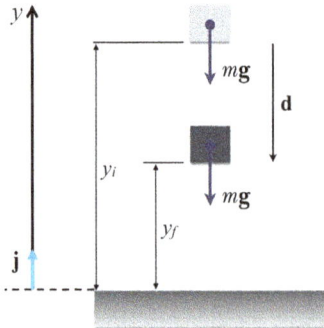

Figure 8.1: The work done on the brick by the gravitational force.

8.1.2 Elastic potential energy

Now, let us consider a composed system consisting of a block attached at the end of a spring with elastic force constant k, as shown in Fig. 8.2. The elastic force gives the force exerted by the spring on the block given as

$$F_s = -kx. \tag{8.6}$$

We showed in Chapter 7 that the work done by the spring's elastic force on a block connected to it is

$$W_s = \frac{1}{2}kx_i^2 - \frac{1}{2}kx_f^2. \tag{8.7}$$

Here, it is assumed that the initial, x_i, and final, x_f, positions of the block are measured with respect to its equilibrium position, $x = 0$.

Again, we see that W_s depends only on the initial and final positions, respectively, x_i and x_f, of the object, but not on the path. Besides, it vanishes if the path is a closed loop.

Definition 8.2 (Elastic potential energy). By definition, the elastic potential energy function of the object–spring system is given as

$$U_s = \frac{1}{2}kx^2. \tag{8.8}$$

The potential energy, U_s, of the system can be considered as the energy associated with the deformed spring, which can either be compressed or stretched from its equilibrium position.

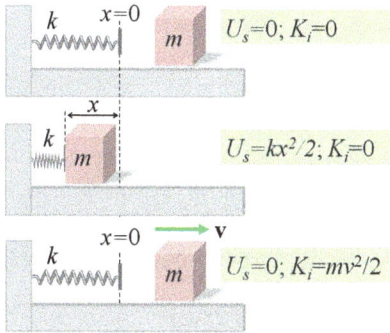

Figure 8.2: (a) An equilibrium state spring on a frictionless horizontal surface. (b) A spring compressed at distance x from its equilibrium position due to the block of mass m. (c) After the block is released from rest state, the elastic potential energy stored in the compressed spring transfers into the kinetic energy of the block.

Combining eq. (8.7) and eq. (8.8), we find that

$$W_s = U_{s,i} - U_{s,f} = -(U_{s,f} - U_{s,i}) = -\Delta U_s. \tag{8.9}$$

Equation (8.9) indicates that the work done on the block by the elastic force equals minus the change in the system (block–spring) elastic potential energy, or decrease in elastic potential energy of the system.

8.2 Conservative and non-conservative forces

It can be shown that the work of gravitational force on a falling object does not depend on whether an object falls vertically or slides down an inclined surface. It does indeed depend on the height of the object from the ground surface. In other words, it depends on the change of the object's elevation. In contrast, the energy loss due to the friction of the surface depends on the magnitude of the distance that the object moves. That is, the work done by the gravitational force does not depend on the path; however, the work done by the friction forces does depend on the path. That difference in the dependence on the path can be used to classify forces in terms of *conservative* and *non-conservative* forces.

For example, the gravitational force is conservative, and the frictional force is non-conservative.

8.2.1 Conservative forces

Two important properties characterize conservative forces:
1. The work a conservative force does on a particle moving between any two points is independent of the path taken by the particle.
2. The work done by that conservative force on a particle moving along any closed path equals zero. (By definition, a path is called a closed path if it begins and ends at the same point.)

The gravitational force and the elastic force exerted by a spring are two examples of the conservative forces. As we showed above, as the object moves between any two points near the Earth's surface, y_i and y_f, the work done by the gravitational force on that object is

$$W_g = mgy_i - mgy_f. \tag{8.10}$$

It can clearly be seen that W_g depends on the initial, y_i, and final, y_f, positions of the object, only; hence, it is independent of the path, and furthermore, W_g equals zero, if the object moves through any closed path (i. e., $y_f = y_i$).

In the case of the object–spring system, the work W_s done by the spring force is given by

$$W_s = \frac{1}{2}kx_i^2 - \frac{1}{2}kx_f^2. \tag{8.11}$$

Similarly, the elastic spring force is conservative because W_s depends again on the initial x_i and final x_f positions of the object, only, and in addition, it is zero for any closed path (that is, $x_f = x_i$).

Usually, the potential energy is associated with only conservative forces. For example, the potential energy of the object–Earth system associated with the gravitational force is

$$U_g = mgy \tag{8.12}$$

and the potential energy of the object–spring system associated with the elastic spring force is

$$U_s = \frac{1}{2}kx^2. \tag{8.13}$$

Definition 8.3 (Work of conservative forces). In general, the work of a conservative force, W_c, on an object equals the decrease in the potential energy associated with the object:

$$W_c = U_i - U_f = -\Delta U \tag{8.14}$$

where U_i is the initial potential energy, and U_f is the final potential energy associated with the object.

8.2.2 Non-conservative forces

The non-conservative forces cause changes in mechanical energy E, defined as the sum of kinetic and potential energies.

For example, if an object slides on a horizontal surface that is not frictionless, the friction force of the surface decreases the kinetic energy of the object. Furthermore, because of the frictional force, the temperature of the object and the surface increases. We can say that heat transfers to the object and the surface. This type of energy associated with temperature (or heat) is called *internal energy*.

Based on experience, the internal energy (or heat) cannot be transferred back to the kinetic energy of the object. In other words, the energy transformation is not reversible. Because the friction force changes the mechanical energy of a system, it is a *non-conservative force*.

The change in kinetic energy of the object because of the friction force relates to the work done by the friction force as

$$W_f = \Delta K_{\text{friction}}. \tag{8.15}$$

But

$$W_f = -f_k d. \tag{8.16}$$

Here, d is the length of the path over which the friction force f_k acts.

Let us consider that the table slides from A to B over the straight-line path of length $d = \sqrt{a^2 + b^2}$ (the red line) in Fig. 8.3. The change in kinetic energy is

$$\Delta K_{\text{friction}}^{(1)} = -f_k \sqrt{a^2 + b^2}. \tag{8.17}$$

Now, suppose the table slides through the path ACB with length $a + b$, from initial point A to the final point B (the blue line). The path length is longer $(a + b > \sqrt{a^2 + b^2})$, and hence the change in kinetic energy is larger in magnitude than that in the case of the motion along a straight line. Along this path, the change in kinetic energy is

$$\Delta K_{\text{friction}}^{(2)} = -f_k(a + b). \tag{8.18}$$

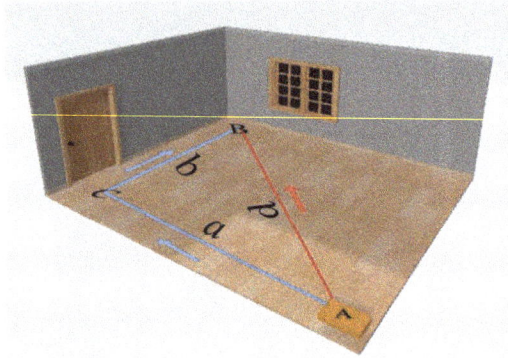

Figure 8.3: The decrease in total mechanical energy due to the friction force depends on the path taken as the table moves from initial point A to the final point B. The reduction in total mechanical energy is larger along the blue path than along the red path.

Definition 8.4 (Work of non-conservative forces). In general, the work done by non-conservative forces equals the change in the kinetic energy, as given by eq. (8.15).

Therefore, the change in kinetic energy for non-conservative forces depends on the path taken by the object between the initial and final points. If the potential energy is involved, then the change in the total mechanical energy depends on the path followed.

8.3 Conservative forces and potential energy

We showed that the work done on a particle by a conservative force does not depend on the path taken by the particle, but the work depends only on the particle's initial and final coordinates. The potential energy function U is defined in terms of the work done by a conservative force:

The work done by a conservative force exerted on a system equals the decrease in the potential energy of the system.

The work done by a conservative force $\mathbf{F}$ as a particle moves along the x-axis is

$$W_c = \int_{x_i}^{x_f} F_x \, dx = -\Delta U \tag{8.19}$$

where F_x is the component of $\mathbf{F}$ along the direction of the displacement.

Therefore, the work done by a conservative force acting on a system equals the negative of the potential energy change associated with that force. Here, the change in the potential energy is defined as

$$\Delta U = U_f - U_i \tag{8.20}$$

or

$$\Delta U = U_f - U_i = -\int_{x_i}^{x_f} F_x \, dx. \tag{8.21}$$

Thus, ΔU is negative when F_x and dx are in the same direction, as when an object is falling in a gravitational field or when a spring pulls an object toward the equilibrium position.

The potential energy indicates the potential, or capability, of an object for either increasing the kinetic energy or doing work as it starts being under the influence of a conservative force acting on the object as exerted by some other member of the composed system.

Often a particular position, namely x_i, is defined as a reference point, and the potential energy differences are measured to that point. Thus, the potential energy function is defined as

$$U_f(x) = -\int_{x_i}^{x_f} F_x \, dx + U_i. \tag{8.22}$$

Note that the value of U_i is often taken to be zero at the reference point. Furthermore, the sign of U_i for nonzero values does not have any significance because it just shifts $U_f(x)$ by a constant value, and only the change in potential energy is physically meaningful.

8.4 Conservation of mechanical energy

Consider an object fixed at some height h above the floor at rest. Therefore, the kinetic energy is zero. On the other hand, the gravitational potential energy of the object–Earth system is

$$U = mgh. \tag{8.23}$$

If the object is free to fall to the floor, then the speed of the object and thus its kinetic energy increase.

In contrast, the potential energy of the system (object–Earth) decreases. Ignoring the air resistance, the amount of potential energy the system loses as the object moves downward transfers to the kinetic energy of the object. Therefore, we can say that the sum of the kinetic and potential energy, that is, the total mechanical energy E remains constant. This is the well-known conservation law of mechanical energy. For the case of an object in free fall, this principle tells us that any increase (or decrease) in potential energy is accompanied by an equal decrease (or increase) of the kinetic energy. Note that the total mechanical energy of a system remains constant in an isolated system of objects that interact only through conservative forces.

Total energy of system

The total mechanical energy E of a system is defined as the sum of the kinetic and potential energies:

$$E = K + U. \tag{8.24}$$

From the principle of conservation of energy, initial mechanical energy, E_i, is equal to the final mechanical energy, E_f:

$$E_i = E_f. \tag{8.25}$$

Therefore,

$$K_i + U_i = K_f + U_f. \tag{8.26}$$

If more than one conservative force acts on an object within a system, a potential energy function is associated with each force. In such a case, we can apply the principle of conservation of mechanical energy for the system,

$$K_i + \sum U_i = K_f + \sum U_f \tag{8.27}$$

where the number of terms in the sums equals the number of conservative forces present.

8.5 Work done by an applied force

When you lift a book through some distance by applying a force to it, the force you apply does work W_p on the book, while the gravitational force does work W_g on the book. If we treat the book as a particle, then the total work done on the book is related to the change in its kinetic energy as described by the work–kinetic energy theorem given by

$$\sum_i W_i = W_p + W_g = \Delta K. \tag{8.28}$$

Also, the gravitational force is a conservative force, and hence

$$W_g = -\Delta U. \tag{8.29}$$

Combining eq. (8.28) and eq. (8.29), we obtain

$$W_p = \Delta E. \tag{8.30}$$

Note that the right side in eq. (8.30) represents the change in the mechanical energy of the book–Earth system. This result indicates that the applied force transfers energy to the system in the form of the kinetic energy of the book and the gravitational potential energy of the book–Earth system.

Thus, we conclude that, if an object is part of a system, then an applied force can transfer energy into or out of the system.

8.6 Exercises

Exercise 8.1. Two blocks are connected by a light string that passes over a frictionless pulley, as shown in Fig. 8.4. The block of mass m_1 lies on a horizontal surface and is connected to a spring of force constant k. The system is released from rest when the spring is unstretched. If the hanging block of mass m_2 falls a distance h before coming to rest, calculate the coefficient of kinetic friction between the block of mass m_1 and the surface.

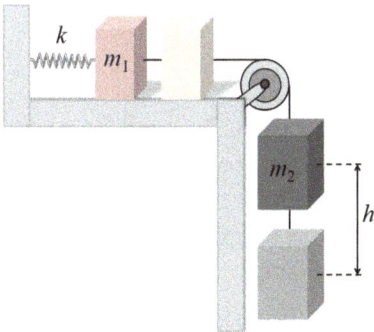

Figure 8.4: As the hanging block moves from its highest elevation to its lowest, the system loses gravitational potential energy but gains elastic potential energy in the spring. Some mechanical energy is lost because of friction between the sliding block and the surface.

Exercise 8.2. A pendulum consists of a sphere of mass m attached to a light cord of length L, as shown in Fig. 8.5. The sphere is released from rest when the cord makes an angle θ_A with the vertical, and the pivot at P is frictionless. (a) Find the speed of the sphere when it is at the lowest point B. (b) What is the tension T_B in the cord at B?

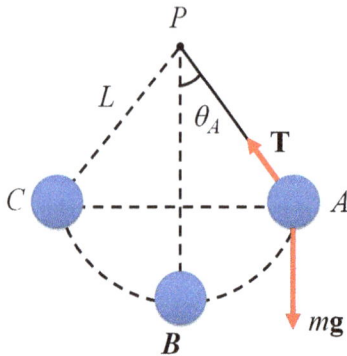

Figure 8.5: A pendulum.

Exercise 8.3. A ball of mass m drops from a height h above the ground, as shown in Fig. 8.6. (a) Neglecting air resistance, determine the speed of the ball when it is at a height y above the ground. (b) Determine the speed of the ball at y if, at the instant of release, it already has an initial speed v_i at the initial altitude h.

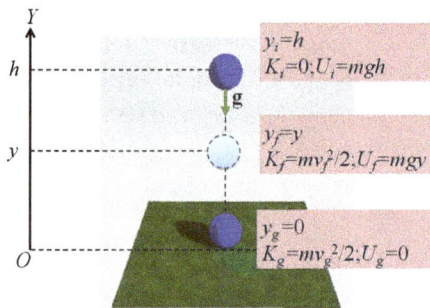

Figure 8.6: A ball is dropped from a height h above the ground.

Exercise 8.4. A crate with mass of 3.00 kg slides down a ramp. The ramp is 1.00 m in length and inclined at an angle of 30.0°, as shown in Fig. 8.7. The crate starts from rest at the top, experiences a constant frictional force of magnitude 5.00 N, and continues to move a short distance on the flat floor after it leaves the ramp. Use energy methods to determine the speed of the crate at the bottom of the ramp.

Exercise 8.5. An object with mass m is released from a height h in a loop-the-loop shown in Fig. 8.8. Use conservation of energy to find the ratio of h/R if the object was to reach point A in Fig. 8.8. What should this ratio be if it reaches point B? Neglect friction.

Exercise 8.6. The Atwood machine shown in Figure 8.9 is released when the block m_2 is 2.0 m above the floor. Find the velocity of the system when m_2 hits the ground. Using the conservation law of energy, calculate the acceleration of the system.

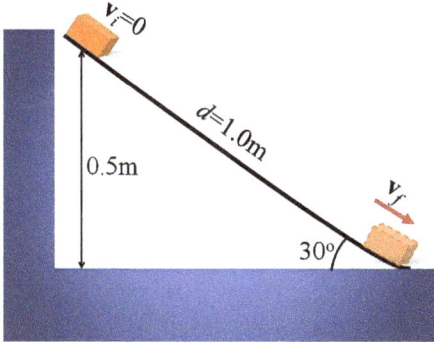

Figure 8.7: A crate slides down a ramp under the influence of gravity.

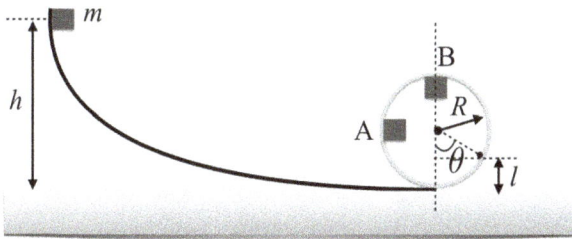

Figure 8.8: Loop-the-loop problem demonstrating the use of conservation of energy.

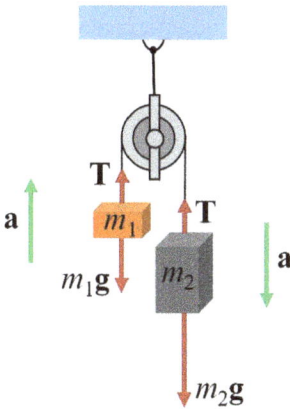

Figure 8.9: Atwood machine.

Exercise 8.7. Calculate the elastic potential energy stored in a spring with a spring constant $k = 800\,\text{N/m}$ and stretched by $10.0\,\text{cm}$.

Exercise 8.8. Find the potential energy of a block of mass $100\,\text{kg}$, which is moved $10.0\,\text{m}$ along a $30°$ inclined plane.

Exercise 8.9. Show that the work of gravitational force on a falling object does not depend on whether an object falls vertically or slides down an inclined surface. For this purpose, consider a block with mass m at height h above the ground, which can take two possible paths, namely path (1) corresponding to a free fall and path (2) corresponding to the movement along an inclined surface of length d. Calculate the work done by the gravitation force along each path. Assume that the inclined surface is frictionless.

Exercise 8.10. What is the work done by the gravitational force if you lift an object of mass m to a height h and then walk a distance s horizontally, as shown in Fig. 8.10?

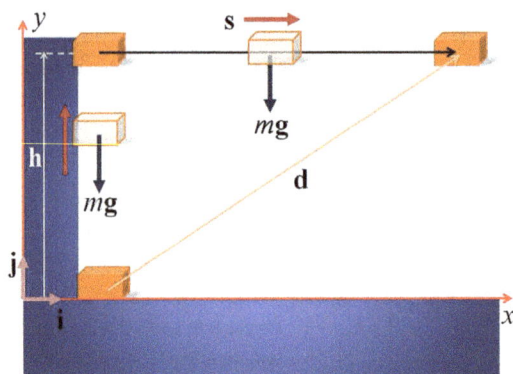

Figure 8.10: Illustration of the movement of an object lifted to a height h by a person who then walked a distance s horizontally.

Exercise 8.11. What is the work done by an external force applied by a person on an object of mass m to lift it to a height h and then walking a distance s horizontally, as shown in Fig. 8.11?

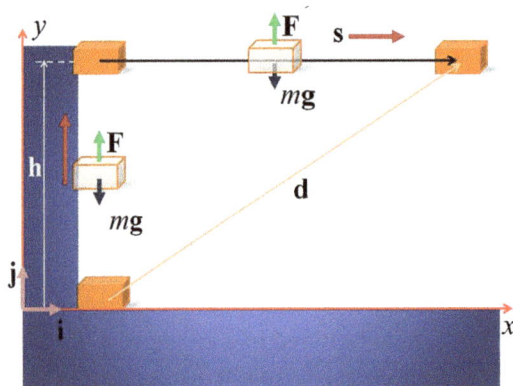

Figure 8.11: Illustration of the movement of an object lifted to a height h by the external force **F** applied by someone who then walked a distance s horizontally.

9 Linear momentum and collisions

So far, in our study of classical mechanics, we have studied the motion of a single particle or body primarily. To extend our comprehension of mechanics, we must now begin to examine the interactions of particles. To start this study, we define and explore a new concept, the center of mass, which will allow us to make mechanical calculations for a system of particles. First, we introduce linear momentum and its conservation law, then we discuss the collisions.

9.1 Linear momentum and Newton's second law

Linear momentum

For a particle of mass m moving with a velocity $\mathbf{v}$ the linear momentum is given as the product of the mass and velocity:

$$\mathbf{p} = m\mathbf{v}. \tag{9.1}$$

Equation (9.1) indicates that the linear momentum is a vector quantity because it equals the product of a scalar quantity, m, and a vector quantity, $\mathbf{v}$. The direction of $\mathbf{p}$ is along $\mathbf{v}$, and the SI unit is

$$\mathrm{kg \cdot m/s}. \tag{9.2}$$

For a particle moving in an arbitrary direction in three-dimensional space, $\mathbf{p}$ has three components, namely (p_x, p_y, p_z). Equation (9.1) can also be written as follows for each component:

$$p_x = mv_x \tag{9.3}$$
$$p_y = mv_y \tag{9.4}$$
$$p_z = mv_z. \tag{9.5}$$

According to Newton, $m\mathbf{v}$ is called *quantity of motion*, which is a graphic description of another term used, *momentum*.[1]

A relationship between the linear momentum of a particle and the total force acting on a particle can be established using Newton's second law of motion:

$$\mathbf{F}_R = \sum_i \mathbf{F}_i = m\frac{d\mathbf{v}}{dt} = \frac{d(m\mathbf{v})}{dt} = \frac{d\mathbf{p}}{dt}. \tag{9.6}$$

---

1 The word *momentum* originates from the Latin word for *movement*.

https://doi.org/10.1515/9783110755824-009

Newton's second law for a particle

The rate change on time of the linear momentum of a particle equals the resultant force acting on the particle:

$$\frac{d\mathbf{p}}{dt} = \mathbf{F}_R. \tag{9.7}$$

9.2 Conservation of momentum for a two-particle system

Consider a two-particle system isolated from their surroundings composed of particles 1 and 2, which interact with each other (see Fig. 9.1). In particular, the particles exert a force on each other, obeying Newton's third law, and there are no external forces present. If a force from particle 1 (such as the gravitational force) acts on particle 2, then, based on Newton's third law, there is a second internal force equal in magnitude but opposite in direction that particle 2 exerts on particle 1.

Denoting by $\mathbf{p}_1$ the momentum of particle 1 and that of particle 2 by $\mathbf{p}_2$ at some instant as shown in Fig. 9.1. Newton's second law as applied to each particle can be written as

$$\mathbf{F}_{21} = \frac{d\mathbf{p}_1}{dt} \tag{9.8}$$

$$\mathbf{F}_{12} = \frac{d\mathbf{p}_2}{dt} \tag{9.9}$$

where $\mathbf{F}_{21}$ is the force acting on particle 1 by particle 2 and $\mathbf{F}_{12}$ is the force exerted by particle 1 on particle 2. Applying Newton's third law, $\mathbf{F}_{12} = -\mathbf{F}_{21}$, or alternatively

$$\mathbf{F}_{12} + \mathbf{F}_{21} = 0. \tag{9.10}$$

Using eq. (9.6), we write

$$\frac{d\mathbf{p}_1}{dt} + \frac{d\mathbf{p}_2}{dt} = \frac{d(\mathbf{p}_1 + \mathbf{p}_2)}{dt} = 0. \tag{9.11}$$

Denoting the total momentum of the system, $\mathbf{p}_{tot}$ as $\mathbf{p}_1 + \mathbf{p}_2 = \mathbf{p}_{tot}$, we obtain

$$\frac{d\mathbf{p}_{tot}}{dt} = 0 \tag{9.12}$$

which indicates that

$$\mathbf{p}_{tot} = \sum_{system} \mathbf{p} = \text{constant}. \tag{9.13}$$

Denoting by $(\mathbf{p}_{1i}, \mathbf{p}_{2i})$ and $(\mathbf{p}_{1f}, \mathbf{p}_{2f})$, respectively, the initial and the final values of the momenta of particles 1 and 2, eq. (9.13) can alternatively be written as

$$\mathbf{p}_{1i} + \mathbf{p}_{2i} = \mathbf{p}_{1f} + \mathbf{p}_{2f}. \tag{9.14}$$

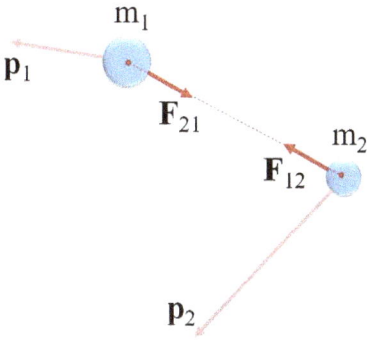

Figure 9.1: The momentum of particle 1 and 2 at some instant.

Equation (9.13) and eq. (9.14) are known as the *law of conservation of linear momentum.* This law, which is considered one of the most important laws of mechanics, can also be written for any number of particles in an isolated system. It can be stated as follows.

If two or more particles in an isolated system interact, then the total momentum of the system remains constant.

This law indicates that, for an isolated system, the total momentum at any instant of time is equal to its initial value.

Here, no assumption was made regarding the natural origin of the forces acting on each particle of the system. However, it is required that these forces must be internal to the system.

9.3 Impulse and momentum

We found above that the second law of Newton can also be written as

$$\frac{d\mathbf{p}}{dt} = \mathbf{F}. \tag{9.15}$$

Multiplying both sides by dt, we obtain

$$d\mathbf{p} = \mathbf{F}dt. \tag{9.16}$$

Integrating eq. (9.16) over some time interval, say from t_i to t_f, we obtain the change on the momentum of a particle due to the force $\mathbf{F}$ acting on the time interval $\Delta t = t_f - t_i$:

$$\Delta \mathbf{p} = \mathbf{p}_f - \mathbf{p}_i = \int_{t_i}^{t_f} \mathbf{F}\, dt. \tag{9.17}$$

Here, $\mathbf{p}_i$ is the momentum at time t_i and $\mathbf{p}_f$ is the momentum at time t_f.

Knowing the function of force with time, one can evaluate the integral in eq. (9.17). The time integral of the force in eq. (9.17) is called the *impulse* of the force $\mathbf{F}$ acting on a particle over the time interval Δt. Therefore, the impulse is a vector quantity given as

$$\mathbf{I} \equiv \int_{t_i}^{t_f} \mathbf{F}\, dt = \Delta \mathbf{p}. \tag{9.18}$$

Impulse–momentum theorem
The impulse of the force $\mathbf{F}$ acting on a particle equals the change in the momentum of the particle caused by that force.

That is also known as *impulse–momentum theorem*, and it is equivalent to New-ton's second law.

Equation (9.18) indicates that impulse is a vector quantity with magnitude equal to the area under the force–time curve. That is also shown in Fig. 9.2, where it is assumed that the force varies in time, and it is nonzero in the time interval $\Delta t = t_f - t_i$.

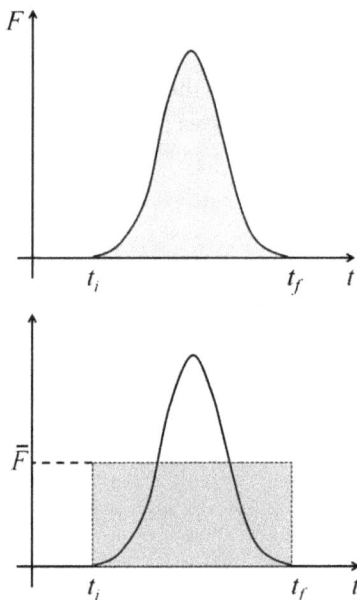

Figure 9.2: A force acting on a particle may vary in time.

Besides, the impulse vector has the same direction as the direction of the change in momentum vector, $\Delta \mathbf{p}$. Moreover, the impulse has the dimensions of momentum, that is, kg m/s, as indicated by eq. (9.18).

It is important to note that impulse is not a property of a particle itself. Indeed, it is a measure of the degree to which an external force exerted on the particle changes its momentum. Therefore, the statement that an impulse is given to a particle indicates that momentum is transferred from an external agent to the particle.

For a time-varying force, we can define the time-average force as

$$\bar{\mathbf{F}} \equiv \frac{1}{\Delta t} \int_{t_i}^{t_f} \mathbf{F}\, dt \qquad (9.19)$$

where $\Delta t = t_f - t_i$. Thus, the impulse in terms of the average force can be determined as

$$\mathbf{I} \equiv \bar{\mathbf{F}}\Delta t. \qquad (9.20)$$

In general, the time-averaged force (see also Fig. 9.2) can be assumed as a constant force that gives to the particle in the time interval Δt the same impulse as the time-varying force gives over this same interval to that particle.

Equation (9.18), in principle, is used to compute the impulse when the force $\mathbf{F}$ is known as a function of time. The estimation of the impulse becomes especially simple if the force exerted on the particle is constant on time. In this case,

$$\mathbf{F} = \bar{\mathbf{F}} \qquad (9.21)$$

and eq. (9.20) becomes

$$\mathbf{I} = \mathbf{F}\Delta t. \qquad (9.22)$$

In many physical situations, the concept of the impulse approximation is used because of the assumption that the resultant force exerted on a particle acts for a short time, and it is much higher than all the other forces present. This approximation is especially useful in treating collisions in which the duration of the crash is concise. When this approximation is made, we refer to the force as an *impulsive force*.

9.4 Collisions

If two particles come together for a short time and producing impulsive forces acting on each other, then we say that a collision occurs. The forces are assumed to be much greater than all other external forces present.

Figure 9.3 shows a simple case of the collision occurring between two macroscopic particles with masses m_1 and m_2, respectively. If two particles collide, the impulsive forces acting on each particle vary in time in complicated ways, for example, as shown in Fig. 9.3. Besides, they obey the third law of Newton.

If $\mathbf{F}_{21}$ is the average force acting on particle 1 by particle 2 during the time interval Δt (assuming no other forces act on the particles), then the change in the momentum of particle 1 due to the collision with particle 2 is

$$\Delta \mathbf{p}_1 = \mathbf{F}_{21}\Delta t. \qquad (9.23)$$

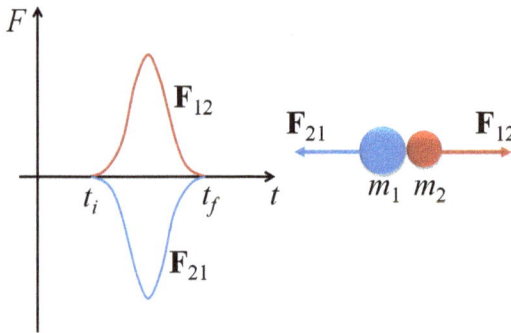

Figure 9.3: The collision between two objects as a result of direct contact.

Similarly, if $\mathbf{F}_{12}$ is the average force acting on particle 2 by particle 1 during the same time interval Δt, then the change in the momentum of particle 2 due to the collision with particle 1 is

$$\Delta\mathbf{p}_2 = \mathbf{F}_{12}\Delta t. \tag{9.24}$$

Using Newton's third law:

$$\mathbf{F}_{21} = -\mathbf{F}_{12}. \tag{9.25}$$

Therefore, it can be found that

$$\Delta\mathbf{p}_1 = -\Delta\mathbf{p}_2 \tag{9.26}$$

or

$$\Delta\mathbf{p}_1 + \Delta\mathbf{p}_2 = 0. \tag{9.27}$$

Conservation law of momentum for any collision
Denoting the total momentum of the system of the two masses as

$$\mathbf{p}_{\text{system}} = \mathbf{p}_1 + \mathbf{p}_2 \tag{9.28}$$

then

$$\mathbf{p}_{\text{before,system}} = \mathbf{p}_{\text{before,1}} + \mathbf{p}_{\text{before,2}} \tag{9.29}$$
$$= \mathbf{p}_{\text{after,system}} = \mathbf{p}_{\text{after,1}} + \mathbf{p}_{\text{after,2}}.$$

Therefore, it can be stated that the total linear momentum of the system of particles is constant in time:

$$\mathbf{p}_{\text{system}} = \mathbf{p}_1 + \mathbf{p}_2 = \text{constant.} \tag{9.30}$$

The total momentum before the collision of an isolated system is equal to the total momentum of the system after the collision. That is the conservation law of the momentum of an isolated system (i. e., the external forces are negligible) for any collision.

We showed that the momentum of an isolated system is conserved in any collision in which external forces are negligible. On the other hand, kinetic energy may not necessarily be constant. That depends on the type of collision. There exist two types of collisions, namely elastic collisions and inelastic collisions; they are discussed in the following.

9.4.1 Elastic collisions

An elastic collision occurs between two objects in which total kinetic energy and the total momentum are conserved before and after the collision.

Mathematically, we can write for any elastic collision

$$K_i = K_f \tag{9.31}$$

$$\mathbf{p}_i = \mathbf{p}_f. \tag{9.32}$$

For example, the billiard-ball collisions and the collisions of air molecules with the walls of a container at ordinary temperatures are approximately elastic. However, in reality, there is always some loss of energy in any collision process. In general, if the objects are rigid, then the process can be considered as an almost elastic collision. Pure elastic collisions occur, in general, between atomic and subatomic particles.

In Fig. 9.4 a particle of mass m_1 and initial velocity $\mathbf{v}_{1i}$ is making a very short elastic collision with another particle, which is assumed initially at rest. After the collision, the final velocity of the projectile is $\mathbf{v}_{1f}$, and its scattering angle is θ to its initial direction. The final velocity of the second particle is $\mathbf{v}_{2f}$, and the scattering angle is ϕ with respect to the initial direction of the first particle.

From the conservation law of the momentum, we can write

$$\mathbf{p}_{1i} + \mathbf{p}_{2i} = \mathbf{p}_{1f} + \mathbf{p}_{2f}. \tag{9.33}$$

The projection along the vertical (y-axis) and horizontal (x-axis) gives the following two equations:

$$m_1 v_{1ix} = m_1 v_{1fx} + m_2 v_{2fx} \tag{9.34}$$

$$0 = m_1 v_{1fy} + m_2 v_{2fy} \tag{9.35}$$

or

$$m_1 v_{1i} = m_1 v_{1f} \cos\theta + m_2 v_{2f} \cos\phi \tag{9.36}$$

$$0 = m_1 v_{1f} \sin\theta - m_2 v_{2f} \sin\phi. \tag{9.37}$$

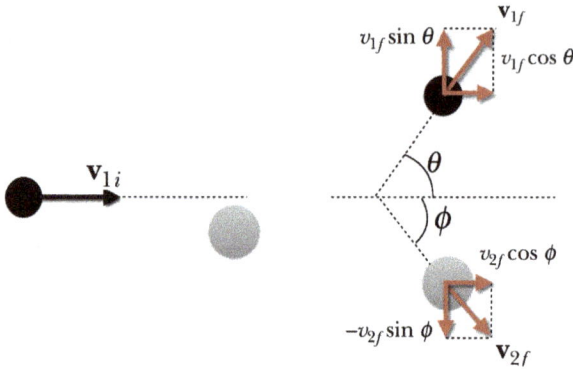

Figure 9.4: An elastic glancing collision between two particles: (Left) Before the collision and (Right) after the collision.

Here, the minus sign in the second equation is because after the collision, particle 2 has a y component of velocity that is downward. The problem can be solved analytically, if no more than two of the seven quantities in two equations are unknown.

If the collision is elastic, conservation of kinetic energy, with $v_{2i} = 0$, gives

$$\frac{1}{2}m_1 v_{1i}^2 = \frac{1}{2}m_1 v_{1f}^2 + \frac{1}{2}m_2 v_{2f}^2. \tag{9.38}$$

Knowing the initial speed of particle 1 and both masses, we are left with four unknowns ($v_{1f}, v_{2f}, \phi, \theta$). Because we have only three equations, one of the four remaining quantities must be given if we are to determine the motion after the collision from conservation principles alone. For example, suppose we want to express all these quantities in terms of the scattering angle θ.

Using eq. (9.36), we get

$$\tan\phi = \frac{v_{1f} \sin\theta}{v_{1i} - v_{1f}\cos\theta}. \tag{9.39}$$

From eq. (9.38), we get

$$v_{2f} = \sqrt{\frac{m_1}{m_2}(v_{1i}^2 - v_{1f}^2)}. \tag{9.40}$$

9.4.2 Inelastic collision

In an inelastic collision, the momentum is conserved before and after the collision, but not the total kinetic energy.

There are two types of inelastic collisions. The *perfectly inelastic* collision occurs when the particles hit each other and stick together after the collision. For example, it happens when a meteorite collides with the Earth.

The *inelastic* collision occurs if the colliding particles do not stick together after the collision. In these collisions, some part of the kinetic energy is lost. For example, if a rubber ball hits a hard surface, an inelastic collision occurs because a part of the kinetic energy of the ball is lost due to the deformation of the ball while it is in contact with the surface.

In general, in most collisions, kinetic energy is not conserved before and after the collision. That is because some of the kinetic energy is transferred into internal energy, elastic potential energy when the objects are deformed, or into rotational energy. Elastic and perfectly inelastic collisions are limiting cases; most collisions fall somewhere between them.

In the case of perfectly inelastic collision, consider two particles of masses m_1 and m_2 moving with initial velocities $\mathbf{v}_{1i}$ and $\mathbf{v}_{2i}$ along a straight line, as shown in Fig. 9.5. The two particles collide head-on, stick together, and then move with some common velocity $\mathbf{v}_f$ after the collision. Because momentum is conserved in any collision, we can say that the total momentum before the collision equals the total momentum of the composite system after the collision:

$$m_1\mathbf{v}_{1i} + m_2\mathbf{v}_{2i} = (m_1 + m_2)\mathbf{v}_f \tag{9.41}$$

or

$$\mathbf{v}_f = \frac{m_1\mathbf{v}_{1i} + m_2\mathbf{v}_{2i}}{m_1 + m_2}. \tag{9.42}$$

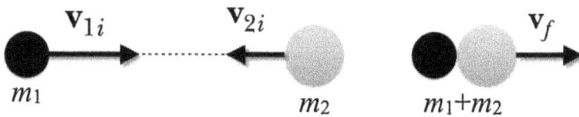

Figure 9.5: An inelastic glancing collision between two particles: (Left) Before the collision and (Right) after the collision.

9.5 Center of mass

To describe the overall motion of a mechanical system in terms of a single particular point, the so-called *center of mass* of the system is determined. Here, the mechanical system is a system of N particles, such as a system of (interacting) atoms in a box, or an extended object.

In the following we show that the center of mass of the system moves as if all the mass of the system is concentrated at that position. Also, if the resultant external force acting on the system is $\mathbf{F}^e$ and the total mass of the system is $M = \sum_{i=1}^{N} m_i$, the acceleration of the center of mass, $\mathbf{a}_{c.\,m.}$, is

$$\mathbf{a}_{c.\,m.} = \frac{\mathbf{F}^e}{M}. \tag{9.43}$$

Equation (9.43) indicates that the system moves as if the resultant external force acts on a single unique particle of mass M located at the center of mass. Note that this behavior is independent of other motion included in the system, such as the internal motion, for example, rotation or vibration of the system. This result is implicitly assumed in the derivations above, in this chapter, because many cases referred to the motion of extended objects are treated as particles.

First, we can consider a mechanical system of two discrete particles with masses m_1 and m_2, respectively. The masses are connected by a rigid rod with negligible mass, as shown in Fig. 9.6. For a given system, the center of mass is defined below as the *average position* of the system's mass. It locates somewhere on the line joining the particles, and we expect to be closer to the particle having the larger mass.

If a single external force acts on the system at some point on the rod somewhere between the center of mass and the smallest particle, the system rotates clockwise (see Fig. 9.6(top)). Also, if the force acts on the system at a point on the rod located between the center of mass and the more massive particle, the system rotates counterclockwise, as shown in Fig. 9.6(middle). On the other hand, if the force is applied on the system locates at the center of mass, the entire system moves in the direction of the force, and no rotation occurs (see Fig. 9.6(bottom)). That experiment can in practice be used to determine the center of mass of the system.

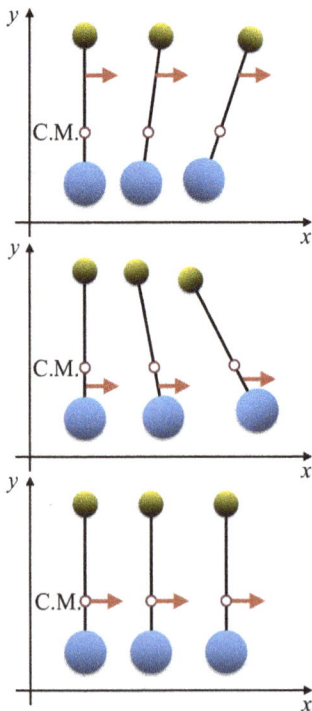

Figure 9.6: A light, rigid rod connects two particles of unequal mass. (Top) The system rotates clockwise when the applied force acts between the less massive particle and the center of mass. (Middle) The system rotates counterclockwise when a force is applied between the more massive particle and the center of mass. (Bottom) The entire system moves in the direction of the force, and no rotation occurs if the applied force locates at the center of mass.

Center of mass for a system of two particles

For a system composed of two particles having masses m_1 and m_2, respectively, at the positions $\mathbf{r}_1$ and $\mathbf{r}_2$, respectively, the center of mass is defined mathematically as

$$\mathbf{r}_{\text{c.m.}} = \frac{m_1\mathbf{r}_1 + m_2\mathbf{r}_2}{m_1 + m_2}. \tag{9.44}$$

$\mathbf{r}_{\text{c.m.}}$ is a point in space that locates somewhere between the line joining the two masses and it is closer to the particle with larger mass. If the system is composed of two particles with equal masses, the center of mass is in the midway between the particles.

That concept can be extended to a system of many particles in three dimensions. Any extended irregular shaped object, as shown in Fig. 9.7, can always be approximated as a collection of *lumped masses*.

Center of mass for a system of particles

For a system of N particles having masses m_i (for $i = 1, 2, \ldots, N$) in three dimensions at the positions $\mathbf{r}_i = (x_i, y_i, z_i)$ (for $i = 1, 2, \ldots, N$), respectively, the center of mass is defined as

$$\mathbf{r}_{\text{c.m.}} = \frac{\sum_{i=1}^{N} m_i\mathbf{r}_i}{\sum_{i=1}^{N} m_i}. \tag{9.45}$$

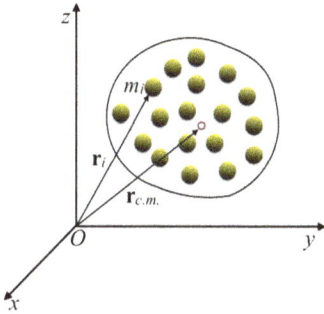

Figure 9.7: A lumped-mass parameter depiction of an irregular shaped object.

The Cartesian coordinates of this vector are $x_{\text{c.m.}}$, $y_{\text{c.m.}}$, and $z_{\text{c.m.}}$ defined as

$$x_{\text{c.m.}} = \mathbf{r}_{\text{c.m.}} \cdot \mathbf{i} = \frac{\sum_{i=1}^{N} m_i x_i}{\sum_{i=1}^{N} m_i} \tag{9.46}$$

$$y_{\text{c.m.}} = \mathbf{r}_{\text{c.m.}} \cdot \mathbf{j} = \frac{\sum_{i=1}^{N} m_i y_i}{\sum_{i=1}^{N} m_i}$$

$$z_{\text{c.m.}} = \mathbf{r}_{\text{c.m.}} \cdot \mathbf{k} = \frac{\sum_{i=1}^{N} m_i z_i}{\sum_{i=1}^{N} m_i}$$

where $\mathbf{i}$, $\mathbf{j}$ and $\mathbf{k}$ are unit vectors along the x-, y- and z-axes, respectively.

If the number of lumped masses approaches infinity, then the system of discrete masses becomes continuous. For that continuous body the summation converges to the exact value of the vector $\mathbf{r}_{c.m.}$ and the sum converges to an integral given as

$$\mathbf{r}_{c.m.} = \frac{\int_V \mathbf{r}\rho(\mathbf{r})dV}{\int_V \rho(\mathbf{r})dV}.$$

(9.47)

In eq. (9.47), $\rho(\mathbf{r})$ is the density of the object and V is the entire volume. This equation can only be solved if $\mathbf{r}\rho(\mathbf{r})$ can be integrated inside the volume V. That requirement is not always achievable and hence in most of the engineering applications a lumped-mass parameter approach using computers is employed.

For any symmetric object with a uniform mass distribution per unit of volume, the center of mass lies on an axis of symmetry and any plane of symmetry. For example, for a rod, the center of mass is located midway between rod's ends in the axis along the rod. Also, the center of mass of three-dimensional symmetric shapes, such as a sphere or a cube, lies at its geometric center.

An extended object can be seen as a continuous distribution of mass; each small mass element δm is acted upon by the force of gravity: $\delta\mathbf{F} = \delta m\mathbf{g}$. The resultant effect of these forces is equivalent to the effect of a single force, $M\mathbf{g}$, acting on a special point, called the center of gravity, where M is given by

$$M = \int_V \rho(\mathbf{r})dV.$$

(9.48)

If $\mathbf{g}$ is constant over the mass distribution, then the center of gravity coincides with the center of mass. If an extended object is pivoted at its center of gravity, it balances in any orientation.

9.6 Motion of a system of particles

Velocity of the center of mass
The physical significance and utility of the center of mass concept can be observed by taking the time derivative of the position vector given by eq. (9.45), which gives the velocity of the center of mass:

$$\mathbf{v}_{c.m.} = \frac{d\mathbf{r}_{c.m.}}{dt} = \frac{\sum_{i=1}^N m_i \frac{d\mathbf{r}_i}{dt}}{\sum_{i=1}^N m_i} = \frac{\sum_{i=1}^N m_i \mathbf{v}_i}{\sum_{i=1}^N m_i}$$

(9.49)

where $\mathbf{v}_i$ is the velocity of the ith particle.

The total momentum of the system

Equation (9.49) can also be written as

$$M\mathbf{v}_{\text{c.m.}} = \sum_{i=1}^{N} m_i \mathbf{v}_i = \sum_{i=1}^{N} \mathbf{p}_i = \mathbf{p}_{\text{tot}}.$$ (9.50)

Here, M denotes the overall mass of the system of masses m_i for $i = 1, 2, \ldots, N$ defined as

$$M = \sum_{i=1}^{N} m_i$$ (9.51)

and $\mathbf{p}_{\text{tot}}$ is the total momentum of the system, which is equal to the total mass, M, multiplied by the velocity of the center of mass, $\mathbf{v}_{\text{c.m.}}$. In other words, the total linear momentum of the system equals the linear momentum of a single particle of mass M moving with a velocity $\mathbf{v}_{\text{c.m.}}$.

Acceleration of the center of mass

If we take the derivative of eq. (9.49), we obtain the expression for the acceleration of the center of the mass of the system of N particles as

$$\mathbf{a}_{\text{c.m.}} = \frac{d\mathbf{v}_{\text{c.m.}}}{dt} = \frac{\sum_{i=1}^{N} m_i \frac{d\mathbf{v}_i}{dt}}{\sum_{i=1}^{N} m_i} = \frac{\sum_{i=1}^{N} m_i \mathbf{a}_i}{\sum_{i=1}^{N} m_i}$$ (9.52)

where $\mathbf{a}_i$ is the acceleration of the ith particle. Equation (9.52) can be rearranged and, using Newton's second law, we obtain

$$M\mathbf{a}_{\text{c.m.}} = \sum_{i=1}^{N} m_i \mathbf{a}_i = \sum_{i=1}^{N} \mathbf{F}_i$$ (9.53)

where $\mathbf{F}_i$ is the total force acting on the ith particle of the system, which is the sum of the internal and external forces on the system, hence, it can be written as

$$\mathbf{F}_i = \sum_{j=1\neq i}^{N} \mathbf{F}_{ij}^{\text{int}} + \mathbf{F}_i^e$$ (9.54)

where $\mathbf{F}_{ij}^{\text{int}}$ is the internal force of particle j on the particle i and $\mathbf{F}_i^e$ is the external force on the particle i. Based on the third law of the Newton, the internal force exerted by particle i on particle j equals in magnitude and opposite in direction to the internal force exerted by particle j on particle i: $\mathbf{F}_{ij}^{\text{int}} = -\mathbf{F}_{ji}^{\text{int}}$. Therefore,

$$\sum_{i=1}^{N} \sum_{j=1\neq i}^{N} \mathbf{F}_{ij}^{\text{int}} = \cdots + \mathbf{F}_{ij}^{\text{int}} + \mathbf{F}_{ji}^{\text{int}} + \cdots$$ (9.55)

$$= \cdots + \mathbf{F}_{ij}^{\text{int}} - \mathbf{F}_{ij}^{\text{int}} + \cdots$$

$$= 0.$$

Newton's second law for a system of particles

Then eq. (9.53) can be written as

$$M\mathbf{a}_{\text{c. m.}} = \sum_{i=1}^{N} m_i \mathbf{a}_i = \sum_{i=1}^{N} \mathbf{F}_i^e = \mathbf{F}^e \tag{9.56}$$

or

$$\frac{d\mathbf{p}_{\text{tot}}}{dt} = \sum_{i=1}^{N} \mathbf{F}_i^e = \mathbf{F}^e \tag{9.57}$$

which is Newton's second law for a system of particles. Comparing that with Newton's second law for a single particle, we say that: *The center of mass of a system of particles of combined mass M moves like an equivalent particle of mass M would move under the influence of the resultant external force on the system.*

It can be seen that if the resultant external force is zero, then from eq. (9.57) we get

$$\frac{d\mathbf{p}_{\text{tot}}}{dt} = 0 \tag{9.58}$$

or

$$\mathbf{p}_{\text{tot}} = \text{constant.} \tag{9.59}$$

That is, the total linear momentum of a system of particles is conserved if no net external force is acting on the system. It follows that, for an isolated system of particles, both the total momentum and the velocity of the center of mass are constant in time. This is a generalization to a many-particle system of the conservation law of total linear momentum discussed for a two-particle system.

9.7 Exercises

Exercise 9.1. A golf ball of mass 50 g is struck with a club (Fig. 9.8). The force exerted on the ball by the club varies from zero, at the instant before contact, up to some maximum value (at which the ball is deformed) and then back to zero when the ball leaves the club. Thus, the force–time curve is qualitatively described by Fig. 9.2. Assuming that the ball travels 200 m, estimate the magnitude of the impulse caused by the collision.

Exercise 9.2. In a particular crash test, an automobile of mass 1500 kg collides with a wall, as shown in Fig. 9.9. The initial and final velocities of the automobile are $\mathbf{v}_i = -15.0\mathbf{i}\,\text{m/s}$ and $\mathbf{v}_f = 2.60\mathbf{i}\,\text{m/s}$, respectively. If the collision lasts for 0.150 s, find the impulse caused by the collision and the average force exerted on the automobile.

Figure 9.8: A golf ball being struck by a club.

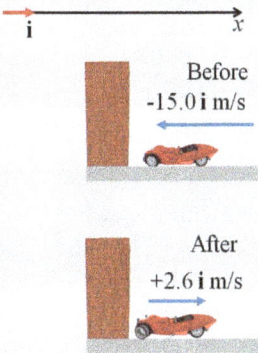

Figure 9.9: The car's momentum changes as a result of its collision with the wall.

Exercise 9.3. A car of mass 1800 kg stopped at a traffic light is struck from the rear by a car with mass of 900 kg, and the two become entangled. If the smaller car was moving at 20.0 m/s before the collision, what is the velocity of the entangled cars after the collision?

Exercise 9.4. A system consists of three particles located as shown in Fig. 9.10. Find the center of mass of the system.

Exercise 9.5. Show that the center of mass of a rod of mass M and length L lies midway between its ends, assuming the rod has a uniform mass per unit length.

Exercise 9.6. A rocket is fired vertically upward. At the instant, it reaches an altitude of 1000 m and speed of 300 m/s, it explodes into three equal fragments. One fragment continues to move upward with a speed of 450 m/s following the explosion. The second fragment has a speed of 240 m/s and is moving east right after the explosion. What is the velocity of the third fragment right after the explosion?

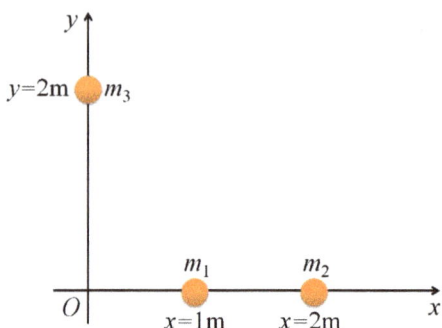

Figure 9.10: A system composed of three masses.

Exercise 9.7. A car is stopped for a traffic signal. When the light turns green, the car accelerates, increasing its speed from zero to 5.20 m/s in 0.832 s. What linear impulse and average force does a passenger of mass 70.0 kg experience in the car?

Exercise 9.8. A tennis player receives a shot with the ball (0.0600 kg) traveling horizontally at 50.0 m/s and returns the shot with the ball traveling horizontally at 40.0 m/s in the opposite direction. What is the impulse delivered to the ball by the racket?

Exercise 9.9. A steel ball of a mass of 3.00 kg strikes a wall with a speed of 10.0 m/s at an angle of 60.0° with the surface. It bounces off with the same speed and angle (Fig. 9.11). If the ball is in contact with the wall for 0.200 s, what is the average force exerted on the ball by the wall?

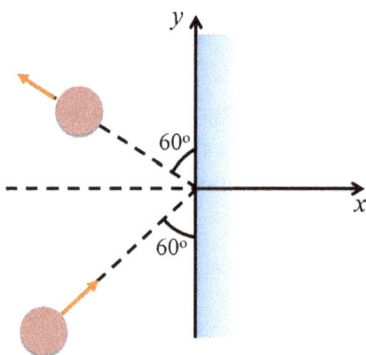

Figure 9.11: A ball striking a wall.

Exercise 9.10. Consider a frictionless track ABC as shown in Fig. 9.12. A block of mass $m_1 = 5.00$ kg is released from A. It makes a head-on elastic collision at B with a block of mass $m_2 = 10.0$ kg that is initially at rest. Calculate the maximum height to which m_1 rises after the collision.

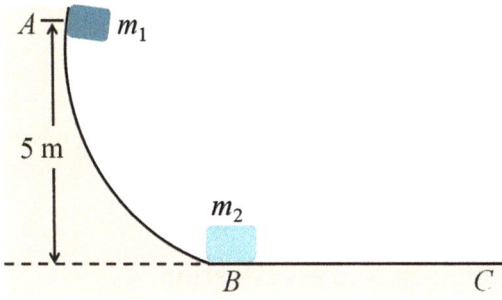

Figure 9.12: Elastic collisions between two objects.

10 Rotation of a rigid body

In this chapter, we introduce the kinematics and dynamics of the motion of a rigid body. An object is rigid, by definition, if it is non-deformable. In other words, it is an object in which distances between all pairs of particles remain constant.

10.1 Angular displacement, velocity, and acceleration

Figure 10.1 shows a planar (flat) rigid body of arbitrary shape confined to the xy plane and rotating about the fixed z-axis through O. The z-axis is perpendicular to the plane of the figure, and O is the origin of a xy coordinate system.

We look at the motion of a particle located at point P that is at a fixed distance r from the origin O and rotates about point O in a circle of radius r. Note that the rigid body is composed of millions of particles, and furthermore every particle on the object undergoes circular motion about O. For convenience, the position of P is represented in polar coordinates, namely (r, θ) with r being the distance from the origin to point P and θ being the polar angle measured counterclockwise from the positive x-axis direction. Clearly, the polar angle θ is the only coordinate that changes in time. On the other hand, r remains unchanged. In contrast, in Cartesian coordinates, both x and y vary in time according to

$$x = r \cos \theta \tag{10.1}$$

$$y = r \sin \theta. \tag{10.2}$$

Assuming that at $t = 0$ the particle lies somewhere at the positive x-axis (i. e., $\theta = 0$), then after some time the particle moves along the circle (see Fig. 10.1) with an angle θ describing an arc of length s. There is a relationship between the arc s and the angular position θ:

$$s = r\theta \tag{10.3}$$

or

$$\theta = \frac{s}{r}. \tag{10.4}$$

From eq. (10.4), the unit of θ is radian (rad). However, because θ is the ratio of an arc length and the radius of the circle, it is a pure number. The full circle as a circumference is equal to $2\pi r$, and thus, an $360°$ angle corresponds to an angle in radians of

$$\theta = \frac{2\pi r}{r} \, \text{rad} = 2\pi \, \text{rad}. \tag{10.5}$$

https://doi.org/10.1515/9783110755824-010

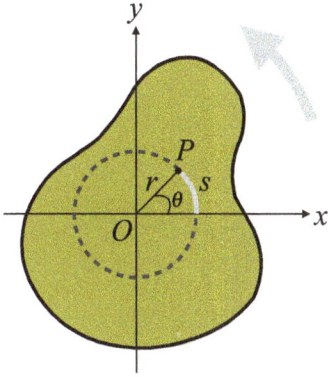

Figure 10.1: A rigid body rotation.

That is also called one revolution. Hence,

$$1\,\text{rad} = \frac{360°}{2\pi} \approx 57.3°. \tag{10.6}$$

In general, any angle in degrees converts to an angle in radians, using the relation $2\pi\,\text{rad} = 360°$:

$$\theta\,(\text{rad}) = \frac{\pi}{180°}\theta\,(\text{deg}). \tag{10.7}$$

Definition 10.1 (Angular displacement). If a particle in a rigid body moves from position P to a position Q in a time Δt along a circle with fixed radius r (see also Fig. 10.2),
the radius vector rotates through an angle

$$\Delta\theta = \theta_f - \theta_i. \tag{10.8}$$

Here, $\Delta\theta$ defines the *angular displacement* of the particle. In eq. (10.8), θ_i is the
polar angle at the initial position P and θ_f is the polar angle at the final position Q.

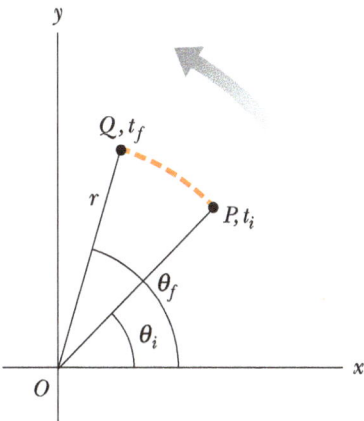

Figure 10.2: A particle in a rigid body rotation.

Definition 10.2 (Average angular velocity). By definition, the average angular velocity, denoted by $\bar{\omega}$, is the ratio of the angular displacement $\Delta\theta$ to the time interval Δt during which that displacement takes place:

$$\bar{\omega} = \frac{\Delta\theta}{\Delta t} = \frac{\theta_f - \theta_i}{t_f - t_i}. \tag{10.9}$$

Similar to linear velocity, the instantaneous angular velocity can be defined as follows.

Definition 10.3 (Instantaneous angular velocity). By definition, the instantaneous angular velocity, denoted by ω, is the limit when $\Delta t \to 0$ of the ratio of the angular displacement $\Delta\theta$ with the time interval Δt during which that displacement takes place:

$$\omega = \lim_{\Delta t \to 0} \frac{\Delta\theta}{\Delta t} = \frac{d\theta}{dt}. \tag{10.10}$$

Note that the units of the angular velocity are radians per second (or shortly rad/s), or rather second^{-1} (s^{-1}) because radians are not dimensional. Here, ω is positive, if θ increases for a counterclockwise motion; otherwise, it is negative if θ decreases when the rotation motion is clockwise.

If the instantaneous angular velocity of an object changes from ω_i at time t_i to ω_f at the time t_f during the time interval $\Delta t = t_f - t_i$, the object gains an angular acceleration.

Definition 10.4 (Average angular acceleration). By definition, the average angular acceleration, denoted by $\bar{\alpha}$, of a rotating object is the ratio of the angular velocity change $\Delta\omega$ with the time interval Δt during which that change takes place:

$$\bar{\alpha} = \frac{\omega_f - \omega_i}{t_f - t_i} = \frac{\Delta\omega}{\Delta t}. \tag{10.11}$$

In analogy to linear acceleration:

Definition 10.5 (Instantaneous angular acceleration). By definition, the instantaneous angular acceleration, denoted by α, of a rotating object is the limit $\Delta t \to 0$ of the ratio of the change in the angular velocity $\Delta\omega$ with the time interval Δt during which that change takes place:

$$\alpha = \lim_{\Delta t \to 0} \frac{\omega_f - \omega_i}{t_f - t_i} = \frac{d\omega}{dt}. \tag{10.12}$$

In either eq. (10.11) or eq. (10.12), the units of the angular acceleration are radians per second squared (or shortly rad/s^2), or just second^{-2} (s^{-2}). Note that α is positive when the angular velocity rate change of counterclockwise rotation is increasing or when the angular velocity rate change of clockwise rotation is decreasing.

When a rigid body rotates about a fixed axis, all particles on the rigid body rotate through the same angle, and they have the same angular speed and angular acceleration. That is, the quantities θ, ω, and α characterize the rotational motion of the entire *rigid body. Using these quantities, we can significantly simplify the analysis of rigid-body rotation.*

We have not specified any direction for ω and α. Strictly speaking, these variables are the magnitudes of the angular velocity and angular acceleration vectors, respectively, $\boldsymbol{\omega}$ and $\boldsymbol{\alpha}$, and hence they are always reported positive.

Here, we consider the rotation about a fixed axis; however, in general, the directions of the vectors can be specified by assigning a positive or negative sign to ω and α. In the case of rotations about the fixed axis, this direction, which uniquely specifies the rotational motion, is along the axis of rotation. Therefore, the directions of $\boldsymbol{\omega}$ and $\boldsymbol{\alpha}$ are both along this axis. When the rigid body rotates in the xy plane (as in Fig. 10.1), the direction of $\boldsymbol{\omega}$ is out of the plane of the diagram perpendicular to it if the rotation motion is counterclockwise and into the plane of the diagram when the rotation is clockwise. The right-hand rule is demonstrated in Fig. 10.3. That is, the extended right thumb points in the direction of $\boldsymbol{\omega}$ if our four fingers of the right hand are wrapped in the direction of rotation. The direction of $\boldsymbol{\alpha}$ follows from its definition,

$$\boldsymbol{\alpha} = \frac{d\boldsymbol{\omega}}{dt}. \tag{10.13}$$

Thus, the direction of $\boldsymbol{\alpha}$ is the same as the direction of $\boldsymbol{\omega}$ if the angular speed increases in time, and it is antiparallel to $\boldsymbol{\omega}$ if the angular speed decreases in time.

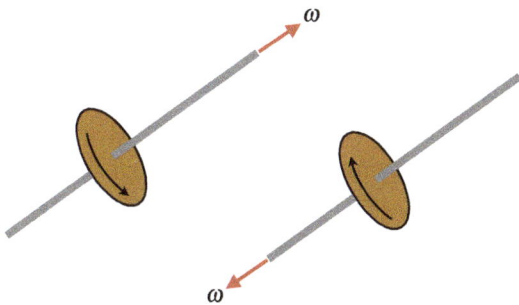

Figure 10.3: The right-hand rule for determining the direction of the angular velocity vector.

10.2 Rotational kinematics

In analogy to linear motion, in the following the rotational motion about a fixed axis is analyzed for the motion under constant angular acceleration. In particular, the kinematic relationships for this type of motion is developed. First, we write

$$\bar{\alpha} = \alpha = \frac{\omega_f - \omega_i}{t_f - t_i}. \tag{10.14}$$

Taking $t_i = 0$ and $t_f = t$, we find

$$\omega_f = \omega_i + \alpha t \tag{10.15}$$

for α constant, which represents the first kinematic equation for the rotational motion. Furthermore, for the rotation motion with α constant, we write

$$\bar{\omega} = \frac{1}{2}(\omega_i + \omega_f) = \frac{1}{2}(\omega_i + \omega_i + \alpha t) = \frac{1}{2}(2\omega_i + \alpha t) \qquad (10.16)$$

which on the other hand is

$$\frac{1}{2}(2\omega_i + \alpha t) = \bar{\omega} = \frac{\theta_f - \theta_i}{t_f - t_i} = \frac{\theta_f - \theta_i}{t} \qquad (10.17)$$

or

$$\theta_f = \theta_i + \omega_i t + \frac{1}{2}\alpha t^2. \qquad (10.18)$$

Equation (10.18) represents the second kinematic equation for rotational motion.
From eq. (10.15), we define the time t as follows:

$$t = (\omega_f - \omega_i)/\alpha \qquad (10.19)$$

and by substituting it into eq. (10.18), we obtain

$$\omega_f^2 - \omega_i^2 = 2\alpha(\theta_f - \theta_i) \qquad (10.20)$$

which is the third kinematic equation of rotational motion with constant α.

10.3 Angular and linear quantities

In this section, we define some useful relationships between the angular speed and acceleration of the rigid-body rotational motion and the linear speed and acceleration of any arbitrary point of this rigid body. We mentioned that when a rigid body rotates about a fixed axis, see Fig. 10.4, every particle of the object moves along a circle with a center located at the axis of rotation.

Since the point P moves in a circle with some radius r, the linear velocity vector **v** is tangent to the circular path, and hence it is a tangential velocity. The magnitude of the tangential velocity at the point P of the rigid body is by definition the so-called tangential speed given as

$$v = \frac{ds}{dt}. \qquad (10.21)$$

Here, s is the arc describing the length of the path (or distance) along the circle moved by the point P. Furthermore, $s = r\theta$ where r is constant for the point P. Therefore, we obtain

$$v = \frac{ds}{dt} = \frac{d(r\theta)}{dt} = r\frac{d\theta}{dt} = r\omega. \qquad (10.22)$$

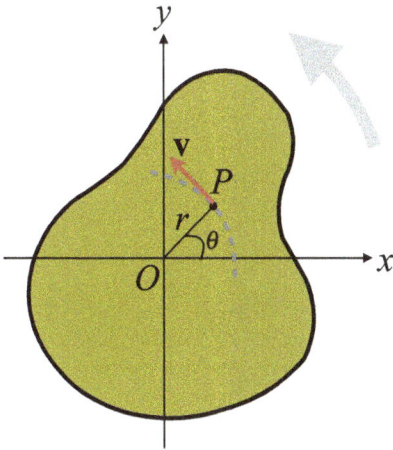

Figure 10.4: Rotation of a rigid body. Vector **v** denotes the linear velocity vector at point P in the rigid body.

Equation (10.22) indicates that the tangential speed of a point in a rotating rigid body equals the product of the perpendicular distance of that point from the axis of rotation r with the angular speed ω. Furthermore, the linear speed is proportional to r. Therefore, every point on the rigid body has the same angular speed; however, not all points of the rigid body have the same linear speed because r is different.

Equation (10.22) shows that the linear speed of a point on the rotating rigid body increases as one moves outward from the center of rotation. For example, as we intuitively expect, the outer end of a swinging baseball bat (which is further from the axis of rotation) moves faster than the handle (which is closer to the axis of rotation).

In the following, we find a relationship between the angular acceleration of the rotating rigid body with the tangential acceleration of the point P. For this purpose, we take the time derivative of v in eq. (10.22):

$$a_t = \frac{dv}{dt} = \frac{d(r\omega)}{dt} \tag{10.23}$$

or

$$a_t = r\frac{d\omega}{dt} = r\alpha \tag{10.24}$$

where α is the angular acceleration.

Equation (10.24) indicates that the tangential component of the linear acceleration, namely a_t, of a point in a rigid body that rotates is given by the product of the distance of the point from the axis of rotation, r, with the angular acceleration α.

Using the relation between the magnitude of the centripetal (or radial) acceleration a_r and the magnitude of the linear velocity as (see eq. (4.53), Chapter 4):

$$a_r = \frac{v^2}{r} \tag{10.25}$$

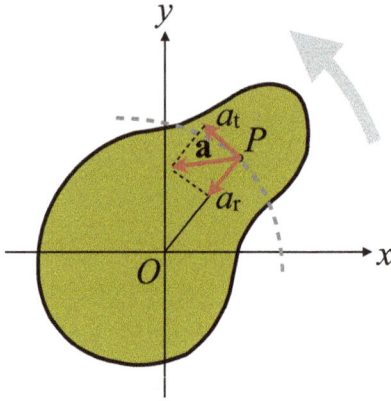

Figure 10.5: Rotation of a rigid body. Vector **a** is the linear acceleration of particle at point P, and a_r and a_t are the radial and tangential components, respectively.

which is directed toward the center of rotation (see Fig. 10.5), we obtain (using eq. (10.22))

$$a_r = \frac{v^2}{r} = r\omega^2. \tag{10.26}$$

We use the definition of the total linear acceleration vector,

$$\mathbf{a} = \mathbf{a}_t + \mathbf{a}_r. \tag{10.27}$$

Combining eq. (10.24) and eq. (10.26), we obtain the total linear acceleration magnitude of the point:

$$a = \sqrt{a_t^2 + a_r^2} = \sqrt{r^2\alpha^2 + r^2\omega^4} = r\sqrt{\alpha^2 + \omega^4}. \tag{10.28}$$

10.4 Rotation energy

To define the kinetic energy of a rigid body that rotates, we first describe it as a collection of particles. Furthermore, we assume that the rigid body rotates about the fixed z-axis with an angular speed ω (Fig. 10.6). To each particle we associate a mass m_i (where the index i runs over all particles of the rigid body), and a linear speed v_i. Therefore, the kinetic energy of the rotating rigid body is determined as

$$K_R = \sum_i K_i = \sum_i \frac{1}{2} m_i v_i^2 = \frac{1}{2} \sum_i m_i r_i^2 \omega^2 \tag{10.29}$$

or

$$K_R = \frac{1}{2} \left(\sum_i m_i r_i^2 \right) \omega^2. \tag{10.30}$$

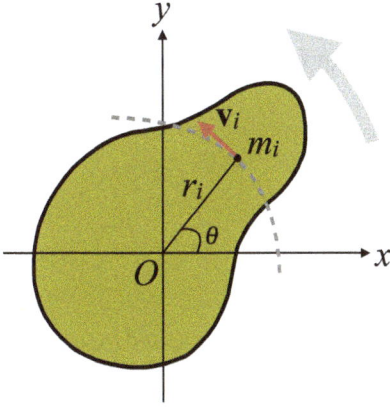

Figure 10.6: Rotation of a rigid body. Vector v_i is the linear velocity of particle m_i at the distance r_i from the center O.

Definition 10.6 (Moment of inertia). By definition, the moment of inertia I is

$$I = \sum_i m_i r_i^2.$$

(10.31)

From the definition of the moment of inertia (see eq. (10.31)), the dimensions of I are ML^2 (kg m^2 in SI units).

Using eq. (10.30) and eq. (10.31), we obtain a simplified expression for the kinetic energy:

$$K_R = \frac{1}{2} I \omega^2.$$

(10.32)

Equation (10.32) indicates that the expression of K_R is similar to that of the linear kinetic energy $K = mv^2/2$ because both depend on the square of speed. Furthermore, the moment of inertia I depends on the mass, as shown in eq. (10.31).

10.5 Calculation of the moments of inertia

To calculate the moment of inertia of an extended rigid body, we assume that the object is divided into many small elementary volumes with masses Δm_i, where the index i runs over all volume elements. From the definition, eq. (10.31), the moment of inertia of the rigid body is approximated as

$$I \approx \sum_i r_i^2 \Delta m_i.$$

(10.33)

Taking the limit of this sum in eq. (10.33) as $\Delta m_i \to 0$, we obtain the exact moment of inertial of the rigid body:

$$I = \lim_{\Delta m_i \to 0} \sum_i r_i^2 \Delta m_i = \int r^2 dm.$$

(10.34)

We can express the moments of inertia in terms of the volume of the elements dV, using $\rho = m/V$ with ρ being the density of the object and V its volume. We can express dm in terms of dV as

$$dm = \rho dV \tag{10.35}$$

then eq. (10.34) can be written as

$$I = \int_V r^2 \rho(r) dV. \tag{10.36}$$

For homogeneous objects, ρ is constant and the integral can be evaluated for a known geometry. For ρ not being constant, its variation with position must be known to complete the integration. The density given by $\rho = m/V$ is called the *volume density*. We will also use other expressions of density, for example, when dealing with a sheet of uniform thickness t, we can use the *surface density* $\sigma = \rho t$, and when mass is distributed uniformly along a rod of cross-sectional area A, we will use the *linear density* $\mu = M/L = \rho A$, that is, the mass per unit length of the rod.

Theorem 10.1. *The moment of inertia of a rigid body about any given axis is expressed as the sum of the moment of inertia of it about the axis passing through its center of mass and Md^2 with d being the distance from the center of mass to the axis of rotation:*

$$I = I_{c.m.} + Md^2. \tag{10.37}$$

Here, M is the entire mass of the rigid body. This is known as the parallel axis theorem *or Huygens–Steiner theorem.*

Proof. In Fig. 10.7, we illustrate the theorem. Suppose that there are two axes of rotation. First rotation axis is passing through the center of mass, and the other is passing through some arbitrary point O'. By definition, eq. (10.34), the moment of inertial about the axis passing through the center of mass is

$$I_{c.m.} = \int r^2 dm \tag{10.38}$$

and the moment of inertial about the axis passing through the point O' is

$$I = \int (r')^2 dm \tag{10.39}$$

where

$$\mathbf{r}' = \mathbf{r} + \mathbf{d}. \tag{10.40}$$

Therefore,

$$I = \int |\mathbf{r} + \mathbf{d}|^2 dm \tag{10.41}$$
$$= \int r^2 dm + 2\mathbf{d} \cdot \int \mathbf{r} dm + d^2 \int dm.$$

The second term is zero because the origin of the coordinate system is at the center of mass of the rigid body. Therefore, we obtain

$$I = I_{c.m.} + Md^2. \tag{10.42}$$

☐

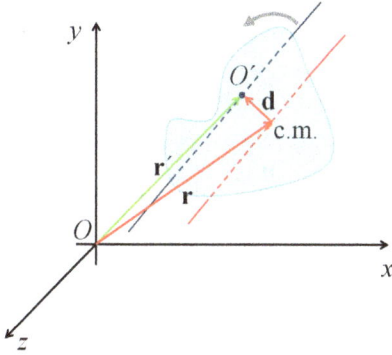

Figure 10.7: Rotation of a rigid body about two axes passing through the center of mass (*c.m.*) and O', respectively.

10.6 Torque

The force exerted on a rigid body, pivoted about an axis, causes the body to rotate about that axis. The capability of a force to rotate a rigid body about an axis of rotation is measured by *torque* τ, which is a vector quantity.

Suppose we have a rod pivoted on the axis through O in Fig. 10.8. We apply a force **F** at an angle ϕ to the horizontal direction (x-axis).

Definition 10.7 (Torque). By definition, the magnitude of the torque associated with the force **F** is as follows:

$$\tau = rF \sin \phi = dF. \tag{10.43}$$

Here, r is the distance between the pivot point and the point where **F** is acting. d is the distance perpendicular to the line of action of **F** from the pivot point O. From the right triangle in Fig. 10.8 that has the x-axis as its hypotenuse, we have

$$d = r \sin \phi. \tag{10.44}$$

Here, d is called the *moment arm* (or *lever arm*) of the applied **F**.

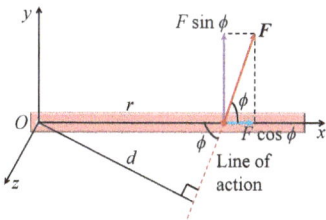

Figure 10.8: The force **F** has a larger rotating tendency about O as both its magnitude F and the moment of arm d increase. It is the component $F \sin \phi$ that tends to rotate the rod about O.

Note that the definition of the torque requires the specification of a reference axis. From eq. (10.43), the torque is the product of the applied force and the moment arm d of that force, which is defined only if an axis of rotation is specified.

In the cases when more than one force is acting on a rigid body, as shown in the example of Fig. 10.9, each force produces a rotation about the pivot at point O. For example, $\mathbf{F}_2$ rotates the body clockwise, and $\mathbf{F}_1$ rotates it counterclockwise.

By convention, the sign of the torque resulting from a force is positive, if the force rotates counterclockwise and is negative if it rotates clockwise. In the example shown in Fig. 10.9, the torque resulting from $\mathbf{F}_1$, which has a moment arm d_1, is positive and equal to $+F_1 d_1$; the torque from $\mathbf{F}_2$ is negative and equal to $-F_2 d_2$. Hence, the resultant torque about O is

$$\sum \tau = \tau_1 + \tau_2 = F_1 d_1 - F_2 d_2. \tag{10.45}$$

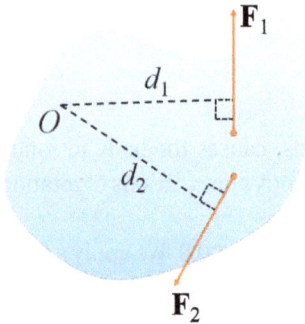

Figure 10.9: The force $\mathbf{F}_1$ tends to rotate the object counter-clockwise about O, and $\mathbf{F}_2$ tends to rotate it clockwise.

Similar to the force, torque is a vector quantity, and hence it has a magnitude and direction. As a vector, the torque is defined as

$$\boldsymbol{\tau} = \mathbf{r} \times \mathbf{F}. \tag{10.46}$$

Equation (10.46) indicates that the direction of the torque changes if the direction of the force $\mathbf{F}$ or its position $\mathbf{r}$ changes. Since ϕ is the angle between $\mathbf{r}$ and $\mathbf{F}$, eq. (10.43) implies that the maximum torque is obtained for an angle 90°. Meanwhile a minimum torque is obtained for $\phi = 0°$; that is, a force perpendicular to the axis of rotation will not cause any rotation.

10.7 Relation between the torque and angular acceleration

Consider first a particle with mass m rotating in a horizontal circle of radius r because of a tangential force $\mathbf{F}_t$ and radial force $\mathbf{F}_r$. Knowing that the tangential force produces a tangential acceleration based on the second law of Newton:

$$a_t = \frac{F_t}{m}.$$

(10.47)

The magnitude of the torque about the center of the circle due to the force $\mathbf{F}_t$ is

$$\tau = F_t r \sin \phi = F_t r \sin 90° = (ma_t)r$$

(10.48)

because the angle between the tangential force $\mathbf{F}_t$ and $\mathbf{r}$ is $\phi = 90°$, and hence $\sin \phi = 1$. Using the relationship between the tangential acceleration and the angular acceleration,

$$a_t = r\alpha$$

(10.49)

we obtain

$$\tau = (mr^2)\alpha = I\alpha.$$

(10.50)

In eq. (10.50), $I = mr^2$ is the moment of inertia of the mass m rotating about the vertical z-axis passing through the origin a circle.

The discussion can be extended for any rigid body of arbitrary shape rotating about a fixed axis, as shown in Fig. 10.10. The body can be composed of an infinite number of mass elements dm of little volume. Considering the Cartesian coordinate system connected to the rigid body, every of those small mass elements rotates about the origin O in a circle centered in O. The external tangential force $d\mathbf{F}_t$ produces a tangential acceleration a_t for every mass element dm. According to Newton's second law:

$$dF_t = (dm)a_t.$$

(10.51)

Then the torque produced by $d\mathbf{F}_t$ for every element about the origin is given by

$$d\tau = (rdm)r\alpha = (r^2 dm)\alpha.$$

(10.52)

Integrating this expression gives the resultant torque as

$$\tau = \int (r^2 dm)\alpha = \alpha \int r^2 dm.$$

(10.53)

Here, α is taken outside the integral because every mass elements have the same α. From the definition of the moment of inertia of a rigid body, we get

$$\tau = I\alpha.$$

(10.54)

That is the same relationship as we found for a particle rotating in a circle. The relation obtained in eq. (10.54) is analogous to the second law of Newton for the linear motion.

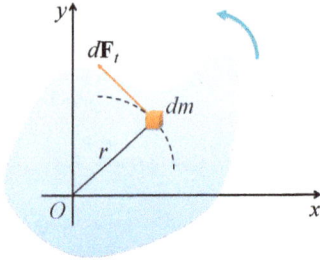

Figure 10.10: A rigid body rotating about an axis through O. Each mass element dm rotates about O with the same angular acceleration α, and the net torque on the object is proportional to α.

10.8 Work, power and energy in rotational motion

Consider a rigid body rotating about an axis through point O, as shown in Fig. 10.11. Assume the external force $\mathbf{F}$ acting on a single point P, lying in the page plane. Furthermore, ds is an infinitesimal distance of the rotation of the body due to the force in an interval of time dt. Then the work done by the force $\mathbf{F}$ on the rigid body is

$$dW = \mathbf{F} \cdot d\mathbf{s}. \tag{10.55}$$

Here, $ds = rd\theta$. Furthermore, this expression is written as

$$dW = Fds\cos(90° - \phi) = Fr\sin\phi d\theta. \tag{10.56}$$

Here, $90° - \phi$ is the angle between $\mathbf{F}$ and $d\mathbf{s}$. On the other hand, the torque due to the force $\mathbf{F}$ is

$$\boldsymbol{\tau} = \mathbf{r} \times \mathbf{F} \tag{10.57}$$

with a magnitude given as

$$\tau = rF\sin\phi. \tag{10.58}$$

Therefore, the work can take the following form:

$$dW = \tau d\theta. \tag{10.59}$$

To determine the rate at which the work is being done by the force $\mathbf{F}$ as the object rotates about a fixed axis, we divide both sides of eq. (10.59) by dt:

$$\frac{dW}{dt} = \tau\frac{d\theta}{dt}. \tag{10.60}$$

Here, $\omega = d\theta/dt$ is the angular velocity and dW/dt is the instantaneous power P. Therefore, we obtain

$$P = \tau\omega. \tag{10.61}$$

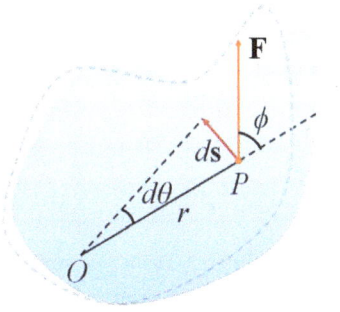

Figure 10.11: A rigid body rotates about an axis through O under the action of an external force **F** applied at P.

Equation (10.61) is analogous to $P = Fv$ in the case of linear motion, and the expression $dW = \tau d\theta$ is analogous to $dW = F_x dx$.

In studying linear motion, we derived the *work–kinetic energy theorem*, very useful in describing the movement of a system. We extend this concept to rotational motion. Similar to linear motion, we expect that when the symmetric object rotates about a fixed axis, then the work done by external forces on the object equals the change in the rotational energy.

For this purpose, using eq. (10.54),

$$\tau = I\alpha = I\frac{d\omega}{dt} \tag{10.62}$$

and applying the chain rule, we write

$$\tau = I\frac{d\omega}{dt} = I\frac{d\omega}{d\theta}\frac{d\theta}{dt} = I\frac{d\omega}{d\theta}\omega. \tag{10.63}$$

After rearranging this expression, we obtain

$$\tau d\theta = I\omega d\omega. \tag{10.64}$$

Integrating this expression, the left-hand side will give the total work:

$$W = \int_{\theta_i}^{\theta_f} \tau d\theta. \tag{10.65}$$

Therefore, we can write

$$W = \int_{\omega_i}^{\omega_f} I\omega d\omega = \frac{1}{2}I\omega_f^2 - \frac{1}{2}I\omega_i^2. \tag{10.66}$$

Work–kinetic energy theorem for rotational motion

Therefore, we obtain the so-called *work–kinetic energy theorem for rotational motion*:

The net work done by external forces in rotating the symmetric rigid body about a fixed axis equal the change in the object's rotational kinetic energy.

In eq. (10.66), the angular speed changes from ω_i to ω_f as the angular position changes from θ_i to θ_f; that is,

$$\omega_f^2 - \omega_i^2 = 2\alpha(\theta_f - \theta_i). \tag{10.67}$$

Combining eq. (10.66) and eq. (10.67), and using the relation in eq. (10.54), we obtain

$$W = \tau(\theta_f - \theta_i). \tag{10.68}$$

10.9 Exercises

Exercise 10.1. Consider a wheel rotating with a constant angular acceleration of $3.50\,\text{rad/s}^2$. Assume that the angular speed of the wheel is $2.00\,\text{rad/s}$ at $t_i = 0$. (a) What is the angle that the wheel rotates in $2.00\,\text{s}$? (b) Find the angular speed at $t = 2.00\,\text{s}$.

Exercise 10.2. Consider the oxygen molecule (O_2), which rotates about the z-axis in the xy plane. Assume that the z-axis passes through the center of the molecule, and it is perpendicular to its length. The mass of each oxygen atom is 2.66×10^{-26} kg, and the average distance between the two atoms is $d = 1.21 \times 10^{-10}$ m (the atoms are treated as point masses) at room temperature. (a) Find the moment of inertia of the molecule about the z-axis. (b) When the angular speed of the molecule about the z-axis is 4.60×10^{12} rad/s, determine its rotational kinetic energy.

Exercise 10.3. Consider four tiny spheres fastened at the corners of a frame of negligible mass laying in the xy plane (Fig. 10.12). Assume that the radii of the spheres are small compared with the dimensions of the frame and that the system rotates about the y-axis with an angular speed of ω. What are the moment of inertia and the rotational kinetic energy about this axis?

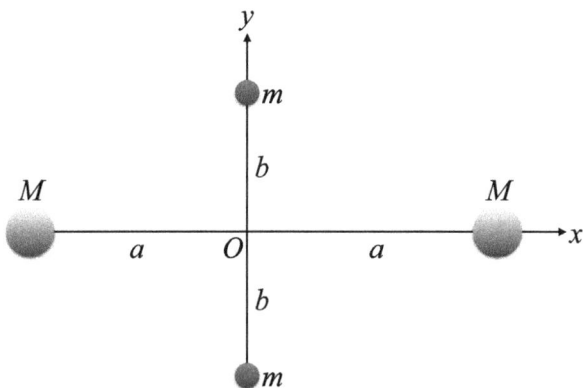

Figure 10.12: A system of four spheres.

Exercise 10.4. Consider the uniform rigid rod of mass M and length L as shown in Fig. 10.13. What is the moment of inertia of the rod about an axis perpendicular to the rod through (a) the center of mass and (b) one end (the y'-axis)?

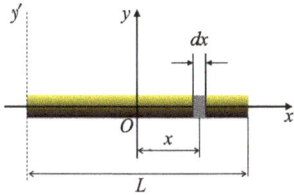

Figure 10.13: A uniform rigid rod of length L with a moment of inertia about the y-axis being less than that about the y'-axis.

Exercise 10.5. Consider a uniform hoop of mass M and radius R that is rotating about the axis perpendicular to the plane of the hoop and passing through its center of mass (see Fig. 10.14). What is the moment of inertia?

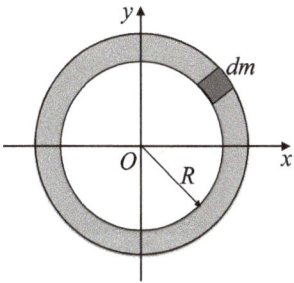

Figure 10.14: The mass elements dm of a uniform hoop are all the same distance from O.

Exercise 10.6. Consider a uniform solid cylinder has a radius R, mass M, and length L. Find the moment of inertia of that cylinder about the central axis (z-axis in Fig. 10.15).

Exercise 10.7. Consider a uniform rod with length L and mass M, which is free to rotate on a frictionless pin passing through one end, as shown in Fig. 10.16. The rod is released from rest in the horizontal position. Find its angular speed when it reaches its lowest position.

Exercise 10.8. Consider two cylinders having masses m_1 and m_2, where $m_1 \neq m_2$, connected by a string passing over a pulley, as shown in Fig. 10.17. The pulley has a radius R and moment of inertia I about its axis of rotation. The string does not slip on the pulley, and the system is released from rest. Find the linear speeds of the cylinders after cylinder 2 ends a distance h and the angular speed of the pulley at this time.

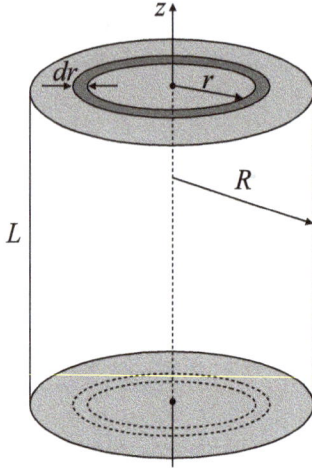

Figure 10.15: Calculating I about the z-axis for a uniform solid cylinder.

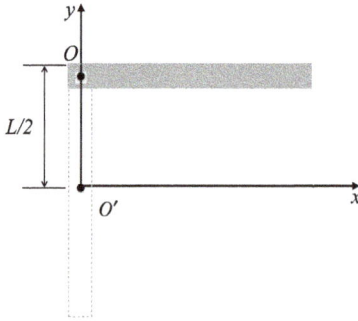

Figure 10.16: Calculating I about the z-axis for a uniform solid cylinder.

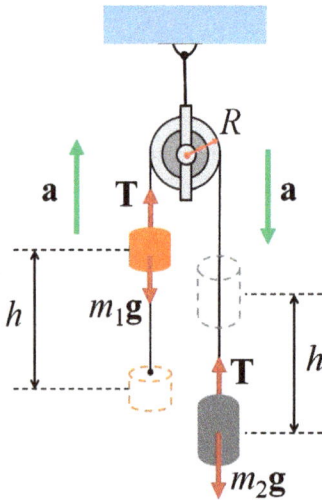

Figure 10.17: Rotating system of two cylinders.

Exercise 10.9. Suppose a wheel starts rotating from rest with constant angular acceleration. It reaches the angular speed of 12.0 rad/s in 3.00 s. (a) Find the magnitude of the angular acceleration. (b) Find the angle (in radians) through which it rotates in this time.

Exercise 10.10. Suppose that the angular position of a swinging door is described by $\theta = (5.00+10.0t+2.00t^2)$ rad. What is the angular position, angular speed, and angular acceleration of the door (a) at $t = 0$ and (b) at $t = 3.00$ s?

11 Rolling motion and angular momentum

In Chapter 10, we discussed the motion of a rigid body rotating about a fixed axis. In this chapter, we introduce a more general motion in which the axis of rotation is not fixed in space. First, we discuss the so-called *rolling motion*. Next, we introduce *angular momentum* as an important quantity in the rotational dynamics. In analogy to the conservation of linear momentum, the angular momentum of a rigid body is always conserved if no external torques act on the object.

11.1 Rolling motion of a rigid body

Assume a cylinder (or a sphere) rolling on a straight line. In particular, the center of mass moves along a straight line, and a point on the circumference moves in a complex path called a *cycloid*. In such motion, the axis of rotation always remains parallel to its initial orientation in space. For that, we consider a uniform cylinder of radius R rolling without slipping on a horizontal surface (see also Fig. 11.1). When the cylinder rotates through an angle θ about the axis of rotation passing through its center, the center of mass of the cylinder moves along a straight line with a distance $l = s = R\theta$. To calculate the linear speed of the center of mass for pure rolling motion, the following formula can be used:

$$v_{c.m.} = \frac{ds}{dt} = R\frac{d\theta}{dt} = R\omega. \tag{11.1}$$

In eq. (11.1), ω is the angular velocity of the cylinder. Note that eq. (11.1) holds if the cylinder or a sphere rolls without slipping, which is the condition for pure rolling motion. In addition, the magnitude of the linear acceleration of the center of mass for that pure rolling motion is

$$a_{c.m.} = \frac{dv_{c.m.}}{dt} = R\frac{d\omega}{dt} = R\alpha. \tag{11.2}$$

Here, α is the angular acceleration of the cylinder.

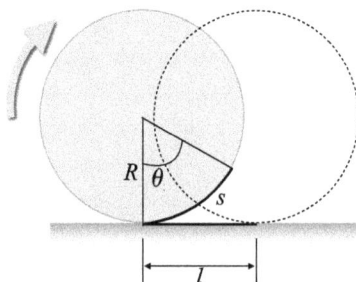

Figure 11.1: In pure rolling motion, as the cylinder rotates through an angle θ, its center of mass moves a linear distance l.

https://doi.org/10.1515/9783110755824-011

The kinetic energy of the rolling cylinder is

$$K = \frac{1}{2}I_p\omega^2. \tag{11.3}$$

In eq. (11.3), I_p is the moment of inertia about the rotation axis through P on the rim. Using the parallel-axis theorem (introduced in Chapter 10), we substitute

$$I_p = I_{c.m.} + MR^2 \tag{11.4}$$

into eq. (11.3), and we obtain

$$K = \frac{1}{2}I_{c.m.}\omega^2 + \frac{1}{2}MR^2\omega^2. \tag{11.5}$$

Using the relationship between the linear and angular speed as $v_{c.m.} = R\omega$, we get

$$K = \frac{1}{2}I_{c.m.}\omega^2 + \frac{1}{2}Mv_{c.m.}^2. \tag{11.6}$$

In eq. (11.6), the first term is the rotational kinetic energy of the cylinder about the center of mass; the second term is the translation kinetic energy the cylinder through space without rotating. Therefore, for a rolling object, the total kinetic energy equals the sum of the rotational kinetic energy about the center of mass and the translation kinetic energy of the center of mass.

11.2 Angular momentum of a particle

In analogy to the linear momentum, used to describe linear motion, the *angular momentum* is used to describe the rotational motion.

For a particle with mass m located at the position $\mathbf{r}$ and moving with a linear velocity $\mathbf{v}$, we can define the angular momentum as follows.

Definition 11.1 (Angular momentum). By definition, the instantaneous angular momentum $\mathbf{L}$ of the particle relative to the origin O of a reference frame is given as

$$\mathbf{L} = \mathbf{r} \times \mathbf{p}. \tag{11.7}$$

The SI unit of angular momentum is $kg \cdot m^2/s$. It is important to note that both the magnitude and the direction of $\mathbf{L}$ depend on the choice of origin. For example, consider a shift of the origin from point O to O' such that $\mathbf{r}_{OO'} = \mathbf{d}$ (which is a vector with tail at point O and tip at O'). Then the position of the particle with respect to the reference frame with origin in O' is

$$\mathbf{r}' = \mathbf{r} + \mathbf{d}. \tag{11.8}$$

The angular momentum $\mathbf{L}'$ of the particle relative to the origin O' is

$$\mathbf{L}' = \mathbf{r}' \times \mathbf{p} \tag{11.9}$$
$$= (\mathbf{r} + \mathbf{d}) \times \mathbf{p}$$
$$= \mathbf{L} + \mathbf{d} \times \mathbf{p}$$

where $\mathbf{L}$ is the angular momentum relative to the origin O. Equation (11.9) indicates that $\mathbf{L}' \neq \mathbf{L}$ because $\mathbf{d} \times \mathbf{p} \neq 0$.

Often, the right-hand rule is used to determine the direction of vector $\mathbf{L}$. Note that the direction of $\mathbf{L}$ is perpendicular to the plane formed by $\mathbf{r}$ and $\mathbf{p}$ (see also Fig. 11.2).

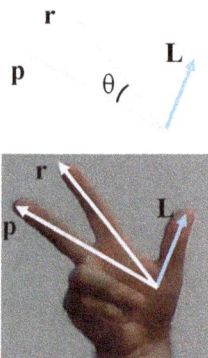

Figure 11.2: The angular momentum $\mathbf{L}$ of a particle of mass m and linear momentum $\mathbf{p}$ located at the vector position $\mathbf{r}$. The value of $\mathbf{L}$ depends on the origin about which it is measured. Also, it is a vector perpendicular to both $\mathbf{r}$ and $\mathbf{p}$.

Because, $\mathbf{p} = m\mathbf{v}$, the magnitude of $\mathbf{L}$ is

$$L = |\mathbf{r} \times \mathbf{p}| \tag{11.10}$$
$$= |\mathbf{r}| \cdot |\mathbf{p}| \sin \theta$$
$$= rmv \sin \theta.$$

In eq. (11.10), θ is the angle between $\mathbf{r}$ and $\mathbf{p}$. Furthermore, eq. (11.10) indicates that L is 0 when $\mathbf{r}$ and $\mathbf{p}$ are co-linear vectors ($\theta = 0°$ or $\theta = \pi$). If $\mathbf{r}$ is perpendicular to $\mathbf{p}$, then $\theta = \pi/2$, and hence we obtain the maximum value, $L = mvr$.

We use Newton's second law in the form

$$\sum_i \mathbf{F}_i = \frac{d\mathbf{p}}{dt} \tag{11.11}$$

with $\sum_i \mathbf{F}_i$ being the resultant force exerted on the particle. Then the resultant torque is

$$\sum_i \boldsymbol{\tau}_i = \mathbf{r} \times \left(\sum_i \mathbf{F}_i \right) = \mathbf{r} \times \frac{d\mathbf{p}}{dt} \tag{11.12}$$

$$= \frac{d}{dt}(\mathbf{r} \times \mathbf{p}) - \left(\frac{d\mathbf{r}}{dt}\right) \times \mathbf{p}$$

$$= \frac{d\mathbf{L}}{dt} - \mathbf{v} \times (m\mathbf{v})$$

$$= \frac{d\mathbf{L}}{dt}$$

because the term $m(\mathbf{v} \times \mathbf{v}) = 0$. Equation (11.12) indicates that the resultant torque is equal to the time derivative of the angular momentum:

$$\sum_i \boldsymbol{\tau}_i = \frac{d\mathbf{L}}{dt}. \tag{11.13}$$

Note that in eq. (11.13), both $\sum_i \boldsymbol{\tau}_i$ and $\mathbf{L}$ are measured about the same origin. Furthermore, eq. (11.13) holds for any origin fixed in an inertial frame.

11.3 Angular momentum of a system of particles

For a system of N particles with angular momenta $\mathbf{L}_i$ for $i = 1, 2, \ldots, N$, the total angular momentum is

$$\mathbf{L} = \sum_{i=1}^{N} \mathbf{L}_i = \sum_{i=1}^{N} \mathbf{r}_i \times \mathbf{p}_i. \tag{11.14}$$

Consider again a linear transformation on the particle coordinates as $\mathbf{r}_i' = \mathbf{r}_i + \mathbf{d}$, where $\mathbf{d}$ is a constant vector. The following transformation of the angular momentum of the system of particles occurs:

$$\mathbf{L}' = \sum_{i=1}^{N} \mathbf{r}_i' \times \mathbf{p}_i \tag{11.15}$$

$$= \sum_{i=1}^{N} [\mathbf{r}_i \times \mathbf{p}_i + \mathbf{d} \times \mathbf{p}_i]$$

$$= \mathbf{L} + \sum_{i=1}^{N} \mathbf{d} \times \mathbf{p}_i$$

$$= \mathbf{L} + \mathbf{d} \times \mathbf{P}.$$

Thus, the total angular momentum on the new coordinates is the angular momentum on the old coordinates, plus the angular momentum of the vector $\mathbf{d}$. Only if total linear momentum $\mathbf{P}$ of system is fixed to 0, $\mathbf{P} = 0$, then $\mathbf{L}' = \mathbf{L}$.

The individual angular momentum $\mathbf{L}_i$ changes with time t, therefore, $\mathbf{L}$ will change with t. The time derivative of the total angular momentum is

$$\frac{d\mathbf{L}}{dt} = \frac{d}{dt} \sum_{i=1}^{N} \mathbf{L}_i = \sum_{i=1}^{N} \frac{d\mathbf{L}_i}{dt}. \tag{11.16}$$

We assume that the net force acting on each particle i is sum of the internal forces of all other particles j ($j \neq i$) and a single external force $\mathbf{F}_i^e$. Using eq. (11.13), we get

$$\frac{d\mathbf{L}}{dt} = \sum_{i=1}^{N} (\boldsymbol{\tau}_i + \boldsymbol{\tau}_i^e) = \sum_{i=1}^{N} \mathbf{r}_i \times \left(\sum_{j=1 \neq i}^{N} \mathbf{F}_{ij} + \mathbf{F}_i^e \right). \tag{11.17}$$

The double sum $\sum_{i=1}^{N} \mathbf{r}_i \times \sum_{j=1 \neq i}^{N} \mathbf{F}_{ij}$ is a combination of the following terms:

$$\mathbf{r}_i \times \mathbf{F}_{ij} + \mathbf{r}_j \times \mathbf{F}_{ji} \tag{11.18}$$

where, based on the third law of Newton: $\mathbf{F}_{ij} = -\mathbf{F}_{ji}$, we get

$$\mathbf{r}_i \times \mathbf{F}_{ij} + \mathbf{r}_j \times \mathbf{F}_{ji} = (\mathbf{r}_i - \mathbf{r}_j) \times \mathbf{F}_{ij} = \mathbf{r}_{ij} \times \mathbf{F}_{ij} = 0 \tag{11.19}$$

because $\mathbf{F}_{ij}$ is parallel to $\mathbf{r}_{ij}$. To prove that, suppose that the mutual forces between two particles i and j can be obtained from a potential function V_{ij} of the distance between the two particles $|\mathbf{r}_i - \mathbf{r}_j|$. For example, the force on the particle i due to particle j is

$$\mathbf{F}_{ji} = -\nabla_i V_{ij}(|\mathbf{r}_i - \mathbf{r}_j|)$$

which satisfies the law of action and reaction (Newton's third law):

$$\mathbf{F}_{ji} = -\nabla_i V_{ij}(|\mathbf{r}_i - \mathbf{r}_j|) = +\nabla_j V_{ij}(|\mathbf{r}_i - \mathbf{r}_j|) = -\mathbf{F}_{ij}$$

where $\mathbf{F}_{ij}$ is the force acting on the particle j due to particle i. Furthermore, we can write

$$\nabla_i V_{ij}(|\mathbf{r}_i - \mathbf{r}_j|) = \frac{\partial |\mathbf{r}_i - \mathbf{r}_j|}{\partial \mathbf{r}_i} \frac{\partial V_{ij}(|\mathbf{r}_i - \mathbf{r}_j|)}{\partial |\mathbf{r}_i - \mathbf{r}_j|} = (\mathbf{r}_i - \mathbf{r}_j) f$$

where f is a scalar function of only the distance $|\mathbf{r}_i - \mathbf{r}_j|$, given as

$$f(|\mathbf{r}_i - \mathbf{r}_j|) = \frac{1}{|\mathbf{r}_i - \mathbf{r}_j|} \frac{\partial V(|\mathbf{r}_i - \mathbf{r}_j|)}{\partial |\mathbf{r}_i - \mathbf{r}_j|}.$$

This indicates that the forces $\mathbf{F}_{ij}$ (or $\mathbf{F}_{ji}$) lie along the direction connecting the two particles.

Note that, if V_{ij} will also depend on other difference vectors associated with the two particles, such as velocities, then the forces would still be equal and opposite, but not necessarily lie along the direction between them.

Therefore, we obtain

$$\frac{d\mathbf{L}}{dt} = \sum_{i=1}^{N} \mathbf{r}_i \times \mathbf{F}_i^e = \boldsymbol{\tau}^e \tag{11.20}$$

where $\boldsymbol{\tau}^e$ is the total external torque on the system due to the external forces. Therefore, the time rate of change of the total angular momentum of a system about some origin in an inertial frame equals the net external torque acting on the system about that origin.

Conservation law of the angular momentum

Furthermore, if there is no external forces acting on the system, then $\tau^e = 0$, and we obtain

$$\frac{d\mathbf{L}}{dt} = 0 \qquad (11.21)$$

or

$$\mathbf{L} = \text{constant} \qquad (11.22)$$

which is the *conservation law of the angular momentum* of the system of particles.

Consider again the linear transformation on the particle coordinates, $\mathbf{r}_i' = \mathbf{r}_i + \mathbf{d}$, as discussed above. The net torque on system for the new coordinates is

$$\sum_i \boldsymbol{\tau}_i' = \sum_{i=1}^{N} \mathbf{r}_i' \times \dot{\mathbf{p}}_i \qquad (11.23)$$

$$= \sum_{i=1}^{N} [\mathbf{r}_i \times \dot{\mathbf{p}}_i + \mathbf{d} \times \dot{\mathbf{p}}_i]$$

$$= \sum_i \boldsymbol{\tau}_i + \sum_{i=1}^{N} \mathbf{d} \times \dot{\mathbf{p}}_i$$

$$= \sum_i \boldsymbol{\tau}_i + \mathbf{d} \times \frac{d\mathbf{P}}{dt}.$$

It can be seen that the net torque is conserved only if the total linear momentum of system is a conserved quantity: $d\mathbf{P}/dt = 0$. Such transformations could correspond to, for example, the position of particles with respect to the center of mass of the system. Or they could correspond to transformations of the coordinates in a system with periodic boundary conditions.

11.4 Angular momentum of a rigid body

Assume a rigid body rotating about a fixed z-axis, as shown in Fig. 11.3. We determine the angular momentum of this body as follows. Each particle with mass m_i at a distance r_i from the origin O of the rigid body rotates in the xy plane about the z-axis with an angular speed ω. The magnitude of the angular momentum of the ith particle about the point O is

$$L_i = m_i v_i r_i = m_i r_i^2 \omega. \qquad (11.24)$$

Here, the relation $v_i = r_i \omega_i$ is used. For the all-rigid body, we can sum over all particles composing the rigid body as

$$L_z = \sum_i m_i r_i^2 \omega = I\omega. \qquad (11.25)$$

I is the moment of inertia of the rigid body about z-axis given as

$$I = \sum_i m_i r_i^2.$$ (11.26)

Here, L_z is the projection of **L** along the z-axis.

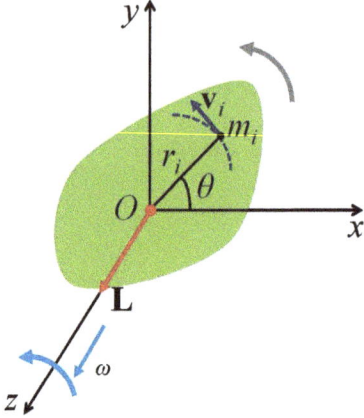

Figure 11.3: Rotation of a rigid body about the z-axis.

The time derivative of L_z where I is constant for a rigid body is

$$\frac{dL_z}{dt} = I\frac{d\omega}{dt} = I\alpha.$$ (11.27)

Here α is the angular acceleration relative to the axis of rotation z. In addition, dL_z/dt is equal to the resultant external torque $\sum_i \tau_i^e$, thus, we get

$$\sum_i \tau_i^e = I\alpha$$ (11.28)

where the sum runs over all particles of the rigid body.

Conservation law of the angular momentum of rigid body

If there are not external forces acting on the system, then $\sum_i \tau_i^e = 0$, and we obtain

$$\frac{dL_z}{dt} = 0$$ (11.29)

or

$$L_z = \text{constant}$$ (11.30)

which is the *conservation law of the angular momentum* of the rigid body rotation.

The effect of conservation of momentum could be understood much better if we consider flight platforms such as helicopters. A helicopter has a primary rotor and a tail rotor and the entire system has a total angular momentum. If the tail rotor fails, the copter can go into a tailspin and crash.

11.5 Precessional motion

Precessional motion is an interesting motion that represents a top spinning motion about its axis of symmetry, as shown in Fig. 11.4. If the spinning is very fast, the symmetry axis of the top spinning object rotates around the z-axis. This type of rotational motion is called *precessional motion*. Note that this motion is slower than the spinning motion around the symmetry axis. In this case (see also Fig. 11.4), the center of mass (c. m.) is shifted causing a torque different from 0:

$$\boldsymbol{\tau} = \mathbf{r} \times (M\mathbf{g}) \tag{11.31}$$

where $\mathbf{F}_g = M\mathbf{g}$ is the gravitational force of the top symmetric object. Since the top is spinning around its own axis, it does not fall down. We denote by $\mathbf{L}$ the angular momentum of the spinning motion, which lie along the axis of the rotation. Motion around the z-axis is precessional motion, which occurs due to the torque $\boldsymbol{\tau}$. Here, $\boldsymbol{\tau}$ changes the direction of the symmetry axis of the top. Actually, there are two forces acting on the top, namely the gravitational force $\mathbf{F}_g$ and normal resistance force $\mathbf{n}$. Therefore, the net torque is

$$\boldsymbol{\tau} = \boldsymbol{\tau}_1 + \boldsymbol{\tau}_2 \tag{11.32}$$

where

$$\boldsymbol{\tau}_1 = \mathbf{r} \times M\mathbf{g} \tag{11.33}$$
$$\boldsymbol{\tau}_2 = \mathbf{r} \times \mathbf{n} = 0$$

where $\boldsymbol{\tau}$ lies on the xy plane.
Furthermore, we can write

$$\boldsymbol{\tau} = \frac{d\mathbf{L}}{dt} \tag{11.34}$$

where the change of $\mathbf{L}$ is due to the precessional motion around the z-axis, as indicated in Fig. 11.4:

$$\Delta\mathbf{L} = \mathbf{L}_f - \mathbf{L}_i. \tag{11.35}$$

From Fig. 11.4, it can be seen that $\Delta\mathbf{L}$ is perpendicular to $\mathbf{L}$, and the magnitude of $\mathbf{L}$ does not change. $\Delta\mathbf{L}$ is also perpendicular to the torque $\boldsymbol{\tau}$, and hence $\Delta\mathbf{L}$ lies on the plane xy. Thus, the magnitude of the torque is

$$\tau = Mgr. \tag{11.36}$$

From Fig. 11.4, we have

$$d\phi = \frac{dL}{L} = \frac{\tau dt}{L} = \frac{Mgr dt}{L}. \tag{11.37}$$

Thus,

$$\omega_p = \frac{d\phi}{dt} = \frac{Mgr}{L} \tag{11.38}$$

where ω_p is the precessional frequency. Using $L = I\omega$, where ω is the frequency of the spinning motion around the symmetry axis of the top, we get

$$\omega_p = \frac{Mgr}{I\omega}. \tag{11.39}$$

Note that in this relation (eq. (11.39)), $\omega \gg \omega_p$, which is a criterion for the precessional motion to be observed.

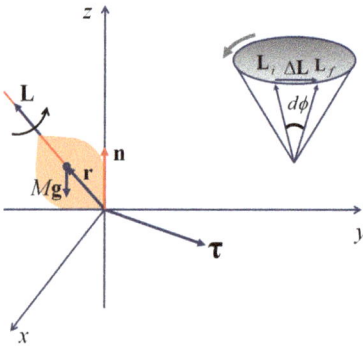

Figure 11.4: Definition of precessional motion.

11.6 Exercises

Exercise 11.1. Consider the rolling motion of a sphere with mass M and radius R down a rough incline, as shown in Fig. 11.5. We assume that the sphere rolls with angular velocity ω without slipping and is released from rest at the top of the incline. Determine $v_{c.m.}$.

Exercise 11.2. Consider a particle moves in the xy plane in a circular path of radius r, as shown in Fig. 11.6. Find the magnitude and direction (a) of its angular momentum relative to O when its linear velocity is $\mathbf{v}$, and (b) of $\mathbf{L}$ in terms of the particle's angular speed ω.

Exercise 11.3. Calculate the magnitude of the angular momentum of a bowling ball spinning at 10 rev/s. Its mass is 6.0 kg and radius 12 cm.

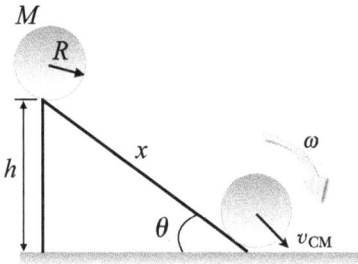

Figure 11.5: A sphere rolling down an incline. Mechanical energy is conserved if no slipping occurs.

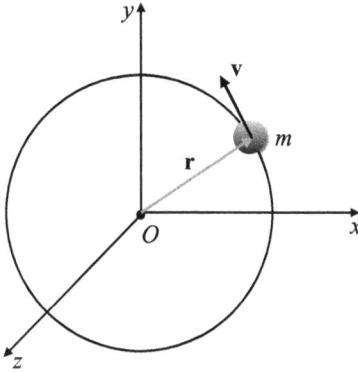

Figure 11.6: A particle moving in a circle of radius r has an angular momentum about O.

Exercise 11.4. A rigid rod of mass M and length d is pivoted without friction about its center, as shown in Fig. 11.7. Two particles of mass m_1 and m_2, respectively, are connected to its ends. The combination rotates in a vertical plane with an angular speed ω. Give an expression for (a) the magnitude of the angular momentum of the system, and (b) the magnitude of the angular acceleration of the system when the rod makes an angle θ with the horizontal direction.

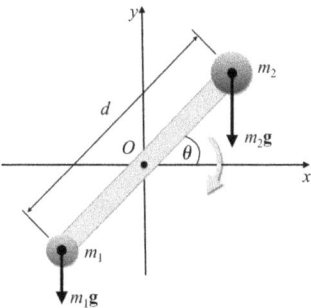

Figure 11.7: A rotating rod about z-axis through O.

Exercise 11.5. Two blocks of masses m_1 and m_2 are attached at the ends of a light cord that passes over a pulley with a radius R, as shown in Fig. 11.8. The moment of inertia

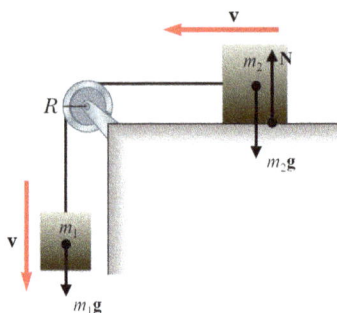

Figure 11.8: A rotating system.

of pulley about its axle is I. The second block slides on a horizontal and frictionless surface. Determine the expression for the linear acceleration of the two objects, using the concepts of angular momentum and torque.

Exercise 11.6. Consider a star rotating with a period of 30 days about an axis through its center, which undergoes a supernova explosion. After the explosion, the stellar core with a radius of 1.0×10^4 km collapses into a neutron star, which has a radius of 3.0 km. Find the period of rotation of the neutron star.

Exercise 11.7. A particle with mass 1.50 kg moves in the xy plane with a velocity of $\mathbf{v} = (4.20\mathbf{i} - 3.60\mathbf{j})$ m/s. Determine the particle's angular momentum when its position vector is $\mathbf{r} = (1.50\mathbf{i} + 2.20\mathbf{j})$ m.

Exercise 11.8. Suppose that the position vector of a particle of mass 2.00 kg versus time is given by the expression $\mathbf{r} = (6.00\mathbf{i} + 5.00t\mathbf{j})$ m. Determine the angular momentum of the particle about the origin as a function of time.

Exercise 11.9. A particle of mass m moves in a circle of radius r at a constant speed v, as shown in Fig. 11.9. If the motion begins at point Q, determine the angular momentum of the particle about point P as a function of time.

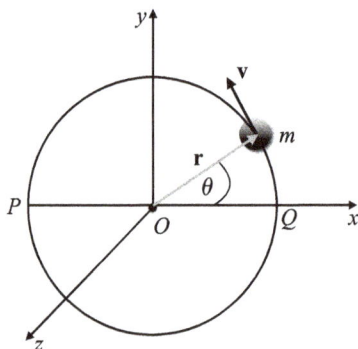

Figure 11.9: A particle moving in a circle.

Exercise 11.10. Consider a uniform solid sphere with a radius of 0.500 m and a mass of 15.0 kg, It rotates counterclockwise about a vertical axis through its center. Determine its vector angular momentum when its angular speed is 3.00 rad/s.

12 Static equilibrium and elasticity

In the previous chapters on classical mechanics, we discussed several quantities, such as displacement, velocity, acceleration, force, work and energy, and momentum in both translational and rotational motion.

Here, we discuss the use of Newton's first law for both linear and rotational motion to analyze the mechanics of a body in *static equilibrium*. In particular, that is very important in continuum mechanics for calculations of forces and torques on elastic bodies. For example, to design the columns of a bridge able to withstand the forces, it is essential to know the forces acting on the columns supporting the bridge.

In our view here, the term *equilibrium* implies either that the body is at rest (i. e., the speed $v = 0$), or that the center of mass moves with constant velocity. Here, we discuss the former case, in which the object is described as being in static equilibrium.

12.1 Conditions for equilibrium

Newton's first law provides the mathematical requirement for static equilibrium as

$$\sum_i \mathbf{F}_i = 0. \tag{12.1}$$

Equation (12.1) indicates that the necessary condition for a static equilibrium is that the resultant force applied to an object is zero. Besides, for the real (extended) objects (rigid bodies), which cannot be seen as particles, a second condition is required for static equilibrium. That is, the resultant torque acting on the extended object should be zero. Note that the equilibrium does not require that the object is at rest; for example, a rigid body rotates about an axis through a fixed point O with an angular velocity ω constant but still be in equilibrium.

Conditions for equilibrium
In general, an object is considered to be in a rotational equilibrium only if the angular acceleration of the object is $\alpha = 0$. For rotation about a fixed axis the following law applies:

$$\sum_i \tau_i = I\alpha. \tag{12.2}$$

Then the second necessary condition for equilibrium is that the resultant torque about any axis must be zero:

$$\sum_i \tau_i = 0. \tag{12.3}$$

https://doi.org/10.1515/9783110755824-012

Equation (12.1) and eq. (12.3) give the two necessary conditions for the equilibrium of an object.

The first condition (see eq. (12.1)) is a statement of *translational equilibrium*, which indicates that for an object, the linear acceleration of the center of mass at any inertial reference frame must be zero. The second condition (see eq. (12.3)) is a statement of *rotational equilibrium*, which indicates that an object is in equilibrium if the angular acceleration about an axis is zero.

The expressions entering in eq. (12.1) and eq. (12.3) are vectors, and hence these two equations are equivalent, in general, to six scalar equations. Namely, three from the first condition of equilibrium given by eq. (12.1), and three from the second condition of equilibrium given by eq. (12.3). These correspond to x, y, and z components, respectively, given by

$$\sum_i F_{ix} = 0 \qquad (12.4)$$

$$\sum_i F_{iy} = 0 \qquad (12.5)$$

$$\sum_i F_{iz} = 0 \qquad (12.6)$$

and

$$\sum_i \tau_{ix} = 0 \qquad (12.7)$$

$$\sum_i \tau_{iy} = 0 \qquad (12.8)$$

$$\sum_i \tau_{iz} = 0. \qquad (12.9)$$

12.2 Elastic properties of solids

In our discussions so far, the objects are considered as rigid, that is, they do not change shape when external forces act on them. However, in general, all objects are deformable, and hence, they may change the shape or the size when external forces are applied. The objects have the tendency using internal forces to resist the deformations as these changes take place.

In our discussion that follows, we introduce two terms: *stress* and *strain*.

12.2.1 Stress

Definition 12.1 (Stress definition). By definition, the stress is proportional to the external force causing a deformation. Mathematically, stress is given as the ratio of the external force acting on an object with its cross-sectional area:

$$\sigma = \frac{\mathbf{F}}{\mathbf{A}}. \tag{12.10}$$

The SI units of stress are N/m^2 or *Pascal*. Equation (12.10) indicates that the stress is inversely proportional to the cross-sectional area of the object. Therefore, to reduce stress, we increase the area, if weight and cost considerations are not consequential for the object.

In eq. (12.10), σ is physically referred to as the *normal stress*. However, forces applied to a body do not, in general, produce normal stress. In particular, the mechanical failures of materials show that the failure angle is not along a cross-sectional area perpendicular to the direction of the exerted force. Therefore, the stress is usually partitioned into two components; namely, the normal stress σ_n, which normal to the failure plane as shown in Fig. 12.1, and the *shear stress*, denoted by σ_t along the plane of failure:

$$\sigma_n = \frac{F_n}{A} \tag{12.11}$$

$$\sigma_t = \frac{F_t}{A} \tag{12.12}$$

where A is the magnitude of the cross-section surface area vector perpendicular to the normal force vector $\mathbf{F}_n$, and $\mathbf{F}_t$ is tangent to that surface, as shown in Fig. 12.1.

The condition for static equilibrium along the normal and the shear directions are as follows:

$$\sum \mathbf{F}_n = 0 \tag{12.13}$$

and

$$\sum \mathbf{F}_t = 0 \tag{12.14}$$

where the sums run over all applied forces.

Assuming the failure angle is θ, see also Fig. 12.1, we obtain

$$F_n = F \cos \theta = \frac{\sigma_n A}{\cos \theta}. \tag{12.15}$$

The normal stress is then

$$\sigma_n = \frac{F \cos^2 \theta}{A} = \frac{F(1 + \cos 2\theta)}{2A}. \tag{12.16}$$

Similarly, the shear part is

$$F_t = F \sin \theta = \frac{\sigma_t A}{\cos \theta}. \tag{12.17}$$

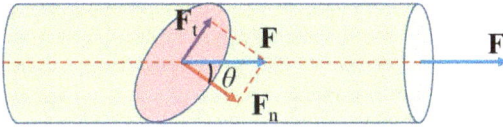

Figure 12.1: Normal and shear stresses along a surface due to the external force **F**. The failure angle is θ.

The shear stress is then

$$\sigma_t = \frac{F \cos\theta \sin\theta}{A} = \frac{F \sin 2\theta}{2A}. \tag{12.18}$$

Note that the stress is not a vector, and hence we cannot assign directions to stress. Furthermore, the stress is also not a scalar, but stress belongs to a class of physical quantities called tensors.[1]

12.2.2 Strain

The degree of deformation is measured by the strain, which is relative elongation or compression of an elastic body under stress.

Consider the rod in Fig. 12.2 under the external force **F** applied along the main axis of the rod. Suppose that the original length of the rod is l, and under the influence of the external force, it becomes $l + \Delta l$.

Definition 12.2 (Strain). By definition, mathematically, the strain is the fractional change in the length of the rod:

$$\varepsilon_n = \frac{\Delta l}{l}. \tag{12.19}$$

Equation (12.19) indicates that the strain is dimensionless. The observations show that, for sufficiently small stresses, the normal stress σ_n is proportional to the strain ε_n. (Note that subscript n at the definition of the strain is used to indicate the relationship observed between ε_n and σ_n for small stresses.)

Definition 12.3 (Elastic modulus). The constant of proportionality depends on the deformed material and the nature of the deformation, and it is called elastic modulus. Thus, the elastic modulus is defined as the ratio of the stress to the resulting strain:

$$E = \frac{\sigma_n}{\varepsilon_n}. \tag{12.20}$$

E (eq. (12.20)) is also called *Young's modulus*. From the definition, the units of E in the SI system are N/m^2 or *Pascal*.

---

1 A tensor is a set of numbers that transform under the rotations according to the rule: $T'_{ij} = R_{ik} R_{jl} T_{kl}$, where **R** is the rotation matrix and $i,j = 1, 2, 3$. Here, T_{ij} is a second rank tensor. Einstein's summation notation is used such that the repeated indices are summed, with $k, l = 1, 2, 3$.

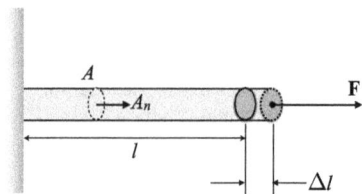

Figure 12.2: A rod subject to a force **F** along it.

In a shear strain, the deformation increases as one moves in a direction perpendicular to the deformation. Examples of shear stress and shear strain are shown in Fig. 12.3.

Definition 12.4 (Shear strain). By definition, the shear strain is

$$\varepsilon_t = \frac{\Delta}{h}. \tag{12.21}$$

Definition 12.5 (Shear modulus). We define a shear modulus G analogous to Young's modulus for sufficiently small shear stresses as follows:

$$G = \frac{\sigma_t}{\varepsilon_t}. \tag{12.22}$$

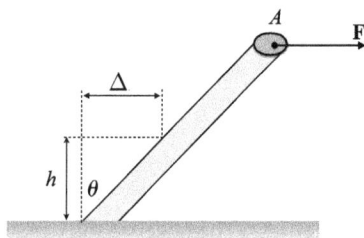

Figure 12.3: Shear strain.

12.2.3 Bulk modulus

The *bulk modulus* characterizes the resistance of a substance to uniform squeezing or to a reduction in pressure when the object is placed in a partial vacuum. For that case, we assume that the external forces acting on an object are perpendicular to all its faces in all directions, as shown in Fig. 12.4. Furthermore, they are distributed uniformly over all the faces. The object changes volume, keeping the same shape subject to this type of deformation.

Definition 12.6 (Volume stress). By definition, the volume stress is given as the ratio of the magnitude of the normal force F with the area A:

$$P = \frac{F}{A}. \tag{12.23}$$

In eq. (12.23), P is also called the *pressure*. When the pressure P of the substance changes by an amount of $\Delta P = \Delta F/A$, then the substance will experience a volume change ΔV.

Definition 12.7 (Volume strain). The *volume strain* is equal to the change in volume ΔV divided by the initial volume of V_i.

Definition 12.8 (Bulk modulus). Thus, we can characterize a volume bulk compression in terms of the bulk modulus, which is defined as

$$B = \frac{\text{Volume stress}}{\text{Volume strain}} = -\frac{\Delta F/A}{\Delta V/V_i} = -\frac{\Delta P}{\Delta V/V_i}. \qquad (12.24)$$

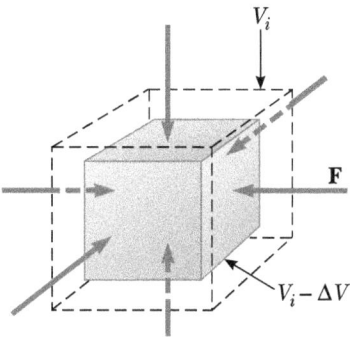

Figure 12.4: A solid is under uniform pressure.

The negative sign is added in eq. (12.24) such that B becomes a positive number. That is necessary because an increase in pressure (positive ΔP) causes a decrease in volume (negative ΔV) and vice versa. The inverse of the bulk modulus, namely $1/B$, is called the *compressibility* of the material.

12.3 Elasticity and plasticity

Equation (12.20) tells us that E is the slope of the line in the stress versus strain plot. Figure 12.5 indicates two distinct regions. We have a linear region, where the material undergoes elastic deformation, and after the applied load is removed, the material relaxes to its original shape. The red region is the plastic region, and, as shown, it propagates with very little increase in the stress, and if not stopped by the removal of the force, it ends up with catastrophic failure. Note that once the material undergoes plastic deformation, it would not retain its original shape.

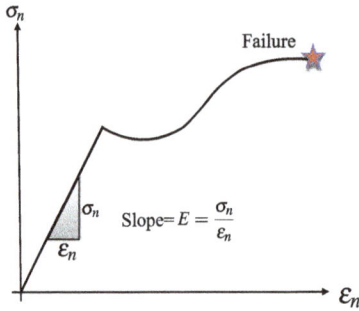

Figure 12.5: Stress–strain curve indicating the elastic and the plastic regions of a given material experiencing mechanical stress.

12.4 Exercises

Exercise 12.1. Find the tension in the cable supporting the weight of hanging mass m as shown in Fig. 12.6.

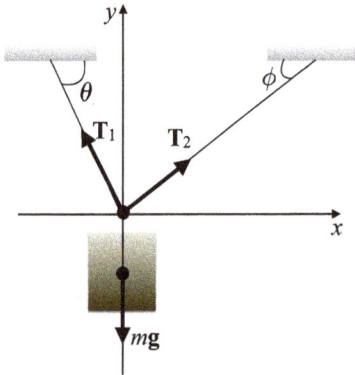

Figure 12.6: An object with a hanging mass in static equilibrium.

Exercise 12.2. An object is stationary and three forces are acting on it. If two of the forces are $F_1 = 5i - 2j$ and $F_2 = 9i + j$, find the third force.

Exercise 12.3. A man of 80 kg is standing 1.5 m from one end of a small 4.0 m long bridge. If the mass of the bridge is 500 kg, calculate the reaction forces at the two ends of the bridge.

Exercise 12.4. Fig. 12.7 shows a person with mass m climbing a ladder with a mass M. Suppose the length of the ladder is l and the person is halfway up the ladder, which makes an angle θ with the floor. The static friction coefficient between the ladder and the ground is μ_s. Find the reaction forces exerted by the floor and by the wall. Find the minimum angle θ_{min} at which the ladder does not slip.

Exercise 12.5. Suppose a uniform horizontal beam with a length of 8.00 m and weight of 200 N is attached to a wall by a pin connection. Assume that its far end is supported

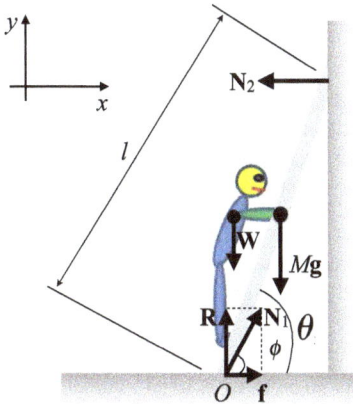

Figure 12.7: A person is climbing a ladder.

by a cable making an angle of 53.0° with horizontal direction. If a 600 N person stands 2.00 m from the wall, find the tension in the cable, the magnitude, and direction of the force exerted by the wall on the beam.

Exercise 12.6. Consider a cable used to support an actor as he swung onto the stage. The tension in the cable is 940 N. What diameter should a 10 m long steel wire have if we do not want it to stretch more than 0.5 cm under these conditions? The Young modulus is $E = 20 \times 10^{10}$ N/m^2.

Exercise 12.7. Consider a solid brass sphere is initially surrounded by air, and the air pressure exerted on it is 1.0×10^5 N/m^2 (normal atmospheric pressure). Then the sphere is lowered into an ocean to a depth at which the pressure is 2.0×10^7 N/m^2. Suppose the volume of the sphere in the air is 0.50 m^3. What is the change in volume once the sphere is submerged if the bulk modulus is 6.1×10^{10} N/m^2?

Exercise 12.8. Assume that Young's modulus of the bone is 1.50×10^{10} N/m^2 and that a bone will fracture if a stress more than 1.50×10^8 N/m^2 is exerted. (a) What is the maximum force that can be exerted on the femur bone in the leg if it has a minimum effective diameter of 2.50 cm? (b) If a compressible force of this magnitude is applied, by how much does the 25.0 cm long bone shorten?

Exercise 12.9. A 200 kg load is hung on a wire with a length of 4.00 m, a cross-sectional area of 0.200×10^{-4} m^2, and a Young's modulus of 8.00×10^{10} N/m^2. What is its increase in length?

Exercise 12.10. When water freezes, it expands by about 9.00 %. What would be the pressure increase inside your automobile's engine block if the water in it froze? The bulk modulus of ice is 2.00×10^9 N/m^2.

13 Oscillatory motion

In this chapter, we introduce the oscillatory motion. In general, if the force applied to a body is proportional to the displacement from a predefined equilibrium position of the body, then this motion is called a *periodic motion*. Other terms are also used for a periodic motion, such as the *harmonic motion, oscillation motion*, or *vibration motion*. The applied force vector has a direction toward the equilibrium position; therefore, it applies a back-and-forth motion about that position periodically.

13.1 Simple harmonic motion

Assume a physical system, such as a block of mass m, connected to the end of a spring. The block is allowed to move on a horizontal direction (chosen as x-axis), on a frictionless surface, as also shown in Fig. 13.1. When the spring is in equilibrium state (*i. e.*, neither stretched nor compressed), the block is at the so-called equilibrium position of the system, $x = 0$.

If the system is disturbed from its equilibrium position, then the block oscillates back and forth about that position. For example, if the block is displaced a small distance $x > 0$ from equilibrium position at $x = 0$, the spring acts on the block with force proportional to that displacement as given by Hooke's law:

$$F_s = -kx. \tag{13.1}$$

F_s is a *restoring force* or *elastic force* with direction toward the equilibrium position. The minus sign in eq. (13.1) indicates that F_s has an opposite direction to the displacement of the object. For instance, when the displacement is on the right of $x = 0$, then $\mathbf{F}_s$ is directed to the left, and vice versa, as illustrated in Fig. 13.1.

Harmonic motion
Applying Newton's second law, we obtain

$$F_s = -kx = ma \tag{13.2}$$

or

$$a = -\frac{k}{m}x. \tag{13.3}$$

Equation (13.3) shows that the acceleration is proportional to the displacement of the block x, and its direction is opposite the direction of the displacement as indicated by the minus sign in front. In general, a system that exhibits this behavior follows a simple harmonic motion.

https://doi.org/10.1515/9783110755824-013

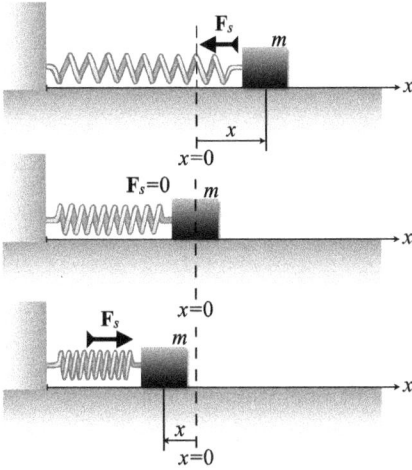

Figure 13.1: A block of mass m attached to a spring moving on a frictionless surface.

By definition, any object moves according to a simple harmonic motion, if the acceleration of the object is proportional to its displacement from an equilibrium position. The acceleration is in the opposite direction to the displacement.

Using the relation between the acceleration a and the time derivatives of x as

$$a = \frac{d^2x}{dt^2} \tag{13.4}$$

and substituting it in eq. (13.3), we get a second order differential equation

$$\frac{d^2x}{dt^2} + \omega^2 x = 0 \tag{13.5}$$

where

$$\omega = \sqrt{\frac{k}{m}}. \tag{13.6}$$

Here, ω is called *angular frequency*. The general form of the solution of this differential equation is as follows:

$$x(t) = A \cos(\omega t + \phi). \tag{13.7}$$

The constants A and ϕ are, respectively, the *amplitude* and *phase angle*. Figure 13.2 shows graphically as a plot of x versus t the physical meanings of the constants A and ϕ. In particular, A characterizes the maximum displacement of the particle from the equilibrium position, and ϕ represents the phase at $t = 0$, defined as

$$\text{phase} = \omega t + \phi. \tag{13.8}$$

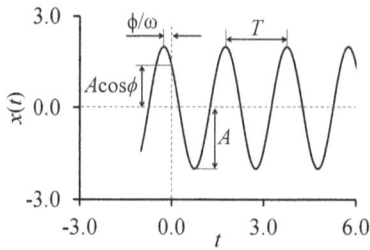

Figure 13.2: x–t graph for a particle undergoing simple harmonic motion.

Initial conditions

These constants are determined from the initial values of the displacement x_0 and velocity v_0, which are known as the initial conditions. For instance, from $x_0 = x(t = 0)$, we obtain

$$x_0 = x(t = 0) = A \cos \phi. \tag{13.9}$$

The particle's velocity at any time t is defined from eq. (13.7) as

$$v(t) = \frac{dx}{dt} = -A\omega \sin(\omega t + \phi). \tag{13.10}$$

Using the initial condition $v_0 = v(t = 0)$, we find

$$v_0 = -A\omega \sin \phi. \tag{13.11}$$

Using eq. (13.9) and eq. (13.11), we get

$$\tan \phi = -\frac{v_0}{\omega x_0} \tag{13.12}$$

or

$$\phi = \tan^{-1}\left(-\frac{v_0}{\omega x_0}\right). \tag{13.13}$$

Also, using eq. (13.9) and eq. (13.11), we determine A in terms of x_0 and v_0 as

$$A = \sqrt{x_0^2 + \left(\frac{v_0}{\omega}\right)^2}. \tag{13.14}$$

Period

From eq. (13.7), the displacement function of time $x(t)$ is periodic with period 2π rad. That is,

$$A \cos(\omega(t + T) + \phi) = A \cos((\omega t + 2\pi) + \phi) \tag{13.15}$$

where T is the period. Furthermore, we have

$$w(t + T) + \phi = (\omega t + 2\pi) + \phi. \tag{13.16}$$

Solving it for T, we get

$$T = \frac{2\pi}{\omega}. \tag{13.17}$$

Therefore, the period T of the motion defines the time, it takes to the particle, to go through one full cycle. We say that the particle has made one oscillation.

Frequency
The inverse of the period T is called *frequency* of the motion. It is usually denoted by the letter f and it represents the number of oscillations that the particle makes per unit of time. Mathematically, it is expressed as

$$f = \frac{1}{T} = \frac{\omega}{2\pi}. \tag{13.18}$$

The SI units of the frequency are cycles per second: s^{-1}, or Hz (hertz).

Angular frequency
The *angular frequency* ω can also be obtained from eq. (13.18):

$$\omega = 2\pi f = \frac{2\pi}{T}. \tag{13.19}$$

The SI units of the angular frequency are radians per second: rad/s
We can also obtain the linear acceleration by taking the first time derivative of the velocity v in eq. (13.10):

$$a(t) = \frac{dv}{dt} = -A\omega^2 \cos(\omega t + \phi). \tag{13.20}$$

Combining eq. (13.7) and eq. (13.20), we get

$$a(t) = -\omega^2 x(t). \tag{13.21}$$

From eq. (13.10), we can see that velocity fluctuates between $\pm A\omega$. Thus, the maximum speed is

$$v_{max} = A\omega. \tag{13.22}$$

From eq. (13.20), the acceleration fluctuates between $\pm A\omega^2$. Thus, the maximum acceleration is

$$a_{max} = A\omega^2. \tag{13.23}$$

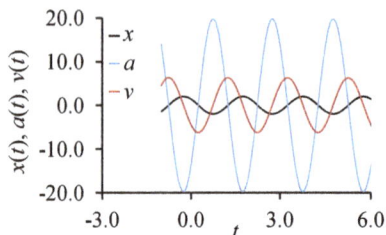

Figure 13.3: Graphical representation of simple harmonic motion.

In Fig. 13.3, we have plotted the displacement, velocity, and acceleration versus time for an arbitrary value of the phase constant. The graphs show that the phase of the velocity differs from the phase of the displacement by $\pi/2$ rad (that is, if x is a maximum or a minimum, the velocity is zero, and if x is zero, the speed is a maximum). Comparing the phase shift between the acceleration and displacement, we see a phase shift of π rad (that is, if x is a maximum, a is a maximum in the opposite direction).

Properties of simple harmonic motion
The following properties of a particle moving in simple harmonic motion are important:
1. The particle's acceleration is proportional to the displacement but is in the opposite direction. That is a necessary and sufficient condition for simple harmonic motion, as opposed to all other kinds of vibration.
2. The particle's displacement from the equilibrium position, its velocity, and its acceleration all vary sinusoidal with time but are not in phase.
3. Both frequency and the period of the motion are independent on the amplitude, but they dependent on the mass and force constant k of the spring:

$$T = 2\pi\sqrt{\frac{m}{k}} \tag{13.24}$$

$$f = \frac{1}{2\pi}\sqrt{\frac{k}{m}}. \tag{13.25}$$

13.2 Energy of the simple harmonic oscillator

We consider the block–spring system shown in Fig. 13.1. Let us calculate the mechanical energy of this system. The kinetic energy of the system is

$$K = \frac{1}{2}mv^2 \tag{13.26}$$

where the velocity v is given by the expression in eq. (13.10). Substituting v into the expression for the kinetic energy, we get

$$K = \frac{1}{2}mA^2\omega^2\sin^2(\omega t + \phi). \tag{13.27}$$

The elastic potential energy, which is stored in the spring due to its deformation x, is given by

$$U = \frac{1}{2}kx^2. \tag{13.28}$$

Replacing the expression for the deformation x, we get

$$U = \frac{1}{2}mA^2\omega^2 \cos^2(\omega t + \phi). \tag{13.29}$$

Then the total mechanical energy of the system is

$$E = K + U \tag{13.30}$$

or

$$E = \frac{1}{2}mA^2\omega^2 \sin^2(\omega t + \phi) + \frac{1}{2}mA^2\omega^2 \cos^2(\omega t + \phi). \tag{13.31}$$

Knowing that

$$\omega = \sqrt{\frac{k}{m}} \tag{13.32}$$

we get

$$E = \frac{1}{2}kA^2 = \frac{1}{2}m\omega^2 A^2. \tag{13.33}$$

Equation (13.33) indicates that the total mechanical energy of a simple harmonic oscillator is constant of motion and is proportional to the square of the amplitude. In Fig. 13.4, we have plotted the kinetic and potential energy of a simple harmonic oscillator. It can be seen that when K increases, U decreases, and vice versa, and hence their sum, $K + U$, is constant.

Note that the total mechanical energy is equal to the maximum potential energy stored in the spring when displacement is maximum. That is, if $x = \pm A$, then $v = 0$ at these points, and hence, there is no kinetic energy. At the equilibrium position there is no displacement $x = 0$, hence $U = 0$, the total energy, all in the form of kinetic energy, is again $(1/2)kA^2$. Thus, we can write

$$E = \frac{1}{2}mv_{max}^2 = \frac{1}{2}m\omega^2 A^2 = \frac{1}{2}m\frac{k}{m}A^2 = \frac{1}{2}kA^2 \tag{13.34}$$

at $x = 0$.

The conservation law of mechanical energy can be used to obtain the velocity for an arbitrary displacement x by expressing the total energy at some arbitrary position x as

$$E = \frac{1}{2}mv^2 + \frac{1}{2}kx^2 = \frac{1}{2}kA^2. \tag{13.35}$$

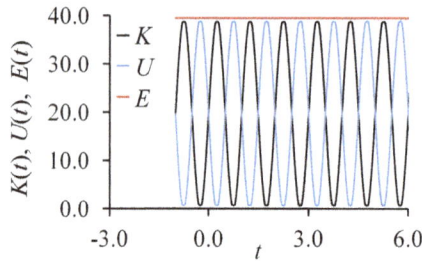

Figure 13.4: Time dependence of the kinetic and potential energy of simple harmonic motion: $A = 3$, $\omega = 2$, $m = 1$ and $\phi = \pi/4$.

Solving it for v, we get

$$v = \pm\sqrt{\frac{k}{m}(A^2 - x^2)} = \pm\omega\sqrt{A^2 - x^2}. \tag{13.36}$$

This equation indicates that $v = 0$ if $x = \pm A$ and the velocity is maximum at equilibrium position: $x = 0$.

13.3 Simple pendulum

The *simple pendulum* is a mechanical system that exhibits periodic motion. It consists of point-like mass m at the end of a light string of length L fixed at the other side, as shown in Fig. 13.5. The motion is due to the gravitational force and occurs in the vertical plane. We will assume that the angle θ is small.

The forces acting on the mass m are the force **T** exerted by the string and the gravitational force **mg**. The gravitational force can be expressed into two components:

$$\text{Tangential component:} \quad F_{gt} = mg\sin\theta \tag{13.37}$$
$$\text{Radial component:} \quad F_{gr} = mg\cos\theta. \tag{13.38}$$

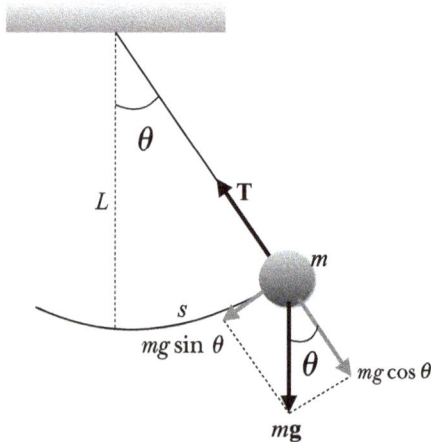

Figure 13.5: The simple pendulum.

The tangential component always acts toward $x = 0$ and is opposite the displacement playing thus the role of a restoring force. We can apply Newton's second law for motion along the tangential direction to get

$$\sum F_t = F_{gt} = -mg \sin \theta = m \frac{d^2s}{dt^2} \qquad (13.39)$$

where the minus sign indicates that the tangential force is opposite of the displacement, and s is the displacement measured along the arc. Knowing that

$$s = L\theta \qquad (13.40)$$

we get

$$\frac{d^2\theta}{dt^2} = -\frac{g}{L} \sin \theta. \qquad (13.41)$$

For small angle approximations we can write

$$\sin \theta \approx \theta. \qquad (13.42)$$

Therefore, we get

$$\frac{d^2\theta}{dt^2} = -\frac{g}{L}\theta \qquad (13.43)$$

which represents a simple harmonic motion with respect to coordinate θ. Denoting ω with

$$\omega = \sqrt{\frac{g}{L}} \qquad (13.44)$$

we get the general solution in the form of

$$\theta(t) = A_\theta \cos(\omega t + \phi) \qquad (13.45)$$

where A_θ is the amplitude (maximum) angular displacement and ω is the angular frequency. ϕ is the initial phase, which depends on the initial conditions:

$$\theta_0 = \theta(t = 0); \quad \dot{\theta}_0 = \left(\frac{d\theta}{dt}\right)_{t=0}. \qquad (13.46)$$

The period of motion is

$$T = \frac{2\pi}{\omega} = 2\pi\sqrt{\frac{L}{g}} \qquad (13.47)$$

and the frequency f is

$$f = \frac{1}{T} = \frac{1}{2\pi}\sqrt{\frac{g}{L}}. \tag{13.48}$$

Thus, the period and frequency of a simple pendulum depend only on the length of the string L and the acceleration due to gravity g, and they are independent on the mass m.

The total mechanical energy of the pendulum–Earth system is composed of the kinetic energy of the pendulum and the gravitational potential energy:

$$E = K + U = \frac{mv^2}{2} + mgh. \tag{13.49}$$

The linear velocity is given as

$$v = L\omega = L\frac{d\theta}{dt} \tag{13.50}$$
$$= -LA_\theta\omega\sin(\omega t + \phi).$$

Taking the y-axis pointing vertically up, and its origin at the position of the mass when $\theta = 0$, we see that the potential energy is

$$U = mgL(1 - \cos\theta). \tag{13.51}$$

Thus, the total mechanical energy is pendulum is given as

$$E = \frac{mL^2 A_\theta^2 \omega^2}{2}\sin^2(\omega t + \phi) + mgL(1 - \cos\theta) \tag{13.52}$$
$$= \frac{mL^2 A_\theta^2 (g/L)}{2}\sin^2(\omega t + \phi) + mgL(1 - \cos\theta)$$
$$= mgL\frac{A_\theta^2}{2}\sin^2(\omega t + \phi) + mgL(1 - \cos\theta).$$

In Fig. 13.6, the variations of the kinetic and potential energy of the simple pendulum are shown. It can be seen that when K increases then U decreases by the same amount, and vice versa, and hence the total energy E is constant.

13.4 Physical pendulum

The *physical pendulum* is an object that oscillates about a fixed axis that does not pass through its center of mass. Also, the object cannot be approximated by a point mass, as shown in Fig. 13.7. In general, we cannot treat the system as a simple pendulum. For the physical pendulum in Fig. 13.7, the rigid body rotates about of fixed axis through

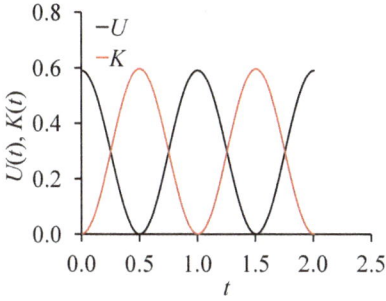

Figure 13.6: The simple pendulum kinetic and potential energy as a function of the time for one period. Arbitrary units are used.

point O, which is at distance L from the center of mass. The magnitude of the torque is

$$\tau = L(mg) \sin \theta. \tag{13.53}$$

Here, θ denotes the angle between the force of gravity and the direction along L. From the law of rotation motion

$$\tau = L(mg) \sin \theta = I\alpha. \tag{13.54}$$

Here, I denotes the moment of inertia about the axis through O and α is the angular acceleration. Knowing that

$$\alpha = \frac{d^2\theta}{dt^2} \tag{13.55}$$

we get

$$\frac{d^2\theta}{dt^2} = -\frac{mgL}{I} \sin \theta. \tag{13.56}$$

Under the small angle approximation, $\sin \theta \approx \theta$, we get

$$\frac{d^2\theta}{dt^2} = -\omega^2\theta \tag{13.57}$$

where

$$\omega^2 = \frac{mgL}{I}. \tag{13.58}$$

The above second order differential equation represents the motion of a simple harmonic oscillator, therefore the solution is

$$\theta(t) = A_\theta \cos(\omega t + \phi) \tag{13.59}$$

where A_θ is the maximum angular displacement. The period is

$$T = \frac{2\pi}{\omega} = 2\pi \sqrt{\frac{I}{mgL}}. \tag{13.60}$$

The frequency f is given as

$$f = \frac{1}{T} = \frac{1}{2\pi}\sqrt{\frac{mgL}{I}}.$$ (13.61)

It can be seen that for the physical pendulum both the period T and frequency f depend explicitly on the mass m of the pendulum.

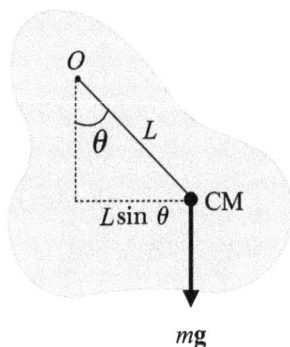

Figure 13.7: The physical pendulum.

13.5 Exercises

Exercise 13.1. Consider an object oscillating with simple harmonic motion along the x-axis. The displacement from the origin varies with time according to the equation

$$x(t) = (4.00\,\text{m}) \cos\left(\pi t + \frac{\pi}{4}\right).$$ (13.62)

t is measured in seconds and the angles in the parentheses are in radians. (a) Find the amplitude, frequency, and period of the motion. (b) What are the velocity and acceleration of the object at any time t? (c) Using the results of part (b), what are the position, velocity, and acceleration of the object at $t = 1.00\,\text{s}$? (d) Find the maximum speed and maximum acceleration of the object. (e) What is the displacement of the object between $t = 0$ and $t = 1.00\,\text{s}$?

Exercise 13.2. Consider a car with a mass of 1300 kg constructed such that its frame is supported by four springs. Each spring has a force constant of 20000 N/m. If two people riding in the car have a combined mass of 160 kg, find the frequency of vibration of the car after it is driven over a pothole in the road.

Exercise 13.3. Suppose a block with a mass of 200 g is attached to the end of a light spring with a force constant of 5.00 N/m, which is free to oscillate on a horizontal, frictionless surface. Initially, the block is displaced 5.00 cm from equilibrium and released from rest. (a) Find the period of its motion. (b) Determine the maximum speed of the block. (c) What is the maximum acceleration of the block? (d) Express the displacement, speed, and acceleration as functions of time.

Exercise 13.4. A 0.500 kg cube connected to a light spring with the force constant of 20.0 N/m. The cube oscillates on a horizontal, frictionless track. (a) Determine the total energy of the system and the maximum speed of the cube when the amplitude of the motion is 3.00 cm. (b) Calculate the velocity of the cube when the displacement is 2.00 cm. (c) Find the kinetic and potential energies of the system when the displacement is 2.00 cm.

Exercise 13.5. Christiaan Huygens suggested that an international unit of length could be defined as the length of a simple pendulum having a period of exactly 1 s. How much shorter would our length unit have been if his suggestion would have been followed?

Exercise 13.6. Suppose that a uniform rod of mass M and length L is pivoted about one end and oscillates in a vertical plane (Fig. 13.8). Find the period of oscillation if the amplitude of the motion is small.

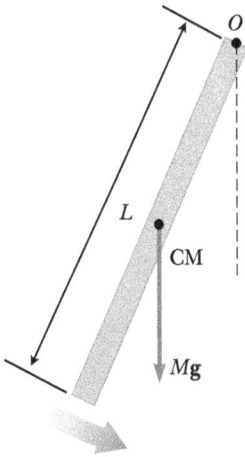

Figure 13.8: The oscillating rod.

Exercise 13.7. Consider a spring stretches by 3.90 cm when a 10.0 g mass is hung from it. If a 25.0 g mass attached to this spring oscillates in simple harmonic motion, calculate the motion period.

Exercise 13.8. Suppose that the coordinates of a particle varies as $x = -A \cos \omega t$. What is the phase constant comparing it with the equation $x = A \cos(\omega t + \phi)$?

Exercise 13.9. The expression gives the displacement of a particle at $t = 0.250$ s

$$x = (4.00 \text{ m}) \cos(3.00\pi t + \pi)$$

where x is in meters and t in seconds. Find (a) the frequency, (b) period, (c) the amplitude, (d) phase constant, and (e) the displacement of the particle at $t = 0.250\,\text{s}$.

Exercise 13.10. The angular displacement of a pendulum is given as

$$\theta = (0.500\,\text{rad})\cos(2\pi t).$$

Find the period and the length of the pendulum.

14 Gravity

Before the 17th century, the laws describing the motion of the objects near the Earth, such as falling objects, the Moon, the planets, and stars, were utterly different. Newton was the first to unify the Moon's motion around the Earth with the attraction of the objects to Earth. Also, he showed only one universal law governing the attractive forces between bodies, known as *Universal Law of Gravitation*.

14.1 Universal law of gravitation

According to the universal law of gravitation of Newton, two bodies are attracted to each other by force proportional to their masses' product and inversely proportional to the square of the distance between the two bodies. Thus, the same force as the force that gives weight to an object on Earth and makes them fall makes the planets move around the Sun.

Newton's law of gravitation
Mathematically, the Universal Law of Gravitation for every two bodies with masses m_1 and m_2 separated by the distance r is given as the following:

$$\mathbf{F}_{12} = G\frac{m_1 m_2}{r^2}\hat{\mathbf{r}}_{12} \tag{14.1}$$

where G is the proportionality constant, known as the *Universal Gravitational Constant*, which is equal to

$$G = 6.67384 \times 10^{-11}\,\mathrm{N} \cdot \frac{\mathrm{m}^2}{\mathrm{kg}^2}. \tag{14.2}$$

In eq. (14.1), $\hat{\mathbf{r}}_{12}$ is a unit vector pointing from the body 1 to 2, and $\mathbf{F}_{12}$ represents the force acting on the body 1 that is exerted by the body 2, as shown in Fig. 14.1. Note that in eq. (14.1), r is the distance between the center of masses of the two bodies. Based on the third law of Newton, the attractive force exerted on the mass m_2 by the mass m_1, $\mathbf{F}_{21}$, is given as

$$\mathbf{F}_{21} = -\mathbf{F}_{12} \tag{14.3}$$
$$|\mathbf{F}_{21}| = |\mathbf{F}_{12}|.$$

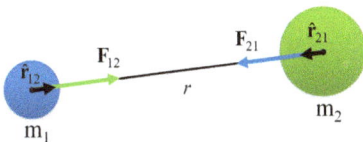

Figure 14.1: Attractive forces between two bodies 1 and 2 with masses m_1 and m_2, respectively.

https://doi.org/10.1515/9783110755824-014

14.2 Second law of Newton and gravity

According to Newton's second law, the force acting on a body of mass m and having acceleration $\mathbf{a}$ is $\mathbf{F} = m\mathbf{a}$. Also, we defined the weight of the body m as a product of its mass and the acceleration of gravity $\mathbf{g}$ due to the force exerted on the object by another body with mass M (such that $m \ll M$). Therefore, it is possible to unify the second law of Newton with Newton's universal law of gravitation by equating the weight of the mass m with the force of gravity exerted on mass m by the mass M, as follows:

$$mg = G\frac{mM}{R^2}\hat{\mathbf{r}}_{12} \tag{14.4}$$

where the unit vector $\hat{\mathbf{r}}_{12}$ points toward the center of mass M, and R is the radius of the body M. Here, we assume that the mass m is on the larger body's surface M or close to its surface.

Therefore, dividing both sides of eq. (14.4) by m, we obtain the vector of the gravitational acceleration as

$$\mathbf{g} = G\frac{M}{R^2}\hat{\mathbf{r}}_{12} \tag{14.5}$$

and its magnitude as

$$g = G\frac{M}{R^2}. \tag{14.6}$$

Equation (14.6) indicates that g is proportional to the mass M and inversely proportional to the square of its radius, and it is intrinsic property of the mass M, and it is independent on the mass m. Therefore, every object close to the surface of the body M will experience the same gravitational acceleration $\mathbf{g}$. Using eq. (14.5) (or eq. (14.6)), the acceleration of gravity can be calculated of a celestial body, if we know the mass and the radius of a given body.

Since gravitational forces are non-contact forces, they exist even when the masses m_1 and m_2 are not in contact. These are the field forces, and hence the gravitational forces are field forces. In analogy to electrostatics, we can define the gravitational field created by the object with mass M at the point in space where the test object with mass m is placed as

$$\mathbf{E}_G = \frac{\mathbf{F}}{m} \tag{14.7}$$

where $\mathbf{F}$ is the gravity force exerted by the object M on the test mass m, given by eq. (14.1). Therefore, we obtain

$$\mathbf{E}_G = G\frac{M}{r^2}\hat{\mathbf{r}}. \tag{14.8}$$

Here, by r we denote the distance from the center of the mass M and the unit vector $\hat{r}$ points toward the center of the object M. Comparing eq. (14.5) and eq. (14.8), we can see that

$$\mathbf{E}_G \equiv \mathbf{g}. \tag{14.9}$$

Figure 14.2 indicates the directions of the gravitational field $\mathbf{g}$ lines pointing toward the center of the object. Therefore, the gravitational field is attractive.

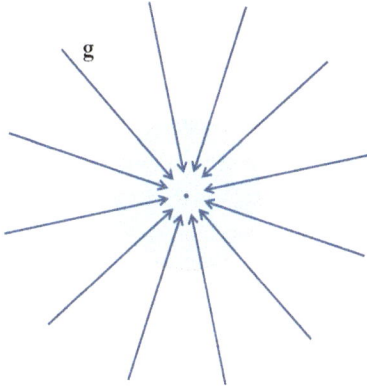

Figure 14.2: Direction of the gravitational field lines pointing toward the center of the body M.

14.3 Gauss's law of gravity

Gauss's law of gravity is defined as the total gravitation flux through a closed surface proportional to the total mass enclosed within that surface. Mathematically, Gauss's law of gravity is written as

$$\Phi_G = CM \tag{14.10}$$

where C is a proportionality constant. Note that Gauss's law of gravity is in analogy to Gauss's law for electrostatics. An integration that determines the gravitational flux over the whole closed surface must be performed of the gravity field. Figure 14.3 illustrates a closed surface (also called Gaussian surface) enclosing an object creating the field of gravity $\mathbf{g}$ in space. Therefore, we can write for the gravitational flux that

$$\Phi_G = N_+ - N_- = -\oint_A \mathbf{g} \cdot d\mathbf{A} \tag{14.11}$$

where $d\mathbf{A} = \hat{n} \cdot dA$, $N_+ = 0$ denotes the number of the gravitational field lines leaving the closed surface and N_- number of gravitational field lines entering the that surface.

Combining eq. (14.10) and eq. (14.11), we obtain

$$CM = -\oint_A \mathbf{g} \cdot d\mathbf{A} \tag{14.12}$$

$$= -\oint_A g\hat{\mathbf{r}} \cdot \hat{\mathbf{n}} dA$$

$$= g\oint_A dA = 4\pi R^2 g$$

where $4\pi R^2$ gives the area of the closed surface, assumed to coincide with the surface of the object. Here, we have assumed that mass M is a spherical object, which is not always the case, and 4π is result of the spherical geometry. Thus, we obtain that

$$g = \frac{CM}{4\pi R^2}. \tag{14.13}$$

Comparing the expressions given by eq. (14.8) and eq. (14.13), we get

$$C = 4\pi G. \tag{14.14}$$

Note that Gauss's law of gravity is a particular case of the gravitational law, and it is used to derive many physical aspects of the gravitational force that can not be explained using the Newtonian approach.

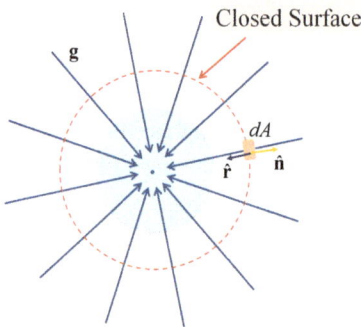

Figure 14.3: A closed surface (Gaussian surface) around the body M. $\hat{\mathbf{r}}$ denotes a unit vector toward the center of M, dA is a surface element of the closed surface and $\hat{\mathbf{n}}$ denotes a unit vector perpendicular to the surface element pointing outward to the closed surface.

14.4 Gravitational potential

We mentioned that the gravitational force is conservative, and hence it can be expressed as minus the gradient of a scalar potential.

Also, in analogy with electric potential, the gravitational field $\mathbf{g}$ is given as

$$\mathbf{g} = -\nabla V \tag{14.15}$$

where V is the gravitational potential.

From eq. (14.15), the gravitational potential is found as

$$V = -\int \mathbf{g} \cdot d\mathbf{r}.$$
(14.16)

The gravitational field $\mathbf{g}$ of the body with mass M, as shown in Fig. 14.4, is written as follows:

$$\mathbf{g} = -G\frac{M}{|\mathbf{r} - \mathbf{r}'|^3}(\mathbf{r} - \mathbf{r}').$$
(14.17)

The minus sign indicates that the gravitational field $\mathbf{g}$ is towards the center of the body M. Substituting eq. (14.17) into eq. (14.16), we obtain

$$V = \int G\frac{M}{|\mathbf{r} - \mathbf{r}'|^3}(\mathbf{r} - \mathbf{r}') \cdot d(\mathbf{r} - \mathbf{r}').$$
(14.18)

After integrating eq. (14.18), we get

$$V(r) = -G\frac{M}{|\mathbf{r} - \mathbf{r}'|}.$$
(14.19)

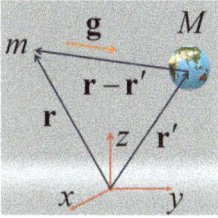

Figure 14.4: Gravitational potential created at position $\mathbf{r}$, with respect to a reference frame, from the body with mass M positioned at $\mathbf{r}'$.

To find the gravitational potential energy U, we consider a test mass m placed in the gravitational field $\mathbf{g}$, which is given as

$$U = mV = -G\frac{mM}{|\mathbf{r} - \mathbf{r}'|}.$$
(14.20)

Assuming a body with spherical mass distribution M, then the change in gravitational potential energy of a mass m that is brought from infinity to height h above the surface of the object M of radius R is

$$\Delta U = U_\infty - U_h = 0 - \left(-G\frac{mM}{R+h}\right)$$
(14.21)

$$= G\frac{mM}{R+h}.$$

Using the expression of the strength of gravitation field g at that position:

$$g = G\frac{M}{(R+h)^2}$$
(14.22)

we get the change in the gravitational potential energy of mass m in the gravitational field:

$$\Delta U = mg(R + h). \tag{14.23}$$

Furthermore, the gravitational potential energy of the mass m at the height h relative to the surface of the object M is

$$U_h = \Delta U(R + h) - \Delta U(R) = mgh \tag{14.24}$$

which is the expression we postulated in the previous chapters.

14.5 Kepler's laws

Three laws describe the orbiting motion of the planets around the Sun, known as Kepler's laws.

Three Kepler laws
1. All the planets move along the elliptical orbits around the Sun. The Sun is located at one of the two foci, as shown in Fig. 14.5.
2. The area swept by the line connecting the Sun and the planet is the same for every equal interval of time, known as the law of conservation of areal velocities.
3. The period of the planet motion is proportional to the square root of the cube of its orbit's semi-major axis, known as the law of periods.

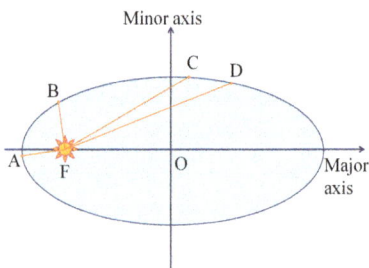

Figure 14.5: The orbiting motion of the planet around the Sun.

14.5.1 First Kepler law

Based on the conservation of mechanical energy, the orbit of the planets moving around the Sun is elliptical. By definition, an ellipse is defined as a set of points P

such that the sum of their distances from two fixed points, namely F and F', called focus points, is constant (see also Fig. 14.6):

$$FP + F'P = C \tag{14.25}$$

where C is a constant.

Based on the conservation law of energy, the system's total mechanical energy is constant, $E = K + U$ is constant, where K is the kinetic energy, and U is the potential energy. During the elliptical motion of the planets around the Sun, or a moon around a planet, the mechanical energy is conserved. Here,

$$\frac{U}{K} = \frac{FP}{F'P}. \tag{14.26}$$

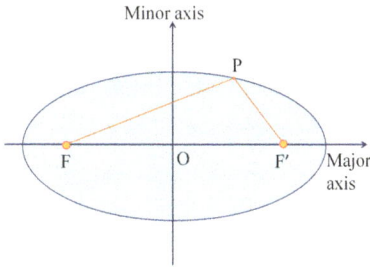

Figure 14.6: Definition of the ellipse.

14.5.2 Kepler's second law

The second law of Kepler holds because of the conservation of angular momentum, which is constant throughout the time of the motion of system. Consider the motion of a planet around the Sun, as shown in Fig. 14.7. The planet moves around the Sun because there is a torque exerted on the planet by the Sun, τ:

$$\tau = \mathbf{r} \times \mathbf{F} = \mathbf{r} \times \left(F \frac{\mathbf{r}}{r} \right) = 0 \tag{14.27}$$

where $\mathbf{F}$ is the attractive force between the Sun and planet. The torque is zero because the force vector $\mathbf{F}$ and $\mathbf{r}$ are parallel, then

$$\tau = \frac{d\mathbf{L}}{dt} = 0 \tag{14.28}$$

and thus $\mathbf{L}$ is a constant vector.

By definition, the angular momentum of the planet is given as $\mathbf{L} = \mathbf{r} \times \mathbf{p}$, where $\mathbf{p} = m\mathbf{v}$. The magnitude of $\mathbf{L}$ is

$$L = |\mathbf{L}| = |\mathbf{r} \times \mathbf{p}| = m|\mathbf{r} \times \mathbf{v}|. \tag{14.29}$$

Furthermore, the linear velocity is given as

$$\mathbf{v} = r\omega = r\dot{\theta} \tag{14.30}$$

where ω is the angular velocity vector, and θ is the angular position vector. Therefore, the torque is

$$\tau = \frac{d\mathbf{L}}{dt} = mr^2\ddot{\theta} = 0 \tag{14.31}$$

which implies that $\ddot{\theta} = 0$, and thus $\frac{d\theta}{dt}$ = constant.

On the other hand, the area dA bound between the two radii and the arc ds along a circle is

$$dA = \frac{1}{2}rvdt = \frac{1}{2}r^2d\theta. \tag{14.32}$$

Thus,

$$\frac{dA}{dt} = \frac{1}{2}r^2\frac{d\theta}{dt} = \text{constant} \tag{14.33}$$

and

$$\frac{dA}{dt} = \text{constant}, \tag{14.34}$$

which is the second law of Kepler.

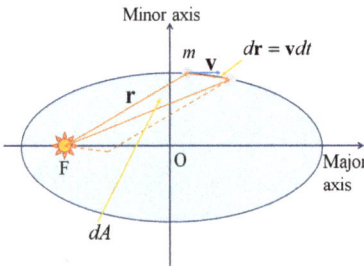

Figure 14.7: Motion of a planet around the Sun.

14.5.3 Kepler's third law

The third law of Kepler for the motion of the planets around the Sun can be derived from the second law. That is,

$$\frac{dA}{dt} = \frac{1}{2}r^2\frac{d\theta}{dt} = \frac{L}{2m}. \tag{14.35}$$

Integrating eq. (14.35) over a period T, we obtain the total area of an ellipse, given as $A = \pi ab$, where a and b are the semi-minor and semi-major axes of the ellipse, respectively. Thus, using eq. (14.35), we get

$$A = \int_0^T \frac{L}{2m}dt = \frac{L}{2m}T = \pi ab. \tag{14.36}$$

Using the following relationships between a and b:

$$b = a\sqrt{1 - e^2} \tag{14.37}$$

$$b = \sqrt{a\frac{L^2}{Gm^2M}}.$$

In eq. (14.37), e denotes the eccentricity of the ellipse, m mass of the planet and M mass of the Sun, we obtain

$$T = \pi\frac{2m}{L}a^{3/2}\sqrt{\frac{L^2}{Gm^2M}}. \tag{14.38}$$

Thus,

$$T = 2\pi a^{3/2}\sqrt{\frac{1}{GM}} \tag{14.39}$$

which is known as the third law of Kepler.

14.6 Orbits of planets, spaceships, and satellites

Suppose a body of mass m moves in a circular orbit around a larger body of mass M. If we assume that the radius is r and $m \ll M$, then the gravitational potential energy of the two-body system is (see eq. (14.20))

$$U = -G\frac{mM}{r}. \tag{14.40}$$

The total mechanical energy of the two-body system can be written as (assuming $m \ll M$)

$$E = \frac{1}{2}mv^2 + U(r) = \frac{1}{2}mv^2 - G\frac{mM}{r}. \tag{14.41}$$

Therefore, from eq. (14.41), we can say that E is either positive, negative or zero, depending on the smaller body's speed, m (such as a planet, spaceship, or satellite). For a system such the Earth–Sun system, $E < 0$ because, by convention, for $r \to \infty$, we obtain $U \to 0$. Then, using eq. (14.41), we obtain the speed of the object in an orbit as

$$v = \sqrt{\frac{2}{m}(E - U(r))}.$$ (14.42)

Using the second law of Newton for the rotational motion, the Newton force is the centripetal force causing a centripetal acceleration $a_r = v^2/r$, then for the motion of the planets, spaceships or satellites with mass m (such that $m \ll M$), we can write

$$m\frac{v^2}{r} = G\frac{mM}{r^2}.$$ (14.43)

From eq. (14.43), the velocity is

$$v = \sqrt{G\frac{M}{r}}.$$ (14.44)

Substituting eq. (14.44) into eq. (14.41), we obtain

$$E = G\frac{mM}{2r} - G\frac{mM}{r} = -G\frac{mM}{2r}$$ (14.45)

which establishes that $E < 0$. Our result shows that the two-body system's mechanical energy is negative for a circular motion such as the motion of a satellite around the Earth. Furthermore, the kinetic energy of the smaller body of mass m, such as a satellite, is positive and equals half of the absolute value of the two-body system. Note that here E represents the system's binding energy (and so it is negative, $E < 0$); that is, the amount of energy necessary to be injected into the system to move the two bodies infinitely far apart.

Combining Kepler's law with that result, we can say that for a two-body system of masses bound by the gravitational forces, both total mechanical energy and the total angular momentum are conserved.

Now, consider an object of mass m, which escapes the Earth of mass M and radius R, as shown in Fig. 14.8. The object starts the escape from the Earth's surface with a velocity v_i. The goal is to calculate the object's initial speed necessary to escape the gravitational field created by the Earth. For that purpose, we use the total mechanical energy conservation law. At the surface of the Earth, the total mechanical energy of the two-body system is

$$E_i = \frac{1}{2}mv_i^2 - G\frac{mM}{R}.$$ (14.46)

At the maximum height h from the surface of the Earth, where the speed of the object becomes zero, $v_f = 0$, the total mechanical energy is

$$E_f = \frac{1}{2}mv_f^2 - G\frac{mM}{R+h}.$$ (14.47)

Using the conservation law of the total mechanical energy, $E_i = E_f$, we have

$$\frac{1}{2}mv_i^2 - G\frac{mM}{R} = -G\frac{mM}{R+h}. \tag{14.48}$$

Solving it for v_i, we obtain

$$v_i = \sqrt{2GM\left(\frac{1}{R} - \frac{1}{R+h}\right)}. \tag{14.49}$$

Escape speed

We denote the minimum speed that the mass m must have when leaving the surface of the Earth to escape the attractive gravitational force of the Earth as the escape speed. That is obtained from eq. (14.49) taking $h \to \infty$:

$$v_{escape} = \sqrt{\frac{2GM}{R}}. \tag{14.50}$$

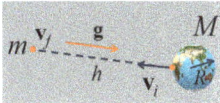

Figure 14.8: An object of mass m escaping the Earth of mass M and radius R.

14.7 Exercises

Exercise 14.1. Mass and radius of Mars are approximately 6.4×10^{23} kg and 3396 km, respectively. What is the acceleration of gravity (g) on Mars in SI units?

Exercise 14.2. Mass and radius of Jupiter are approximately 1898.19×10^{24} kg and 71492 km, respectively. What is the acceleration of gravity (g) on Jupiter in SI units?

Exercise 14.3. Mass and radius of the Mercury are approximately 0.33011×10^{24} kg and 2439.7 km, respectively. What is the acceleration of gravity (g) on Mercury in SI units?

Exercise 14.4. Mass of the Earth is approximately 5.9724×10^{24} kg. Its equatorial and polar radii are 6378.1 km and 6356.8 km, respectively. What is the acceleration of gravity, g, on the equator and poles of Earth in SI units?

Exercise 14.5. Find the velocity of the planet Uranus as it moves around the Sun. Assume the distance between Uranus and the Sun is 19 AU and the Sun's mass is 1.99×10^{30} kg.

Exercise 14.6. At what distance from the Earth is the acceleration of gravity half of that on the Earth?

Exercise 14.7. Find the gravity force between two lead spheres each with a mass of 100.0 kg and 50.0 cm apart.

Exercise 14.8. Calculate the period of Mars assuming the mass of the Sun is M_S = 1.99×10^{30} kg and the distance of Mars from the Sun is 228×10^6 km.

Exercise 14.9. Calculate the escape speed from the Earth for the spacecraft of mass 6000 kg. What is the kinetic energy the spacecraft must have when leaving the surface of the Earth to escape the influence of the Earth's gravitational field?

Exercise 14.10. Calculate the total work necessary for separating three masses m_1, m_2 and m_3 given in the configuration shown in Fig. 14.9.

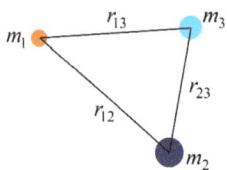

Figure 14.9: The configuration of three masses m_1, m_2, and m_3.

A Solutions

In this appendix, we present the solutions of the problems shown at the end of each chapter.

Solutions Chapter 1

1.1. The final position will be determined by the resultant vector:

$$\mathbf{R} = \mathbf{A} + \mathbf{B} \tag{A.1}$$

with magnitude calculated as

$$R = \sqrt{A^2 + B^2} = \sqrt{(3.0)^2 + (4.0)^2} = 5.0 \text{ m.} \tag{A.2}$$

Therefore, you would find yourself 5.0 m from where you started. The angle with respect to the east cast is

$$\theta = \tan^{-1}\left(\frac{4.0}{3.0}\right) = 53°. \tag{A.3}$$

The total distance traveled is

$$3.0 \text{ m} + 4.0 \text{ m} = 7.0 \text{ m.} \tag{A.4}$$

1.2.
(a) The two vectors are positioning in the xy plane, as shown in Fig. 1.24.
(b) The resultant vector $\mathbf{R} = \mathbf{A} + \mathbf{B}$ has these components:

$$R_x = A_x + B_x = 3.5 + 4.0 = 7.5 \tag{A.5}$$

and

$$R_y = A_y + B_y = (-2.0) + 3.0 = 1.0. \tag{A.6}$$

(c) The magnitude is

$$R = \sqrt{R_x^2 + R_y^2} \tag{A.7}$$
$$= \sqrt{(7.5)^2 + (1.0)^2}$$
$$= \sqrt{56.25 + 1.0} \approx 7.6.$$

https://doi.org/10.1515/9783110755824-015

1.3.

(a) The length of the side opposite the angle $\theta = 30°$ is

$$a = c \sin \theta = (3.0\,\text{m}) \sin 30° = (3.0\,\text{m})0.5 = 1.5\,\text{m}. \tag{A.8}$$

(b) The length of the side adjacent the angle $\theta = 30°$ is

$$b = c \cos \theta = (3.0\,\text{m}) \cos 30° = (3.0\,\text{m})\frac{\sqrt{3}}{2} \approx 2.6\,\text{m}. \tag{A.9}$$

1.4. Using the rules of the derivative, we get

$$\begin{aligned}
\frac{dy}{dx} &= \frac{d}{dx}(ax^5 + bx^3 + c) \\
&= \frac{d}{dx}(ax^5) + \frac{d}{dx}(bx^3) + \frac{d}{dx}(c) \\
&= 5ax^4 + 3bx^2 + 0 \\
&= 5ax^4 + 3bx^2.
\end{aligned} \tag{A.10}$$

1.5. Using the rules of the derivative, we get

$$\begin{aligned}
\frac{dy}{dt} &= \frac{d}{dt}(y_0 + v_0 t + at^2) \\
&= \frac{d}{dt}(y_0) + \frac{d}{dt}(v_0 t) + \frac{d}{dt}(at^2) \\
&= 0 + v_0 + 2at \\
&= v_0 + 2at.
\end{aligned} \tag{A.11}$$

1.6. The resultant vector $\mathbf{R}$ is

$$\mathbf{R} = R_x \mathbf{i} + R_y \mathbf{j} + R_z \mathbf{k} \tag{A.12}$$

where

$$R_x = A_x + B_x = 2 + (-3) = -1 \tag{A.13}$$
$$R_y = A_y + B_y = 5 + (-4) = 1 \tag{A.14}$$
$$R_z = A_z + B_z = 6 + 4 = 10. \tag{A.15}$$

Therefore, we get

$$\mathbf{R} = -\mathbf{i} + \mathbf{j} + 10\mathbf{k}. \tag{A.16}$$

The magnitude of $\mathbf{R}$ is

$$|\mathbf{R}| = \sqrt{R_x^2 + R_y^2 + R_z^2} = \sqrt{(-1)^2 + (1)^2 + (10)^2} \approx 10.1. \tag{A.17}$$

1.7. The dot product is given

$$W = \mathbf{A} \cdot \mathbf{B} = AB \cos \theta = 0.5AB \tag{A.18}$$

since $\cos 60° = 0.5$.

1.8. From the definition, the dot product is given by

$$\mathbf{A} \cdot \mathbf{B} = A_x B_x + A_y B_y + A_z B_z \tag{A.19}$$
$$= (3)(-2) + (2)(2) + (4)(-4)$$
$$= -6 + 4 - 16 = -18.$$

1.9. Figure A.1 gives the direction of the velocity $\mathbf{v}$, where $\theta = 45°$. Then, from Fig. A.1, the components of the velocity are given by

$$v_x = v \cos \theta = (800 \text{ km/h}) \cos 45° \approx 566 \text{ km/h} \tag{A.20}$$
$$v_y = v \sin \theta = (800 \text{ km/h}) \sin 45° \approx 566 \text{ km/h}. \tag{A.21}$$

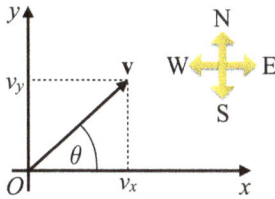

Figure A.1: Direction of the airplane velocity.

1.10. Figure A.2 shows graphically the trajectory of the person, where $\mathbf{r}_1 = (2.0 \text{ km})\mathbf{i}$ and $\mathbf{r}_2 = (1.0 \text{ km})\mathbf{j}$. The resultant displacement is the vector $\mathbf{r}$, given as

$$\mathbf{r} = \mathbf{r}_1 + \mathbf{r}_2 = (2.0 \text{ km})\mathbf{i} + (1.0 \text{ km})\mathbf{j} \tag{A.22}$$

and the magnitude of the resultant displacement is

$$r = \sqrt{(2.0)^2 + (1.0)^2} \approx 2.2 \text{ km}. \tag{A.23}$$

The angle formed by the resultant displacement vector with east direction is

$$\tan \theta = \frac{r_y}{r_x} = \frac{1.0}{2.0} = 0.5 \tag{A.24}$$

or

$$\theta = \tan^{-1}(0.5) \approx 27°. \tag{A.25}$$

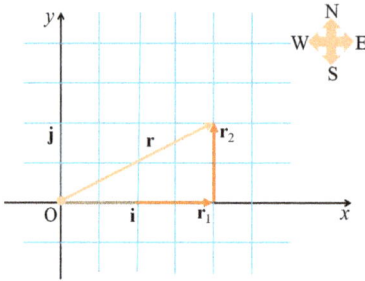

Figure A.2: The trajectory of the person.

1.11. Figure A.3 shows graphically the trajectory of the car, where $v_1 = (-60.0 \text{ km/h})j$ and $v_2 = (-50.0 \text{ km/h})i$. The resultant velocity is the vector v, given as

$$v = v_1 + v_2 = (-50.0 \text{ km/h})i + (-60.0 \text{ km/h})j \qquad (A.26)$$

and the magnitude of the resultant velocity vector is

$$v = \sqrt{(-50.0)^2 + (-60.0)^2} \approx 78.1 \text{ km/h}. \qquad (A.27)$$

The angle formed by the resultant velocity vector with east direction is $\theta = -(90° + \phi)$, where

$$\tan \phi = \frac{v_2}{v_1} = \frac{50.0}{60.0} \approx 0.833 \qquad (A.28)$$

or

$$\phi = \tan^{-1}(0.833) \approx 39.8°. \qquad (A.29)$$

Thus, we get

$$\theta \approx -130°. \qquad (A.30)$$

The minus sign indicates that the angle is measured clockwise.

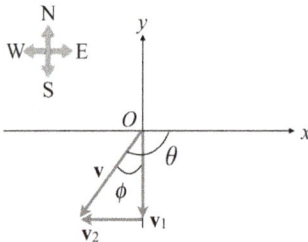

Figure A.3: The trajectory of the car.

1.12. The sum of two vectors is

$$\mathbf{R} = R_x\mathbf{i} + R_y\mathbf{j} + R_z\mathbf{k} \tag{A.31}$$

where

$$R_x = A_x + B_x = 4, \quad R_y = A_y + B_y = 14, \quad R_z = A_z + B_z = -8. \tag{A.32}$$

Therefore, we can write

$$\mathbf{R} = 4\mathbf{i} + 14\mathbf{j} - 8\mathbf{k}. \tag{A.33}$$

Similarly, the difference $\mathbf{R} = \mathbf{A} - \mathbf{B}$ is

$$\mathbf{R} = R_x\mathbf{i} + R_y\mathbf{j} + R_z\mathbf{k} \tag{A.34}$$

where

$$R_x = A_x - B_x = 2, \quad R_y = A_y - B_y = 0, \quad R_z = A_z - B_z = 4. \tag{A.35}$$

Thus, we can write

$$\mathbf{R} = 2\mathbf{i} + 4\mathbf{k}. \tag{A.36}$$

Solutions Chapter 2

2.1. Since the results are needed in km, we first express the length and the width in km:

$$\ell = 100\,\text{m} = 100 \times 10^{-3}\,\text{km}; \quad w = 50.0\,\text{m} = 50.0 \times 10^{-3}\,\text{km}. \tag{A.37}$$

The area of a rectangle is the length multiplied by the width:

$$\begin{aligned} A &= \ell \times w \\ &= \left(100 \times 10^{-3}\,\text{km}\right) \times \left(50.0 \times 10^{-3}\,\text{km}\right) = 5000 \times 10^{-6}\,\text{km}^2 \\ &= 5.00 \times 10^{-3}\,\text{km}^2. \end{aligned} \tag{A.38}$$

2.2. The volume of a sphere is calculated as

$$V = \frac{4}{3}\pi R^3 \tag{A.39}$$

where $R = 2\,\text{mm}$ and $\pi = 3.14159265\ldots$. Thus,

$$V = \frac{4}{3}3.14159265(2.0\,\text{mm})^3 = 33.51\,\text{mm}^3 \approx 34\,\text{mm}^3 \tag{A.40}$$

where the result is approximated in two significant figures.

Now, we can calculate the volume of the tank:

$$V_{tank} = 1\,m^3 = 1 \times (10^3\,mm)^3 = 1 \times 10^9\,mm^3. \qquad (A.41)$$

Then the number of raindrops is

$$N = \frac{V_{tank}}{V} = \frac{1 \times 10^9\,mm^3}{34\,mm^3} \approx 3 \times 10^7. \qquad (A.42)$$

2.3. The mass gain is

$$\Delta m = 70\,\mu g = 70 \times 10^{-6}\,g = 70 \times 10^{-9}\,kg. \qquad (A.43)$$

Therefore, the number of CH molecules, N, is

$$N = \frac{\Delta m}{m_{CH}} = \frac{70 \times 10^{-9}\,kg}{2.2 \times 10^{-26}\,kg} \approx 32 \times 10^{17}. \qquad (A.44)$$

2.4. The mass of the deuteron is

$$M_d = M_p + M_n - BE \qquad (A.45)$$

where $M_p = 1.007276\,amu$ is the mass of proton and $M_n = 1.008665\,amu$ is the mass of neutron. Therefore,

$$M_d = 1.007276\,amu + 1.008665\,amu - 0.002362\,amu \qquad (A.46)$$
$$= 2.015379\,amu.$$

2.5. 1 year is 365.25 days, including the leap year. Then

$$1\,year = 365.25 \times 24 = 8766\,h. \qquad (A.47)$$

We know that 1 hour is 3600 seconds. Therefore,

$$1year = 8766\,h \times 3600\,s/h = 31557600\,s. \qquad (A.48)$$

Since $1\,s = 10^9$ ns, we have

$$1year = 31557600\,s = 31557600 \times 10^9\,ns = 3.15576 \times 10^{16}\,ns. \qquad (A.49)$$

The lower limit of the proton lifetime is

$$\tau_p > 3.15576 \times 10^{16} \times 1.01 \times 10^{34}\,ns \approx 3.18732 \times 10^{50}\,ns. \qquad (A.50)$$

Now we divide this lower limit of the proton lifetime by the mean lifetime of the positive pion:

$$\frac{\tau_p}{\tau_{\pi^+}} > 1.22589 \times 10^{49}. \qquad (A.51)$$

2.6. We know that $1\,\mu m = 10^{-6}\,m$, then

$$\ell = 150\,\mu m = 150 \cdot (10^{-6}\,m) = 150 \times 10^{-6}\,m. \tag{A.52}$$

Since $1\,m = 10^9\,nm$, we have

$$\ell = 150 \times 10^{-6}\,m = (150 \times 10^{-6}) \cdot (10^9\,nm) = 150 \times 10^3\,nm. \tag{A.53}$$

We also know that $1\,m = 10^{10}\,\text{Å}$, thus

$$\ell = 150 \times 10^{-6}\,m = (150 \times 10^{-6}) \cdot (10^{10}\,\text{Å}) = 150 \times 10^4\,\text{Å}. \tag{A.54}$$

2.7. The volume of a sphere is

$$V = \frac{4}{3}\pi R^3. \tag{A.55}$$

The size of the H-atom in m is

$$D_H = 1\,\text{Å} = 1 \times 10^{-10}\,m. \tag{A.56}$$

The radius is

$$R_H = D_H/2 = 5 \times 10^{-11}\,m. \tag{A.57}$$

Therefore, the volume of the H-atom is

$$V_H = \frac{4}{3}\pi R_H^3 \approx 5 \times 10^{-31}\,m^3. \tag{A.58}$$

2.8. Since the density is equal to mass per unit volume, the mass m of the cube is

$$m = \rho V = (2.7\,g/cm^3)(0.20\,cm^3) = 0.54\,g. \tag{A.59}$$

To find the number of atoms N in this mass of aluminum, we can set up a proportion using the fact that one mole of aluminum (27 g) contains 6.02×10^{23} atoms:

$$\frac{N_A}{27\,g} = \frac{N}{0.54\,g} \tag{A.60}$$

$$\frac{6.02 \times 10^{23}\,\text{atoms}}{27\,g} = \frac{N}{0.54\,g}. \tag{A.61}$$

From this,

$$N = \frac{(0.54\,g)(6.02 \times 10^{23}\,\text{atoms})}{27\,g} = 12 \times 10^{21}\,\text{atoms}. \tag{A.62}$$

2.9. For the speed term, we have

$$[v] = \frac{L}{T}. \tag{A.63}$$

For the dimension of the acceleration

$$[a] = \frac{L}{T^2}. \tag{A.64}$$

Therefore, the dimensions of at are

$$[at] = \frac{L}{T^2}T = \frac{L}{T}. \tag{A.65}$$

Hence, the expression is dimensionally correct.

2.10. Since $1\,g = 10^{-3}\,kg$ and $1\,cm = 10^{-2}\,m$, the mass m and volume V in basic SI units are

$$m = 856\,g \times 10^{-3}\,kg/g = 0.856\,kg \tag{A.66}$$

$$V = L^3 = (5.35\,cm \times 10^{-2}\,m/cm)^3 \tag{A.67}$$

$$= (5.35)^3 \times 10^{-6}\,m^3$$

$$= 1.53 \times 10^{-4}\,m^3.$$

Therefore,

$$\rho = \frac{m}{V} = \frac{0.856\,kg}{1.53 \times 10^{-4}\,m^3} \tag{A.68}$$

$$= 5.59 \times 10^3\,kg/m^3.$$

2.11. Suppose that distance between the two cities is 3000 miles. In addition, the length of one step is 2 ft. Then next we can determine the number of steps in one miles:

$$5280\,\frac{ft}{mi} \approx 5000\,\frac{ft}{mi}. \tag{A.69}$$

Therefore,

$$\frac{5000\,ft/mi}{2\,ft/step} = 2500\,step/mi. \tag{A.70}$$

Hence,

$$(3000\,mi) \times (2500\,steps/mi) = 7.5 \times 10^6\,steps \approx 8 \times 10^6\,steps. \tag{A.71}$$

2.12. First, the area A is

$$A = \ell w = 21.3 \times 9.80 = 208.74 \, \text{cm}^2 \approx 209 \, \text{cm}^2. \tag{A.72}$$

The value of A can vary in the range

$$A_{\text{min}} = (21.1 \, \text{cm}) \times (9.70 \, \text{cm}) \approx 205 \, \text{cm}^2 \tag{A.73}$$

$$A_{\text{max}} = (21.5 \, \text{cm}) \times (9.90 \, \text{cm}) \approx 213 \, \text{cm}^2. \tag{A.74}$$

Therefore, the uncertainty is

$$\delta A = 4.00 \, \text{cm}^2 \tag{A.75}$$

and

$$A = (209 \pm 4.00) \, \text{cm}^2. \tag{A.76}$$

2.13.
(a) $100 + 1.51 = 101.51 \approx 102$
(b) $12/3.5 \approx 3.4$
(c) $99 + 13.65 = 112.65 \approx 113$

2.14. We know that

$$1 \, \text{g} = 0.001 \, \text{kg}.$$

Therefore,

$$m = 1234 \, \text{g} = (1234 \times 0.001) \, \text{kg} = 1.234 \, \text{kg}.$$

Using only three significant figures, we can approximate the mass as

$$m = 1.234 \, \text{kg} \approx 1.23 \, \text{kg}.$$

2.15. We know that

$$1 \, \text{cm} = 0.01 \, \text{m}.$$

Therefore, the speed in SI units is

$$v = 2321 \, \frac{\text{cm}}{\text{s}} = 2321 \times 0.01 \, \frac{\text{m}}{\text{s}} = 23.21 \, \frac{\text{m}}{\text{s}}.$$

Using only three significant figures, we can approximate the speed as

$$v = 23.2 \, \frac{\text{m}}{\text{s}}.$$

Solutions Chapter 3

3.1. From the position–time plot (see also Fig. 3.12),

$$x_A = 30\,\text{m} \quad \text{at } t_A = 0\,\text{s} \tag{A.77}$$

and

$$x_F = -50\,\text{m} \quad \text{at } t_F = 50\,\text{s}. \tag{A.78}$$

Using these values

$$\Delta x = x_F - x_A = -50\,\text{m} - 30\,\text{m} = -80\,\text{m}. \tag{A.79}$$

This result indicates that the car ends up 80 m along the negative direction (left in this case) from where it started.

The average velocity is

$$\bar{v}_s = \frac{\Delta x}{\Delta t} = \frac{x_f - x_i}{t_f - t_i} = \frac{x_F - x_A}{t_F - t_A} \tag{A.80}$$

$$= \frac{-80\,\text{m}}{50\,\text{s}} = -1.6\,\text{m/s}.$$

The car's average speed is found by adding the distances traveled and dividing by the total time:

$$\text{Average speed} = \frac{\widehat{AB} + \widehat{BC} + \widehat{CD} + \widehat{DE} + \widehat{EF}}{T} \tag{A.81}$$

$$\approx \frac{|AB| + |BF|}{50\,\text{s}}$$

$$= \frac{\sqrt{(10)^2 + (30)^2} + \sqrt{(40)^2 + (110)^2}}{50\,\text{s}}$$

$$\approx \frac{32\,\text{m} + 117\,\text{m}}{50\,\text{s}}$$

$$\approx 3.0\,\text{m/s}.$$

3.2. Using the position versus time graph for this motion, given in Fig. 3.13, we can find the following.

(a) For the first time interval, we have a negative slope, and hence a negative velocity. Thus, we know that the displacement between A and B must be a negative number having units of meters. Similarly, we expect the displacement between B and D to be positive.

In the first time interval, we set $t_i = t_A = 0$ and $t_f = t_B = 1\,\text{s}$. From this,

$$\Delta x_{A \to B} = x_f - x_i = x_B - x_A \tag{A.82}$$

$$= [-4(1) + 2(1)^2] - [-4(0) + 2(0)^2]$$

$$= -2\,m.$$

For the displacement during the second time interval, we have $t_i = t_B = 1\,s$, $t_f = t_D = 3\,s$, and hence we get

$$\Delta x_{B \to D} = x_f - x_i = x_D - x_B \tag{A.83}$$

$$= [-4(3) + 2(3)^2] - [-4(1) + 2(1)^2]$$

$$= +8\,m.$$

The displacement can also be read directly from the graph in Fig. 3.13.

(b) In the first time interval, $\Delta t = t_f - t_i = t_B - t_A = 1\,s$. Therefore, from the displacement calculated from (a), we find that

$$\bar{v}_{x(A \to B)} = \frac{\Delta x_{A \to B}}{\Delta t} = \frac{-2\,m}{1\,s} = -2\,m/s. \tag{A.84}$$

In the second time interval, $\Delta t = 2\,s$; therefore

$$\bar{v}_{x(B \to D)} = \frac{\Delta x_{B \to D}}{\Delta t} = \frac{8\,m}{2\,s} = +4\,m/s. \tag{A.85}$$

These values agree with the slopes of the lines joining these points in Fig. 3.13.

(c) We can assume that the instantaneous velocity is the same order of magnitude as our previous results, that is, around 4 m/s. From the graph, we find out that the slope of the tangent at position C is greater than the slope of the red line connecting points B and D. Thus, we expect the answer to be greater than 4 m/s. To be accurate, we evaluate the expression of the velocity as a function of time t:

$$v_x(t) = \frac{dx}{dt} = (-4 + 4t)\,m/s. \tag{A.86}$$

Then the instantaneous velocity at $t = 2.5\,s$ is

$$v_x(2.5\,s) = (-4 + 4 \cdot 2.5) = +6\,m/s. \tag{A.87}$$

From the position versus time graph, the slope at $t = 2.5\,s$ gives

$$v_x = +6\,m/s. \tag{A.88}$$

3.3. In Fig. A.4 is shown a $v_x - t$ graph representing the velocity versus time. Because the slope of the entire $v_x - t$ curve is negative, we expect the acceleration to be negative.
(a) We find the velocities at $t_i = t_A = 0$ and $t_f = t_B = 2.0\,s$ by substituting the numerical values of t into the expression for the velocity:

$$v_{xA} = (40 - 5t_A^2)\,m/s = [40 - 5(0)^2]\,m/s = +40\,m/s \tag{A.89}$$

$$v_{xB} = (40 - 5t_B^2)\,m/s = [40 - 5(2.0)^2]\,m/s = +20\,m/s. \tag{A.90}$$

Therefore, the average acceleration in the given interval of time $\Delta t = t_B - t_A = 2.0\,\text{s}$ is

$$\bar{a}_x = \frac{v_{xf} - v_{xi}}{t_f - t_i} = \frac{v_{xB} - v_{xA}}{t_B - t_A} = \frac{(20 - 40)\,\text{m/s}}{(2.0 - 0)\,\text{s}} \tag{A.91}$$
$$= -10\,\text{m/s}^2.$$

The negative sign indicates that the slope of the line joining the initial and final points on the velocity, which represents the acceleration, is negative.

(b) The velocity at any time t is

$$v_{xi} = (40 - 5t^2)\,\text{m/s} \tag{A.92}$$

and the velocity at any time later $t + \Delta t$ is

$$v_{xf} = 40 - 5(t + \Delta t)^2 = 40 - 5t^2 - 10t\Delta t - 5\Delta t^2 \tag{A.93}$$
$$= v_{xi} + \left[-10t\Delta t - 5\Delta t^2\right].$$

Thus, the change in the velocity during the time interval Δt is

$$\Delta v_x = v_{xf} - v_{xi} = \left[-10t\Delta t - 5\Delta t^2\right]\,\text{m/s}. \tag{A.94}$$

Dividing this expression by Δt and taking the limit of the result as $\Delta t \to 0$ gives the acceleration at any time t:

$$a_x = \lim_{\Delta t \to 0} \frac{\Delta v_x}{\Delta t} = \lim_{\Delta t \to 0} (-10t - 5\Delta t) = -10t\,\text{m/s}^2. \tag{A.95}$$

Therefore, at $t = 2.0\,\text{s}$

$$a_x = (-10)(2.0)\,\text{m/s}^2 = -20\,\text{m/s}^2. \tag{A.96}$$

The minus sign indicates that the acceleration is along the negative direction of the x-axis.

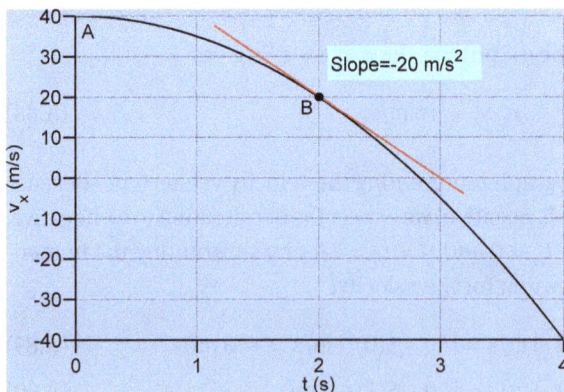

Figure A.4: Velocity versus time graph for a particle moving with a velocity v_x varying with time as $v_x = 40 - 5t^2$.

3.4.

(a) First, we convert velocity into meters per seconds:

$$v_{xf} = 100\,\text{km/h} = 100 \times \frac{10^3}{3600}\,\text{m/s} \approx 27.8\,\text{m/s}. \tag{A.97}$$

Then the initial velocity is

$$v_{xi} = v_{xf}/3 \approx 9.26\,\text{m/s}. \tag{A.98}$$

The acceleration is

$$a_x = \frac{v_{xf} - v_{xi}}{t} = \frac{27.8 - 9.26}{10.0}\,\frac{\text{m}}{\text{s}} \times \frac{1}{\text{s}} = 1.85\,\text{m/s}^2. \tag{A.99}$$

(b) We can calculate the distance traveled during the first 5 s as

$$x_f - x_i = v_{xi}t + \frac{1}{2}a_x t^2 = (9.26\,\text{m/s})(5\,\text{s}) + \frac{1}{2}(1.85\,\text{m/s}^2)(5\,\text{s})^2 \tag{A.100}$$

$$= 46.3\,\text{m} + 23.1\,\text{m} = 69.4\,\text{m}.$$

If we consider no acceleration ($a_x = 0$), then moving with initial velocity of 9.26 m/s will result in a displacement of $(9.26\,\text{m/s})(5\,\text{s}) = 46.3\,\text{m}$ during the first 5 s, indicating that the acceleration of the velocity gives an extra 23.1 m distance traveling.

3.5.

(a) First we convert units from mi/h to m/s:

$$v_{xi} = 140\,\text{mi/h} = 140 \times \frac{1609}{3600}\,\text{m/s} \approx 62.6\,\text{m/s}. \tag{A.101}$$

Since the jet finally stops, the final velocity is

$$v_{xf} = 0.00\,\text{m/s}. \tag{A.102}$$

Therefore, the acceleration is

$$a_x = \frac{v_{xf} - v_{xi}}{t} = \frac{0.00\,\text{m/s} - 62.6\,\text{m/s}}{2.00\,\text{s}} = -31.3\,\text{m/s}^2. \tag{A.103}$$

The negative sign of acceleration indicates that acceleration is along the negative direction of the x-axis, and thus, the jet is slowing down.

(b) We can now use the following equation to solve it for the displacement:

$$x_f - x_i = v_{xi}t + \frac{1}{2}a_x t^2 \tag{A.104}$$

$$= (62.6\,\text{m/s})(2.00\,\text{s}) + \frac{1}{2}(-31.3\,\text{m/s}^2)(2.00\,\text{s})^2$$

$$= (125.2\,\text{m}) - (62.6\,\text{m}) = 62.6\,\text{m}.$$

3.6. We start by writing the equation of the velocity change of the ball:

$$v = v_0 - gt. \tag{A.105}$$

The sign minus indicates that the motion is upward. Here, $g = 9.8\,\text{m/s}^2$. Then

$$v = 25\,(\text{m/s}) - 9.8\,(\text{m/s}^2) \times 1.0\,(\text{s}) \approx 15\,\text{m/s}. \tag{A.106}$$

3.7. Figure 3.14 illustrates the trajectory of the ball.
(a) For the motion from A to B, it is a upward movement. If we consider the y-axis as vertical, then the acceleration along the y-axis is

$$a_y = -g = -9.80\,\text{m/s}^2. \tag{A.107}$$

Using the equation for the velocity change, we have

$$v_{yB} = v_{yA} + a_y t_{AB}. \tag{A.108}$$

Replacing the numerical values:

$$0 = (20.0\,\text{m/s}) - (9.80\,\text{m/s}^2)t_{AB}. \tag{A.109}$$

Solving it for t_{AB}, we find

$$t_{AB} = 20.0/9.80 = 2.04\,\text{s}. \tag{A.110}$$

(b) The displacement along the y direction during that time, t_{AB}, is

$$y_B - y_A = v_{yi}t_{AB} + \frac{1}{2}a_y t_{AB}^2. \tag{A.111}$$

Replacing the numerical values

$$y_B - 0 = (20.0\,\text{m/s})(2.04\,\text{s}) + \frac{1}{2}(-9.80\,\text{m/s}^2)(2.04\,\text{s})^2 \tag{A.112}$$

or

$$y_B = 20.4\,\text{m}. \tag{A.113}$$

(c) To find the time needed for the ball to return to the point where it was thrown, we study the motion from B to C. This is a downward motion with $a_y = g = -9.80\,\text{m/s}^2$. The initial velocity is the velocity at B: $v_{yB} = 0$. The height is y_B, then using the equation for the displacement:

$$y_C - y_B = v_{yB}t_{BC} + \frac{1}{2}a_y t_{BC}^2. \tag{A.114}$$

Replacing the numerical values

$$0 - 20.4 \text{ m} = \frac{1}{2}(-9.80 \text{ m/s}^2)t_{BC}^2. \tag{A.115}$$

Solving it for t_{BC}, we get

$$t_{BC} = 2.04 \text{ s.} \tag{A.116}$$

The total time: $t = t_{AB} + t_{BC} = 4.08$ s.
(d) To determine the velocity of the ball at point C, we use the equation:

$$v_{yC} = v_{yB} + a_y t_{BC}. \tag{A.117}$$

Replacing the numerical values,

$$v_{yC} = 0 + (-9.80 \text{ m/s}^2)(2.04 \text{ s}) \approx -20.0 \text{ m/s.} \tag{A.118}$$

The sign minus indicates that velocity has downward direction.
(e) To find the velocity at time $t_D = 5.00$ s, we use the equation

$$v_{yD} = v_{yC} + a_y t_{CD}. \tag{A.119}$$

After replacing the values and knowing that

$$t_{CD} = 5.00 \text{ s} - 4.08 \text{ s} = 0.920 \text{ s} \tag{A.120}$$

we get

$$v_{yD} = -20.0 \text{ m/s} + (-9.80 \text{ m/s}^2)(0.920 \text{ s}) \tag{A.121}$$
$$\approx -29.0 \text{ m/s.}$$

To find the distance traveled by the ball, we use the following equation:

$$y_D - y_C = v_{yC} t_{CD} + \frac{1}{2} a_y t_{CD}^2. \tag{A.122}$$

Replacing the values, we get

$$y_D - 0 = (-20.0 \text{ m/s})(0.920 \text{ s}) + \frac{1}{2}(-9.80 \text{ m/s}^2)(0.920 \text{ s})^2 \tag{A.123}$$

or

$$y_D = -22.5 \text{ m.} \tag{A.124}$$

The sign minus indicates that the point D is below the zero level horizontal axis.

3.8. The average and instantaneous velocities are equal only for motion with constant velocity, that is, for motion with $a = 0$. To explain this, we can use the two equations for the displacement:

$$x_f - x_i = \bar{v}_x t \tag{A.125}$$

$$x_f - x_i = v_x t + (1/2)at^2 \tag{A.126}$$

where $\bar{v}_x$ is the average velocity and v_x is the instantaneous initial velocity. It can be seen that, for $a = 0$, we have $\bar{v}_x t = v_x t$ or $\bar{v}_x = v_x$.

Alternatively, since graphically the average velocity equals the slope of the line joining the initial and the final position of the body in a position versus time graph, and the instantaneous velocity equals the slope of the tangent at an instant in that graph, then the slopes are equal if the position–time graph is a straight line. That is, $v_x = \bar{v}_x$ if the position versus time is a straight line.

3.9. The answer is no. Even when the average velocity is nonzero for some time interval, the instantaneous velocity may be zero. For example, when there is a change in the velocity direction, the instantaneous velocity becomes zero when the body changes the motion direction. Mathematically, since the instantaneous velocity is given as the derivative of the position with respect to time: $v_x = dx/dt$, at the inflation point, the derivative is zero, and hence the instantaneous velocity becomes zero.

3.10. Starting with the definition of the average velocity and instantaneous velocity as follows:

$$\bar{v}_x = \frac{x_f - x_i}{t} \tag{A.127}$$

$$v_x(t) = \frac{dx}{dt}. \tag{A.128}$$

It can be seen that the average velocity is zero at some time interval if the position of the body does not change, that is, x is constant. Then, from the definition of the $v_x(t)$, the derivative must also be zero.

3.11. The answer is yes. That could happen in the time interval when the velocity is decreasing compared to the initial velocity but remains still positive. In this interval, the acceleration has a negative sign since the motion is slowing down, as also indicated in Fig. A.5.

3.12. The answer is yes. For example, consider the motion with constant velocity, but different from zero. From the definition of the acceleration: $a_x = dv_x/dt$, it turns out that if v_x is constant, then $a_x = 0$.

3.13. The answer is no. From the definition of the acceleration: $a_x = dv_x/dt$, it turns out that if v_x is zero, then a_x must be zero.

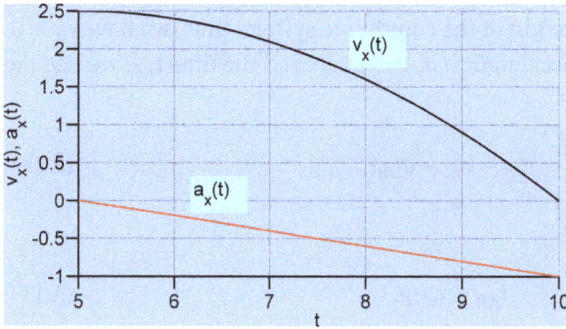

Figure A.5: The velocity $v_x(t)$ and acceleration $a_x(t)$ in some time interval between 5 and 10 seconds.

3.14. The answer is no. This can also be shown from the equations of velocity and displacement:

$$v_x(t) = v_{xi} + a_x t \qquad (A.129)$$

$$x_f - x_i = v_{xi}t + \frac{a_x t^2}{2}. \qquad (A.130)$$

It can be seen that object might stop, but it can never stay stopped for ever. Since, for $v_{xi} = 0$, we have $x_f - x_i = a_x t^2/2 \neq 0$, because $a_x \neq 0$.

3.15. The displacement, $y_f - y_i$ does not depend on the origin of the coordinate system. Neither does the velocity, which depends on the initial velocity.

3.16. To understand the problem, we have sketched the diagram shown in Fig. A.6. In the case when the student throw the ball upward (see Fig. A.6(a)), the trajectory of the ball follows the highest point B, then moves downward until it touches the ground at point C. In the second case, when the student throws the ball down from point A, the movement is down until it touches the ground at point B (see Fig. A.6(b)).

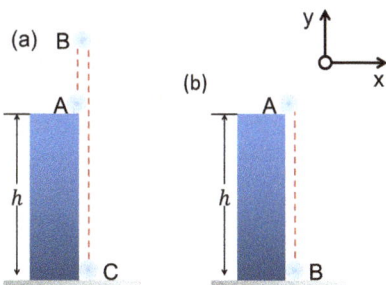

Figure A.6: (a) The ball thrown upward. (b) The ball thrown downward with the same velocity.

For the first ball, the velocity at the point B is $v_{yB} = 0$. The displacement in the upward direction is given as:

$$y_B - y_A = v_{yA}t_{AB} + (1/2)a_y t_{AB}^2. \qquad (A.131)$$

Assume that the point A is the origin of the coordinate system; thus, we have $y_A = 0$. The velocity $v_{yA} = v_{yi}$ and the acceleration $a_y = -g$. To find the time t_{AB}, we use the equation of the velocity:

$$v_{yB} = v_{yA} + at_{AB}.$$ (A.132)

Or

$$t_{AB} = v_{yi}/g.$$ (A.133)

Then

$$y_B = \frac{v_{yi}^2}{g} + \frac{1}{2}(-g)\frac{v_{yi}^2}{g^2} = \frac{v_{yi}^2}{2g}.$$ (A.134)

Then the motion from B to C is downward with initial velocity zero (i. e., free fall) from the height

$$h' = h + \frac{v_{yi}^2}{2g}.$$ (A.135)

To find the velocity at point C, we use the following equation:

$$v_{yC}^2 - v_{yB}^2 = 2a_y(y_C - y_B) = 2gh'$$ (A.136)

or

$$v_{yC}^2 = 2g\left(h + \frac{v_{yi}^2}{2g}\right) = 2hg + 2v_{yi}^2.$$ (A.137)

Hence, the speed at point C is

$$v_{yC} = \sqrt{2hg + 2v_{yi}^2}.$$ (A.138)

For the second ball, the trajectory is shown in Fig. 3.8(b), the motion is downward with initial velocity v_{yi}. Using the equation

$$v_{yB}^2 - v_{yi}^2 = 2a_y(y_B - y_A) = 2gh$$ (A.139)

we get the speed of the second ball as

$$v_{yB} = \sqrt{2hg + v_{yi}^2}.$$ (A.140)

It can be seen that the first ball will touch the ground with higher speed compare to the second ball.

3.17. To understand the problem, we have sketched the diagram shown in Fig. A.7. From the definition of the average velocity, it turns out that the magnitude of the average velocity is

$$|\bar{v}_x| = |\tan(\alpha)|. \tag{A.141}$$

On the other hand, the instantaneous velocity is the derivative of position with respect to time: $v_x(t) = dx/dt$. Therefore, graphically it is characterized by the tangent of the trajectory curve at each instant of time, which is equal to the tangent of the angle β, as indicated in the figure by the green, blue and red lines. Thus,

$$|v_x(t)| = |\tan(\beta)|. \tag{A.142}$$

Since $\alpha > \beta$ always and tangent is a monotonically increasing function in absolute value for angles in the interval $[0, \pi/2]$, we can say that $|\bar{v}_x| > |v_x(t)|$. They are equal only for motion along a straight line in a position–time graph.

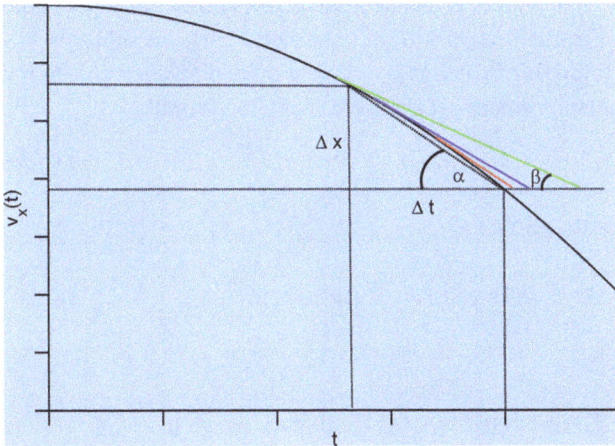

Figure A.7: A diagram of the average and instantaneous velocities.

3.18. Using the definition of the average velocity as

$$\bar{v}_x = \frac{x_f - x_i}{t_f - t_i} \tag{A.143}$$

then if in some interval, let say $t_f - t_i$, the average velocity is zero, then

$$x_f - x_i = \bar{v}_x(t_f - t_i) = 0. \tag{A.144}$$

So, the object does not move or it moves along a closed path, and hence the displacement is zero.

3.19. To answer to the questions, we have to use the following equation:

$$\bar{v}_x = \frac{x_f - x_i}{t_f - t_i}.$$ (A.145)

(a) For the first second:

$$\bar{v}_x = \frac{2.3\,\text{m} - 0.0\,\text{m}}{1.0\,\text{s} - 0.0\,\text{s}} = 2.3\,\text{m/s}.$$ (A.146)

(b) For the last three seconds:

$$\bar{v}_x = \frac{57.5\,\text{m} - 20.7\,\text{m}}{5.0\,\text{s} - 3.0\,\text{s}} = 18.4\,\text{m/s} \approx 18\,\text{m/s}.$$ (A.147)

(c) For the entire period

$$\bar{v}_x = \frac{57.5\,\text{m} - 0.0\,\text{m}}{5.0\,\text{s} - 0.0\,\text{s}} = 11.5\,\text{m/s} \approx 12\,\text{m/s}.$$ (A.148)

3.20.

(a) We can split the total displacement into three parts: in the first part, Δx_1 is the motorist's displacement in the first 35 min; in the second part, the motorist has stopped $\Delta x_2 = 0$; and in the third part, $\Delta x_3 = 130$ km is the displacement for the last 2.00 h. Each of the displacement is calculated using the formula

$$\Delta x = v\Delta t.$$ (A.149)

Therefore, for the first part, we obtain

$$\Delta x_1 = (85\,\text{km/h})\left(\frac{35}{60}\,\text{h}\right) \approx 50\,\text{km}.$$ (A.150)

The total displacement is

$$\Delta x = \Delta x_1 + \Delta x_2 + \Delta x_3 = 180\,\text{km}.$$ (A.151)

(b) The average velocity is calculated as

$$\bar{v}_x = \frac{\Delta x}{\Delta t} = \frac{180\,\text{km}}{\frac{35}{60}\,\text{h} + \frac{15}{60}\,\text{h} + 2.00\,\text{h}} = \frac{180\,\text{km}}{2.83\,\text{h}} \approx 63.6\,\text{km/h}.$$ (A.152)

3.21. For the average velocity we use the formula

$$\bar{v}_x = \frac{x_f - x_i}{t_f - t_i}.$$ (A.153)

(a) From Fig. 3.15, $x_f = 8$ m, $x_i = 0$ m, $t_f = 2$ s and $t_i = 0$ s, thus

$$\bar{v}_x = \frac{8\,\text{m} - 0\,\text{m}}{2\,\text{s} - 0\,\text{s}} = 4\,\text{m/s}.$$ (A.154)

(b) From Fig. 3.15, $x_f = 5\,\text{m}$, $x_i = 0\,\text{m}$, $t_f = 4\,\text{s}$ and $t_i = 0\,\text{s}$. Then we get

$$\bar{v}_x = \frac{5\,\text{m} - 0\,\text{m}}{4\,\text{s} - 0\,\text{s}} = 1.25\,\text{m/s} \approx 1\,\text{m/s}. \tag{A.155}$$

(c) From Fig. 3.15, $x_f = 5\,\text{m}$, $x_i = 8\,\text{m}$, $t_f = 4\,\text{s}$ and $t_i = 2\,\text{s}$. Then we get

$$\bar{v}_x = \frac{5\,\text{m} - 8\,\text{m}}{4\,\text{s} - 2\,\text{s}} = -1.5\,\text{m/s} \approx -2\,\text{m/s}. \tag{A.156}$$

(d) From Fig. 3.15, $x_f = -5\,\text{m}$, $x_i = 5\,\text{m}$, $t_f = 7\,\text{s}$ and $t_i = 4\,\text{s}$. Then we get

$$\bar{v}_x = \frac{-5\,\text{m} - 5\,\text{m}}{7\,\text{s} - 4\,\text{s}} \approx 3\,\text{m/s}. \tag{A.157}$$

(e) From Fig. 3.15, $x_f = 0\,\text{m}$, $x_i = 0\,\text{m}$, $t_f = 8\,\text{s}$ and $t_i = 0\,\text{s}$. Then we get

$$\bar{v}_x = \frac{0\,\text{m} - 0\,\text{m}}{8\,\text{s} - 0\,\text{s}} = 0\,\text{m/s}. \tag{A.158}$$

3.22. To find the average velocity we use the following equation:

$$\bar{v}_x = \frac{x_f - x_i}{t_f - t_i}. \tag{A.159}$$

(a) Here, $t_i = 2.0\,\text{s}$ and $t_f = 3.0\,\text{s}$, and hence

$$x_i = x(t_i) = 10(2.0)^2 = 40\,\text{m} \tag{A.160}$$
$$x_f = x(t_f) = 10(3.0)^2 = 90\,\text{m}. \tag{A.161}$$

Then

$$\bar{v}_x = \frac{90\,\text{m} - 40\,\text{m}}{3.0\,\text{s} - 2.0\,\text{s}} = 50\,\text{m/s}. \tag{A.162}$$

(b) In this case, $t_i = 2.0\,\text{s}$ and $t_f = 2.1\,\text{s}$, and hence

$$x_i = x(t_i) = 10(2.0)^2 = 40\,\text{m} \tag{A.163}$$
$$x_f = x(t_f) = 10(2.1)^2 \approx 44\,\text{m}. \tag{A.164}$$

Then

$$\bar{v}_x = \frac{44\,\text{m} - 40\,\text{m}}{2.1\,\text{s} - 2.0\,\text{s}} = 40\,\text{m/s}. \tag{A.165}$$

3.23. The average velocity is calculated as follows:

$$\bar{v}_x = \frac{x_f - x_i}{t_f - t_i}. \tag{A.166}$$

The average speed is

$$|\bar{v}_x| = \frac{|AB| + |BA|}{t_{AB} + t_{BA}} = \frac{2\ell}{t_{AB} + t_{BA}} \tag{A.167}$$

where $|AB| = |BA| = \ell$ and t_{AB} is the time walking from A to B and t_{BA} is the time walking from B to A.

(a) Let us denote $t_i = 0$ the time when the person starts the walk from point A. Then the time for walking from A to B is

$$t_{AB} = \ell/v_1 \tag{A.168}$$

where ℓ is the distance from A to B. And, the time to walk back from B to A is

$$t_{BA} = \ell/v_2. \tag{A.169}$$

Thus, the final time is

$$t_f = t_{AB} + t_{BA} = \ell\left(\frac{1}{v_1} + \frac{1}{v_2}\right). \tag{A.170}$$

Then the average velocity is

$$\bar{v}_x = \frac{0 - 0}{t_f - 0} = 0\,\text{m/s}. \tag{A.171}$$

(b) The average speed is

$$|\bar{v}_x| = \frac{2\ell}{\ell(\frac{1}{v_1} + \frac{1}{v_2})} = \frac{2}{(\frac{1}{v_1} + \frac{1}{v_2})}. \tag{A.172}$$

3.24.

(a) Using the equations of the kinematics:

$$x_f - x_i = \bar{v}_x \Delta t \tag{A.173}$$

where $\bar{v}_x$ is the average velocity, which is calculated as

$$\bar{v}_x = \frac{\Delta x}{\Delta t} = \frac{5.00\,\text{m} - (-3.00\,\text{m})}{6.00\,\text{s} - 1.00\,\text{s}} = \frac{8.00}{5.00}\,\text{m/s} = 1.60\,\text{m/s}. \tag{A.174}$$

Then

$$x = x_i + 1.60(t_f - t_i) = -3.00 + 1.60(t - 1.00) = -4.60 + 1.60t \tag{A.175}$$

where the units are given in meters. Graphically x versus time t is shown in Fig. A.8.

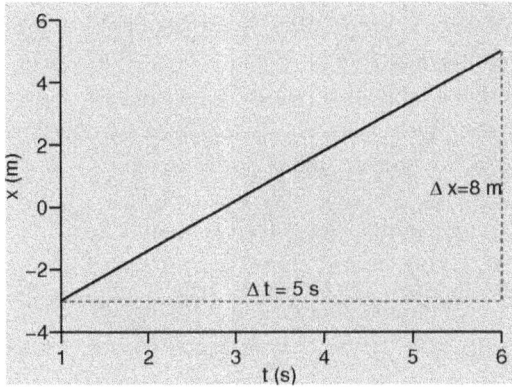

$\Delta x = 8$ m

$\Delta t = 5$ s

t (s)

Figure A.8: The position–time graph.

(b) The instantaneous velocity is given by

$$v_x(t) = \frac{dx}{dt} = 1.60 \text{ m/s}. \qquad (A.176)$$

As expected, since the position–time graph is a straight line, the average velocity and instantaneous velocity are equal.

3.25.

(a) The average velocity is defined as

$$\bar{v}_x = \frac{x_f - x_i}{t_f - t_i}. \qquad (A.177)$$

From the graph in Fig. 3.16, we find that

$$x_f = x(t_f) = x(4.0) = 2.0 \text{ m} \qquad (A.178)$$
$$x_i = x(t_i) = x(2.0) = 6.0 \text{ m}. \qquad (A.179)$$

After replacing the numerical values, we find

$$\bar{v}_x = \frac{2.0 \text{ m} - 6.0 \text{ m}}{4.0 \text{ s} - 2.0 \text{ s}} = -2.0 \text{ m/s}. \qquad (A.180)$$

The negative sign indicates that the displacement is in the negative direction of the x-axis.

(b) The instantaneous velocity is defined as

$$v_x(t) = \frac{dx}{dt} \qquad (A.181)$$

which, graphically, is the slope of the tangent to the curve at time t. In our case, $t = 2.0$ s, then the slope is equal to the tangent of the angle between the tangent and the time axis. That is given by

$$v_x(2.0) = \tan(\alpha) = -\frac{6.0 \text{ m}}{3.5 \text{ s} - 2.0 \text{ s}} = -4.0 \text{ m/s}. \qquad (A.182)$$

The negative sign indicates that the velocity at $t = 2.0$ s is in the negative direction of the x-axis.

(c) The velocity is zero at time t where the tangent is parallel with time axis, since then the tangent of the angle between the tangent at the curve and the time axis is zero. From the graph in Fig. 3.16, it can be seen that this happens at $t = 4$ s.

3.26. We use the following kinematic equation to find the acceleration:

$$v_{xf} = v_{xi} + a_x t \tag{A.183}$$

where $v_{xi} = 60$ m/s and $v_{xf} = 0$. Therefore,

$$a_x = \frac{v_{xf} - v_{xi}}{t} = \frac{0.0 - 60 \text{ m/s}}{15 \text{ s}} = -4.0 \text{ m/s}^2. \tag{A.184}$$

The sign minus indicates that the motion is slowing down or graphically, if we plot velocity versus time, the slope is negative. Also, we can say that the acceleration is along the negative direction of the x-axis.

3.27. The average acceleration is given by

$$\bar{a}_x = \frac{v_{xf} - v_{xi}}{\Delta t} = \frac{22.0 \text{ m/s} - 25.0 \text{ m/s}}{3.50 \times 10^{-3} \text{ s}} \tag{A.185}$$
$$= -857.14 \text{ m/s}$$
$$\approx -0.857 \times 10^3 \text{ m/s} = -0.857 \text{ km/s}.$$

The sign minus indicate that the motion is slowing down, and also, we can say that the acceleration is along the negative direction of the x-axis.

The magnitude of the average acceleration is

$$|\bar{a}_x| = 0.857 \text{ km/s}. \tag{A.186}$$

3.28.

(a) From the equation

$$v_{xf} = v_{xi} + a_x t \tag{A.187}$$

since the particle starts the motion from the rest, then $v_{xi} = 0$. From Fig. 3.17 $a_x = 2.0$ m/s^2 for time from $t = 0$ s to $t = 10$ s. Therefore,

$$v_{xf}(t = 10 \text{ s}) = 0 + (2.0 \text{ m/s}^2)(10 \text{ s}) = 20 \text{ m/s}. \tag{A.188}$$

To find the speed at $t = 20$ s, we determine the acceleration from the time interval 15 s to 20 s.

$$v_{xf}(t = 20 \text{ s}) = v_{xi}(t = 15 \text{ s}) + a_x(20 \text{ s} - 15 \text{ s}). \tag{A.189}$$

To find the $v_{xi}(t = 15\,\text{s})$, we look at the interval of time from 10 s to 15 s:

$$v_x(t = 15\,\text{s}) = v_x(t = 10\,\text{s}) + a_x(15\,\text{s} - 10\,\text{s}) = 20\,\text{m/s} \qquad \text{(A.190)}$$

since $a_x = 0$ from 10 to 15 s time interval. Then

$$v_{xf}(t = 20\,\text{s}) = 20\,\text{m/s} + (-3.0\,\text{m/s}^2)(20\,\text{s} - 15\,\text{s}) \qquad \text{(A.191)}$$
$$= 20\,\text{m/s} - 15\,\text{m/s} = 5.0\,\text{m/s}. \qquad \text{(A.192)}$$

(b) The distance traveled during the first 20 s can be written as

$$x = x_1 + x_2 + x_3 \qquad \text{(A.193)}$$

where x_1 is the distance traveled during the first 10 s, which is calculated as

$$x_1 = a_x t^2/2 = (2.0\,\text{m/s}^2)(10\,\text{s})^2/2 = 100\,\text{m} \qquad \text{(A.194)}$$

x_2 is the distance traveled in the interval from 10 s to 15 s with constant speed:

$$x_2 = v_x(t = 10\,\text{s})(15\,\text{s} - 10\,\text{s}) = (20\,\text{m/s})(5.0\,\text{s}) = 100\,\text{m}. \qquad \text{(A.195)}$$

And, x_3 is the distance traveled in the time interval from 15 s to 20 s, given by

$$x_3 = v_x(t = 15\,\text{s})(20\,\text{s} - 15\,\text{s}) + 0.5(-3.0\,\text{m/s}^2)(20\,\text{s} - 15\,\text{s})^2 \qquad \text{(A.196)}$$
$$= (20\,\text{m/s})(5.0\,\text{s}) - 37.5\,\text{m}$$
$$= 62.5\,\text{m}.$$

The total distance is

$$x = 100\,\text{m} + 100\,\text{m} + 62.5\,\text{m} = 262.5\,\text{m} \approx 2.63 \times 10^2\,\text{m}. \qquad \text{(A.197)}$$

3.29.
(a) In Fig. 3.18 we have plotted acceleration $a_x(t)$ versus time.
(b) The average acceleration of the object in the time interval between $t = 5.00\,\text{s}$ and $t = 15.0\,\text{s}$ is

$$\bar{a}_x = \frac{v_{xf} - v_{xi}}{t_f - t_i} = \frac{8.00\,\text{m/s} - (-8.00\,\text{m/s})}{15.0\,\text{s} - 5.00\,\text{s}} = 1.60\,\text{m/s}^2. \qquad \text{(A.198)}$$

While in the time interval between $t = 0\,\text{s}$ and $t = 20.0\,\text{s}$, the average acceleration is

$$\bar{a}_x = \frac{v_{xf} - v_{xi}}{t_f - t_i} = \frac{8.00\,\text{m/s} - (-8.00\,\text{m/s})}{20.0\,\text{s} - 0.00} = 0.800\,\text{m/s}^2. \qquad \text{(A.199)}$$

3.30.

(a) The position of the particle at time $t = 3.00$ s is

$$x(t = 3.00\,\text{s}) = 2.00 + 3.00(3.00\,\text{s}) - (3.00\,\text{s})^2 \qquad (\text{A.200})$$
$$= 2.00 + 9.00 - 9.00$$
$$= 2.00\,\text{m}.$$

(b) The velocity is calculated as

$$v_x(t) = \frac{dx}{dt} = 3.00 - 2.00t. \qquad (\text{A.201})$$

Then the velocity at time $t = 3.00$ s is

$$v_x(t = 3.00\,\text{s}) = 3.00 - 2.00(3.00\,\text{s}) = -3.00\,\text{m/s}. \qquad (\text{A.202})$$

The sign minus indicates that the particle is moving along the negative direction of the x-axis.

(c) The acceleration is

$$a_x(t) = \frac{dv_x}{dt} = -2.00\,\text{m/s}^2 \qquad (\text{A.203})$$

which is constant acceleration. The minus sign indicates that the particle is slowing down and that the slop of the velocity–time curve is negative. Also, the minus sign indicates that the direction of the acceleration is along the negative direction of the x-axis.

3.31.

(a) First, we find the position of the object at $t = 2$ s:

$$x_i = x(t = 2) = \left(3.00(2.00)^2 - 2.00(2.00) + 3.00\right)\text{m} = 11.0\,\text{m}. \qquad (\text{A.204})$$

Then we find the position at $t = 3.00$ s:

$$x_f = x(t = 3) = \left(3.00(3.00)^2 - 2.00(3.00) + 3.00\right)\text{m} = 24.0\,\text{m}. \qquad (\text{A.205})$$

Then the average velocity is

$$\bar{v}_x = \frac{x_f - x_i}{t_f - t_i} = \frac{24.0\,\text{m} - 11.0\,\text{m}}{3.00\,\text{s} - 2.00\,\text{s}} = 13.0\,\text{m/s}. \qquad (\text{A.206})$$

(b) First we find the instantaneous velocity:

$$v_x(t) = \frac{dx}{dt} = (6.00t - 2.00)\,\text{m/s}. \qquad (\text{A.207})$$

Then the velocity at $t = 2.00$ s

$$v_{xi} = (6.00(2.00) - 2.00)\,\text{m/s} = 10.0\,\text{m/s}. \tag{A.208}$$

The velocity at $t = 3.00$ s is

$$v_{xf} = (6.00(3.00) - 2.00)\,\text{m/s} = 16.0\,\text{m/s}. \tag{A.209}$$

(c) The average acceleration is

$$\bar{a}_x = \frac{v_{xf} - v_{xi}}{t_f - t_i} = \frac{16.0\,\text{m/s} - 10.0\,\text{m/s}}{3.00\,\text{s} - 2.00\,\text{s}} = 6.00\,\text{m/s}^2. \tag{A.210}$$

(d) We calculate first the instantaneous acceleration as

$$a_x = \frac{dv_x}{dt} = 6.00\,\text{m/s}^2 \tag{A.211}$$

which indicates that the acceleration is constant. Therefore, the instantaneous acceleration at both 2.00 s and 3.00 s is equal to $6.00\,\text{m/s}^2$.

3.32.
(a) The following kinematic equation is used to determine acceleration

$$v_{xf} = v_{xi} + a_x t = a_x t \tag{A.212}$$

since $v_{xi} = 0$. From eq. (A.212), you get

$$a_x = v_{xf}/t = (42.0\,\text{m/s})/(8.00\,\text{s}) = 5.25\,\text{m/s}^2. \tag{A.213}$$

(b) The distance is calculated as

$$x_f = x_i + v_{xi}t + a_x t^2/2 \tag{A.214}$$

where $x_i = 0$. After replacing the numerical values

$$x_f = (5.25\,\text{m/s}^2)(8.00\,\text{s})^2/2 = 168\,\text{m}. \tag{A.215}$$

(c) To find the speed after 10 s, assuming the same acceleration, we use the equation

$$v_{xf} = v_{xi} + a_x t = 0 + (5.25\,\text{m/s}^2)(10.0\,\text{s}) = 52.5\,\text{m/s}. \tag{A.216}$$

3.33.

(a) To find the initial speed, we use this equation

$$v_{xf}^2 - v_{xi}^2 = 2a_x(x_f - x_i) \tag{A.217}$$

where $v_{xf} = 2.80\,\text{m/s}$ is the final speed, and $x_f - x_i = 40.0\,\text{m}$ is the displacement. a_x is the acceleration, which must have a negative sign, since the truck is slowing down. To find the acceleration, we use this equation:

$$v_{xf} = v_{xi} + a_x t. \tag{A.218}$$

Thus,

$$a_x = (v_{xf} - v_{xi})/t. \tag{A.219}$$

Replacing the expression for the acceleration (eq. (A.219)) into eq. (A.217), we get

$$v_{xf}^2 - v_{xi}^2 = 2(x_f - x_i)(v_{xf} - v_{xi})/t. \tag{A.220}$$

After replacing the numerical values

$$7.84 - v_{xi}^2 = 9.40(2.80 - v_{xi}). \tag{A.221}$$

Hence, we obtain an equation for v_{xi} as follows:

$$v_{xi}^2 - 9.40 v_{xi} + 18.5 = 0. \tag{A.222}$$

First, we find the determinant:

$$D = (9.40)^2 - 4.00(18.5) = 14.4. \tag{A.223}$$

Then the solutions are

$$v_{xi}^{1,2} = \frac{9.40 \pm \sqrt{14.4}}{2} = 4.70 \pm 1.9. \tag{A.224}$$

Since the v_{xi} must be greater than the final speed, because the truck is slowing down, the physical solution is only

$$v_{xi} = 6.60\,\text{m/s}. \tag{A.225}$$

(b) To find the acceleration, we use

$$a_x = \frac{v_{xf} - v_{xi}}{t} = \frac{2.80\,\text{m/s} - 6.60\,\text{m/s}}{8.50\,\text{s}} = -0.224\,\text{m/s}^2. \tag{A.226}$$

The sign minus indicates that the truck is slowing down, as expected.

3.34. To find the acceleration, we use this equation

$$x_f = x_i + v_{xi}t + \frac{a_x t^2}{2} \qquad \text{(A.227)}$$

where x_i = 3.00 cm is the initial coordinate, v_{xi} = 12.0 cm/s is its initial velocity, x_f = −5.00 cm is the final coordinate, t = 2.00 s is the traveling time, and a_x is the acceleration, which is unknown.

After replacing the numerical values, we get

$$-5.00 \text{ cm} = 3.00 \text{ cm} + (12.0 \text{ cm/s})(2.00 \text{ s}) + a_x \frac{(2.00 \text{ s})^2}{2}. \qquad \text{(A.228)}$$

Solving it for a_x, we obtain:

$$a_x = \frac{-5.00 \text{ cm} - 3.00 \text{ cm} - 24.0 \text{ cm}}{2.00 \text{ s}^2} = -16.0 \text{ cm/s}^2. \qquad \text{(A.229)}$$

The sign minus indicates that the body is slowing down and the acceleration is along the negative direction of the x-axis.

3.35.

(a) To find the position when it changes the direction, we first find the maximum of the $x(t)$ curve. For this, we first calculate the derivative with respect to time:

$$\frac{dx}{dt} = 3.00 - 8.00t \qquad \text{(A.230)}$$

and equalize it to zero, then solve it for t:

$$3.00 - 8.00t = 0 \qquad \text{(A.231)}$$

or t = 3.00/8.00 ≈ 0.375 s. Then the position at this instant of time is

$$x(0.375) = 2.00 + 3.00(0.375) - 4.00(0.375)^2 \qquad \text{(A.232)}$$
$$\approx 2.00 + 1.13 - 0.563 \approx 2.56 \text{ m}.$$

(b) The position of the particle at t = 0 is

$$x(0) = 2.00 \text{ m}. \qquad \text{(A.233)}$$

To find the time that the particle has the same position as it had at t = 0, we solve this equation

$$x(0) = x(t) \qquad \text{(A.234)}$$

or

$$2.00 = 2.00 + 3.00t - 4.00t^2. \qquad \text{(A.235)}$$

The physical solution is

$$t = 3.00/4.00\,\text{s} = 0.750\,\text{s}. \tag{A.236}$$

Then the velocity is

$$v_x = \frac{dx}{dt} = 3.00 - 8.00t \tag{A.237}$$

and

$$v_x(0.750) = 3.00 - 8.00(0.750) = -3.00\,\text{m/s}. \tag{A.238}$$

The sign minus indicates that the particle changes direction into negative x direction.

3.36. To find the velocity of the body, we use this equation

$$v_{xf} = v_{xi} + a_x t \tag{A.239}$$

where v_{xi} is the initial velocity, v_{xf} is the final velocity, a_x is the acceleration, and t is the time of this acceleration. For both cases, $v_{xi} = 5.20\,\text{m/s}$ and $t = 2.50\,\text{s}$.
(a) In this case, $a_x = 3.00\,\text{m/s}^2$, and hence

$$v_{xf} = (5.20\,\text{m/s}) + (3.00\,\text{m/s}^2)(2.50\,\text{s}) = 12.7\,\text{m/s}. \tag{A.240}$$

(b) For $a_x = -3.00\,\text{m/s}^2$, we get

$$v_{xf} = (5.20\,\text{m/s}) + (-3.00\,\text{m/s}^2)(2.50\,\text{s}) = -2.30\,\text{m/s}. \tag{A.241}$$

In the first case, the velocity is positive, so it is along the positive direction of the x-axis. While in the second case, the velocity is negative, and hence it is opposite direction of the positive x-axis (that is, along the negative direction of the x-axis).

3.37.
(a) To find the time, we use this equation

$$x_f - x_i = v_{xi}t + a_x t^2/2 \tag{A.242}$$

where $v_{xi} = 0$ is the initial velocity, and $a_x = 10.0\,\text{m/s}^2$ is the acceleration. Here, $x_f - x_i = 400\,\text{m}$. Then we solve this equation for time t, after we replace the numerical values:

$$400 = 0 + (10.0)t^2/2. \tag{A.243}$$

The physical solution is

$$t = 8.94\,\text{s}. \tag{A.244}$$

(b) To find the speed, we use this equation

$$v_{xf} = v_{xi} + a_x t.$$ (A.245)

Replacing the numerical values

$$v_{xf} = 0 + (10.0)(8.94) = 89.4 \text{ m/s}.$$ (A.246)

3.38.
(a) From the problem, $v_{xi} = 30.0 \text{ m/s}$, $a_x = -2.00 \text{ m/s}^2$. The position is

$$x = x_0 + v_{xi}t + a_x t^2/2$$ (A.247)

where $x_0 = 0$. Thus,

$$x = 30.0t - 2.00t^2/2 = (30.0t - t^2) \text{ m}.$$ (A.248)

The velocity is

$$v_x = \frac{dx}{dt} = (30.0 - 2.00t) \text{ m/s}.$$ (A.249)

Note that the time t is measured in seconds.
(b) The maximum distance the car travels up to the hill can be calculated from the equation

$$v_{xf}^2 - v_{xi}^2 = 2a_x(x_f - x_i)$$ (A.250)

where v_{xf} is the final velocity, since the car stops, then this velocity is zero, $v_{xf} = 0$. From eq. (A.250), we get

$$x_f - x_i = -\frac{v_{xi}^2}{2a_x} = -\frac{900 \text{ m}^2/\text{s}^2}{2(-2.00 \text{ m/s}^2)} = 225 \text{ m}.$$ (A.251)

3.39. Suppose the y positive direction is upward, and the origin is at the top of the building.
(a) Since the ball is released from the rest, $v_{yi} = 0$, and the acceleration is $a_y = -9.80 \text{ m/s}^2$, where the minus sign indicates that acceleration has negative y direction. Then the position is calculated as

$$y_f = y_i + v_{yi}t + a_y t^2/2$$ (A.252)

or

$$y_f = 0 + 0 - (4.90 \text{ m/s}^2)t^2 = -(4.90t^2) \text{ m}.$$ (A.253)

Therefore,

$$y_f = \begin{cases} -4.90\,\text{m}, & t = 1.00\,\text{s} \\ -19.6\,\text{m}, & t = 2.00\,\text{s} \\ -44.1\,\text{m}, & t = 3.00\,\text{s}. \end{cases} \tag{A.254}$$

The minus sign indicates that position is on the negative side of the y-axis.

(b) To find the velocity, we use this equation

$$v_{yf} = v_{yi} + a_y t = -(9.80t)\,\text{m/s}. \tag{A.255}$$

Replacing the numerical values, we get

$$v_{yf} = \begin{cases} -9.80\,\text{m/s}, & t = 1.00\,\text{s} \\ -19.6\,\text{m/s}, & t = 2.00\,\text{s} \\ -29.4\,\text{m/s}, & t = 3.00\,\text{s}. \end{cases} \tag{A.256}$$

The minus sign indicates that the velocity is downward.

3.40. Suppose the y positive direction is upward, and the origin is at the student's position. The acceleration is $a_y = -g = -9.80\,\text{m/s}^2$, where the sign minus indicates that direction is opposite to the positive direction of the y-axis.

(a) To find the initial velocity, we use this equation with $y_i = 0$:

$$y_f = y_i + v_{yi}t + a_y t^2/2 = v_{yi}t - gt^2/2 \tag{A.257}$$

where $y_f = 4.00\,\text{m}$, $t = 1.5\,\text{s}$, then

$$v_{yi} = \frac{y_f + gt^2/2}{t} \tag{A.258}$$

$$= \frac{4.00\,\text{m} + 0.500(9.80\,\text{m/s}^2)(1.50\,\text{s})^2}{1.50\,\text{s}}$$

$$= 10.0\,\text{m/s}.$$

(b) First, let us determine the time until the keys reach maximum height, where $v_{yf} = 0$ using the equation:

$$v_{yf} = v_{yi} - gt \tag{A.259}$$

since $v_{yf} = 0$ at maximum height, you can find

$$t = \frac{v_{yi}}{g} = \frac{10.0\,\text{m/s}}{9.80\,\text{m/s}^2} = 1.02\,\text{s}. \tag{A.260}$$

The roommate caught the keys 1.50 s later, which means that he caught the keys on the way down to the ground, after the keys had reached the maximum height. To find the velocity of the keys, when they are caught, you can use this equation

$$v_{yf} = v_{yi} - gt \qquad (A.261)$$

where v_{yi} = 0 and t = 1.50 s – 1.02 s = 0.480 s. Replacing these values into the above equation, you get

$$v_{yf} = 0 - (9.80 \, m/s^2)(0.480 \, s) = -4.70 \, m/s. \qquad (A.262)$$

The sign minus indicates that the velocity is along the negative y direction; that is, it has the direction downward.

3.41. You can use this equation to first determine the speed with which the ball hits the ground:

$$v_{yf}^2 - v_{yi}^2 = 2a_y h. \qquad (A.263)$$

Suppose the y positive direction is upward, and the origin is at the position where the ball is initially thrown. The acceleration is $a_y = -g = -9.80 \, m/s^2$, where the sign minus indicates that direction is opposite of the positive y direction. Then, from the problem, $h = -30.0 \, m$ and $v_{yi} = -8.00 \, m/s$, where the sign minus indicates that the direction is along the negative y-axis. Replacing the numerical values into the equation, you get

$$v_{yf}^2 = (-8.00 \, m/s)^2 + 2.00(-9.80 \, m/s^2)(-30.0 \, m) = 652 \, m^2/s^2 \qquad (A.264)$$

or $v_{yf} = \pm 25.5 \, m/s$. Physical solution is $v_{yf} = -25.5 \, m/s$, because the direction of the velocity is downward. To find the time, you can use this equation:

$$v_{yf} = v_{yi} - gt \qquad (A.265)$$

from which you can write

$$t = \frac{v_{yf} - v_{yi}}{-g} = \frac{-25.5 \, m/s - (-8.00 \, m/s)}{-9.80 \, m/s^2} \approx 1.79 \, s. \qquad (A.266)$$

3.42. Suppose the y positive direction is upward, and the origin is at the ground position. The acceleration is $a_y = -g = -9.8 \, m/s^2$, where the sign minus indicates that direction is opposite of the positive y direction. Based on the problem, for the first ball, the initial position is $y_{1,i} = h$, and the initial velocity $v_{y1,i} = 0$. For the second ball the initial position is $y_{2,i} = 0$ and the initial velocity $v_{y2,i}$ is unknown. Let us denote with t the time after they meet, then the position of each ball at any time is

$$y_1 = y_{1,i} + v_{y1,i}t - g\frac{t^2}{2} = h - g\frac{t^2}{2} \qquad (A.267)$$

$$y_2 = y_{2,i} + v_{y2,i}t - g\frac{t^2}{2} = v_{y2,i}t - g\frac{t^2}{2} \tag{A.268}$$

where $y_1 = y_2 = h/2$. Then

$$\frac{h}{2} = h - g\frac{t^2}{2} \tag{A.269}$$

$$\frac{h}{2} = v_{y2,i}t - g\frac{t^2}{2}. \tag{A.270}$$

From the first equation, $t = \sqrt{h/g}$. After replacing it into the second equation, you get

$$\frac{h}{2} = v_{y2,i}\sqrt{\frac{h}{g}} - \frac{g(\frac{h}{g})}{2}. \tag{A.271}$$

Solving it for $v_{y2,i}$, you obtain

$$v_{y2,i} = \sqrt{hg}. \tag{A.272}$$

Solutions Chapter 4

4.1. Figure A.9 illustrates the trajectory of the motorist.

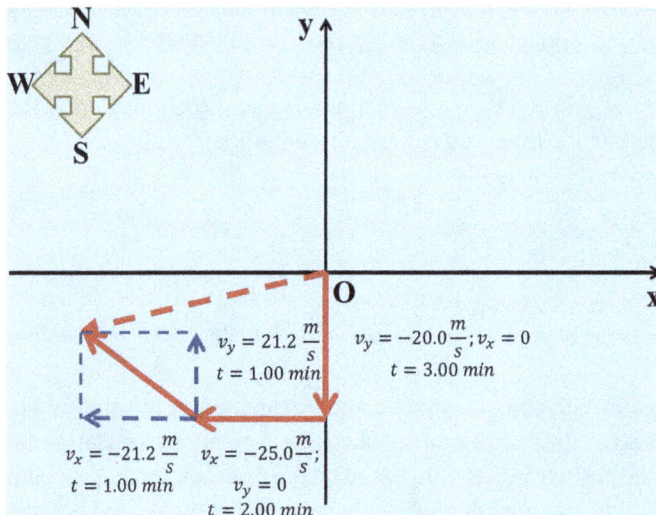

Figure A.9: Illustration of the trajectory of the motorist.

(a) The coordinate system is chosen such that positive x direction is east and positive y direction is north. A graphical illustration of the trajectory of the motorist is shown in Fig. A.9. As it can be seen, first the motorist travels south with velocity along the negative y direction, hence $v_y = -20\,\text{m/s}$ and $v_x = 0$ and time is $t = 3.00\,\text{min} = 180\,\text{s}$. Therefore, the displacement according to y-axis is

$$y_1 = v_y t = (-20\,\text{m/s})(180.0\,\text{s}) = -3600\,\text{m} = -3.60 \times 10^3\,\text{m} \qquad \text{(A.273)}$$

where the minus sign indicates that displacement is in negative y direction. During this time displacement along the x-axis is $x_1 = 0$.

Then, during the next interval of time, $t = 2.00\,\text{min} = 120\,\text{s}$, the motorist travels in the negative x direction. Thus, $v_x = -25.0\,\text{m/s}$, and $v_y = 0$. Therefore, the displacement along the x direction is

$$x_2 = v_x t = (-25.0\,\text{m/s})(120\,\text{s}) = -3000\,\text{m} = -3.00 \times 10^3\,\text{m}. \qquad \text{(A.274)}$$

The sign minus indicates that the displacement is along negative x direction. The displacement along the y direction is $y_2 = 0$.

During the last interval of time, $t = 1.00\,\text{min} = 60.0\,\text{s}$, displacement is in the direction northwest. Therefore, velocity can be written as

$$\mathbf{v} = v_x \mathbf{i} + v_y \mathbf{j} \qquad \text{(A.275)}$$

where

$$v_x = -v\cos(45°) \approx -(30.0\,\text{m/s})(0.707) \approx -21.2\,\text{m/s} \qquad \text{(A.276)}$$
$$v_y = v\sin(45°) \approx (30.0\,\text{m/s})(0.707) \approx 21.2\,\text{m/s} \qquad \text{(A.277)}$$

where the sign minus indicates that x component has the negative x direction. From this, you can find that

$$x_3 = v_x t = (-21.2\,\text{m/s})(60.0\,\text{s}) = -1272\,\text{m} \approx -1.27 \times 10^3\,\text{m} \qquad \text{(A.278)}$$
$$y_3 = v_y t = (21.2\,\text{m/s})(60.0\,\text{s}) = 1272\,\text{m} \approx 1.27 \times 10^3\,\text{m}. \qquad \text{(A.279)}$$

The total displacement vector is

$$\Delta \mathbf{r} = \Delta x \mathbf{i} + \Delta y \mathbf{j} \qquad \text{(A.280)}$$

where

$$\Delta x = x_1 + x_2 + x_3 = 0 - 3.00 \times 10^3\,\text{m} - 1.27 \times 10^3\,\text{m} \qquad \text{(A.281)}$$
$$= -4.27 \times 10^3\,\text{m}$$
$$\Delta y = y_1 + y_2 + y_3 = -3.60 \times 10^3\,\text{m} + 0 + 1.27 \times 10^3\,\text{m} \qquad \text{(A.282)}$$
$$= -2.33 \times 10^3\,\text{m}.$$

Hence,

$$\Delta \mathbf{r} = (-4.27 \times 10^3 \mathbf{i} - 2.33 \times 10^3 \mathbf{j})\,\text{m}. \qquad \text{(A.283)}$$

(b) The average velocity is

$$\bar{\mathbf{v}} = \frac{\Delta \mathbf{r}}{\Delta t} = \bar{v}_x \mathbf{i} + \bar{v}_y \mathbf{j} \tag{A.284}$$

where the interval of time is $\Delta t = 6.00\,\text{min} = 360\,\text{s}$.

$$\bar{v}_x = \frac{\Delta x}{\Delta t} = \frac{-4.27 \times 10^3\,\text{m}}{360\,\text{s}} \tag{A.285}$$
$$= -11.9\,\text{m/s}$$

and

$$\bar{v}_y = \frac{\Delta y}{\Delta t} = \frac{-2.33 \times 10^3\,\text{m}}{360\,\text{s}} \tag{A.286}$$
$$= -6.47\,\text{m/s}.$$

(c) The average speed is

$$\bar{v} = \sqrt{\bar{v}_x^2 + \bar{v}_y^2} = \sqrt{(-11.9)^2 + (-6.47)^2} \tag{A.287}$$
$$= \sqrt{141.61 + 41.86} \approx 13.5\,\text{m/s}.$$

4.2.
(a) The average velocity vector is

$$\bar{\mathbf{v}} = \bar{v}_x \mathbf{i} + \bar{v}_y \mathbf{j} \tag{A.288}$$

where

$$\bar{v}_x = \frac{x_f - x_i}{t_f - t_i} \tag{A.289}$$
$$\bar{v}_y = \frac{y_f - y_i}{t_f - t_i}. \tag{A.290}$$

After replacing the numerical values, you can get

$$x_i = (1.00\,\text{m/s})(2.00\,\text{s}) + 1.00\,\text{m} = 3.00\,\text{m} \tag{A.291}$$
$$x_f = (1.00\,\text{m/s})(4.00\,\text{s}) + 1.00\,\text{m} = 5.00\,\text{m} \tag{A.292}$$
$$y_i = (0.125\,\text{m/s}^2)(2.00\,\text{s})^2 + 1.00\,\text{m} = 1.50\,\text{m} \tag{A.293}$$
$$y_f = (0.125\,\text{m/s})(4.00\,\text{s})^2 + 1.00\,\text{m} = 3.00\,\text{m}. \tag{A.294}$$

Hence,

$$\bar{v}_x = \frac{5.00\,\text{m} - 3.00\,\text{m}}{4.00\,\text{s} - 2.00\,\text{s}} = 1.00\,\text{m/s} \tag{A.295}$$

$$\bar{v}_y = \frac{3.00 \text{ m} - 1.50 \text{ m}}{4.00 \text{ s} - 2.00 \text{ s}} = 0.750 \text{ m/s}. \tag{A.296}$$

Thus, the average velocity vector is

$$\bar{v} = (1.00\mathbf{i} + 0.750\mathbf{j}) \text{ m/s}. \tag{A.297}$$

(b) First, you can calculate the instantaneous velocity

$$\mathbf{v} = \frac{d\mathbf{r}}{dt} \tag{A.298}$$

$$= \frac{dx}{dt}\mathbf{i} + \frac{dy}{dt}\mathbf{j}$$

$$= a\mathbf{i} + 2ct\mathbf{j}.$$

Then the velocity at $t = 2.00$ s is

$$\mathbf{v} = (1.00 \text{ m/s})\mathbf{i} + 2(0.125 \text{ m/s}^2)(2.00 \text{ s})\mathbf{j} = (1.00\mathbf{i} + 0.500\mathbf{j}) \text{ m/s}. \tag{A.299}$$

To calculate the speed at $t = 2.00$ s, we calculate the magnitude of the vector given by eq. (A.299):

$$v = \sqrt{(1.00)^2 + (0.500)^2} \approx 1.12 \text{ m/s}. \tag{A.300}$$

4.3.

(a) The vector of position is

$$\mathbf{r} = x\mathbf{i} + y\mathbf{j} = [(18.0 \text{ m/s})t]\mathbf{i} + [(4.00 \text{ m/s})t - (4.90 \text{ m/s}^2)t^2]\mathbf{j}. \tag{A.301}$$

(b) To calculate the vector of velocity, you can take the derivative of the above expression with respect to time t:

$$\mathbf{v} = \frac{d\mathbf{r}}{dt} = (18.0 \text{ m/s})\mathbf{i} + [(4.00 \text{ m/s}) - (9.80 \text{ m/s}^2)t]\mathbf{j}. \tag{A.302}$$

(c) The vector of acceleration is calculated as derivative of the expression for velocity with respect to time:

$$\mathbf{a} = \frac{d\mathbf{v}}{dt} = (-9.80\mathbf{j}) \text{ m/s}^2. \tag{A.303}$$

(d) The vector position at $t = 3.00$ s is

$$\mathbf{r} = [(18.0 \text{ m/s})(3.00 \text{ s})]\mathbf{i} \tag{A.304}$$

$$+ [(4.00 \text{ m/s})(3.00 \text{ s}) - (4.90 \text{ m/s}^2)(3.00 \text{ s})^2]\mathbf{j}$$

$$= (54.0\mathbf{i} - 32.1\mathbf{j}) \text{ m}.$$

(e) The velocity vector at $t = 3.00$ s is

$$\mathbf{v} = (18.0\mathbf{i} - 25.4\mathbf{j}) \, \text{m/s}. \tag{A.305}$$

(f) While the acceleration is constant, hence

$$\mathbf{a} = (-9.80\mathbf{j}) \, \text{m/s}^2. \tag{A.306}$$

4.4. Figure A.10 presents the graph of the x–y path.

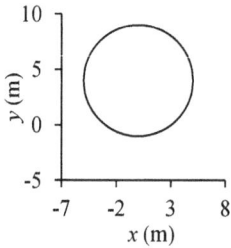

Figure A.10: A plot of the x–y path.

(a) The velocity can be written as

$$\mathbf{v} = v_x\mathbf{i} + v_y\mathbf{j} \tag{A.307}$$

where

$$v_x = \frac{dx}{dt} = -(5.00 \, \text{m})\omega \cos \omega t \tag{A.308}$$

$$v_y = \frac{dy}{dt} = (5.00 \, \text{m})\omega \sin \omega t. \tag{A.309}$$

At time $t = 0$, these components are given by

$$v_x = -(5.00 \, \text{m})\omega \tag{A.310}$$

$$v_y = 0. \tag{A.311}$$

The acceleration vector is

$$\mathbf{a} = a_x\mathbf{i} + a_y\mathbf{j} \tag{A.312}$$

where

$$a_x = \frac{dv_x}{dt} = (5.00 \, \text{m})\omega^2 \sin \omega t \tag{A.313}$$

$$a_y = \frac{dv_y}{dt} = (5.00 \, \text{m})\omega^2 \cos \omega t. \tag{A.314}$$

At time $t = 0$, you can write

$$a_x = 0 \tag{A.315}$$

$$a_y = (5.00 \, \text{m})\omega^2. \tag{A.316}$$

(b) The position, velocity and acceleration vectors can be written as

$$\mathbf{r} = [-(5.00\text{ m})\sin \omega t]\mathbf{i} + [(4.00\text{ m}) - (5.00\text{ m})\cos \omega t]\mathbf{j} \qquad \text{(A.317)}$$
$$\mathbf{v} = [-(5.00\text{ m})\omega \cos \omega t]\mathbf{i} + [(5.00\text{ m})\omega \sin \omega t]\mathbf{j} \qquad \text{(A.318)}$$
$$\mathbf{a} = [(5.00\text{ m})\omega^2 \sin \omega t]\mathbf{i} + [(5.00\text{ m})\omega^2 \cos \omega t]\mathbf{j}. \qquad \text{(A.319)}$$

(c) From the information of the problem, you can, for instance, calculate x^2:

$$x^2 = 25.0 \sin^2 \omega t \qquad \text{(A.320)}$$

and

$$(y - 4.00)^2 = 25.0 \cos^2 \omega t. \qquad \text{(A.321)}$$

Then, add side by side these two equations, to get

$$x^2 + (y - 4.00)^2 = 25.0 \qquad \text{(A.322)}$$

which represents the equation of a circle centered at $(0, 4)$ and radius $R = 5\,\text{m}$. Graphically this is shown in Fig. A.10.

4.5.

(a) To find the acceleration, you can use this equation

$$\mathbf{v} = \mathbf{v}_i + \mathbf{a}(t - t_i) \qquad \text{(A.323)}$$

where $t = 3.00$ s and $t_i = 0$. From this,

$$\mathbf{a} = \frac{\mathbf{v} - \mathbf{v}_i}{t - t_i} = (2.00\mathbf{i} + 3.00\mathbf{j})\,\text{m/s}^2. \qquad \text{(A.324)}$$

Thus,

$$a_x = 2.00\,\text{m/s}^2, \quad a_y = 3.00\,\text{m/s}^2. \qquad \text{(A.325)}$$

(b) To find the coordinates you can use the equation for displacement:

$$\mathbf{r} - \mathbf{r}_i = \mathbf{v}_i t + \mathbf{a}\frac{t^2}{2} \qquad \text{(A.326)}$$

where $\mathbf{r}_i = 0$. Replacing the expressions for $\mathbf{v}_i$ and $\mathbf{a}$, you get

$$\mathbf{r} = [3.00\mathbf{i} - 2.00\mathbf{j}]t + [2.00\mathbf{i} + 3.00\mathbf{j}]\frac{t^2}{2} \qquad \text{(A.327)}$$
$$= ([3.00t + 1.00t^2]\mathbf{i} + [-2.00t + 1.50t^2]\mathbf{j})\,\text{m}.$$

4.6.

(a) To find the acceleration, you can use the equation

$$\mathbf{v} = \mathbf{v}_i + \mathbf{a}t. \tag{A.328}$$

From this

$$\mathbf{a} = \frac{\mathbf{v} - \mathbf{v}_i}{t} = (0.800 \text{ m/s}^2)\mathbf{i} - (0.300 \text{ m/s}^2)\mathbf{j}. \tag{A.329}$$

(b) To find the direction of the acceleration with respect to unit vector $\mathbf{i}$, you can calculate the component of acceleration along this vector:

$$a_x = \mathbf{a} \cdot \mathbf{i} = 0.800 \text{ m/s}^2. \tag{A.330}$$

The sign plus indicates that direction of the acceleration is along the positive $\mathbf{i}$ direction.

(c) To find the position of the fish at $t = 25.0$ s, you can use the equation

$$
\begin{aligned}
\mathbf{r} = \mathbf{r}_i + \mathbf{v}_i t + \mathbf{a}\frac{t^2}{2} \qquad\qquad\qquad\qquad\qquad & \tag{A.331}\\
= (10.0\mathbf{i} - 4.00\mathbf{j}) + (4.00\mathbf{i} + 1.00\mathbf{j})t + (0.800\mathbf{i} - 0.300\mathbf{j})\frac{t^2}{2} &\\
= (10.0 + 4.00t + 0.400t^2)\mathbf{i} + (-4.00 + 1.00t - 0.150t^2)\mathbf{j} &\\
= (10.0 + 4.00(25.0) + 0.400(25.0)^2)\mathbf{i} &\\
\quad + (-4.00 + 1.00(25.0) - 0.150(25.0)^2)\mathbf{j} &\\
= 360\mathbf{i} - 72.75\mathbf{j} &\\
\approx 360\mathbf{i} - 72.8\mathbf{j}. &
\end{aligned}
$$

The components of the position are

$$x = 360 \text{ m}, \quad y = -72.8 \text{ m}. \tag{A.332}$$

Therefore, the fish is moving in the positive x direction and negative y direction.

4.7.

(a) The vector position can be calculated as

$$\mathbf{r} = \mathbf{r}_i + \mathbf{v}_i t + \mathbf{a}t^2/2 \tag{A.333}$$

where $\mathbf{r}_i = 0$ m. After replacing the values from the problem, you get

$$\mathbf{r} = ((5.00t)\mathbf{i} + (1.50t^2)\mathbf{j}) \text{ m}. \tag{A.334}$$

The vector velocity can be calculated by

$$\mathbf{v} = \mathbf{v}_i + \mathbf{a}t. \tag{A.335}$$

You can now replace the expressions of $\mathbf{v}_i$ and $\mathbf{a}$, and get

$$\mathbf{v} = (5.00\mathbf{i} + (3.00t)\mathbf{j})\,\text{m/s}. \tag{A.336}$$

(b) The coordinates of the particle at $t = 2.00$ s are

$$x = (5.00\,\text{m/s})(2.00\,\text{s}) = 10.0\,\text{m} \tag{A.337}$$

and

$$y = (1.50\,\text{m/s}^2)(2.00\,\text{s})^2 = (1.50\,\text{m/s}^2)(4.00\,\text{s}^2) = 6.00\,\text{m}. \tag{A.338}$$

The components of the velocity are

$$v_x = 5.00\,\text{m/s} \tag{A.339}$$

and

$$v_y = (3.00\,\text{m/s}^2)(2.00\,\text{s}) = 6.00\,\text{m/s}. \tag{A.340}$$

The speed is calculated as

$$v = \sqrt{v_x^2 + v_y^2} = \sqrt{25.0 + 36.0} \approx 7.81\,\text{m/s}. \tag{A.341}$$

4.8. To understand the problem, we have sketched the possible trajectory of the mug in Fig. A.11. The origin of the coordinate system is taken at the top of the counter and the axes x and y directions are indicated in the figure.

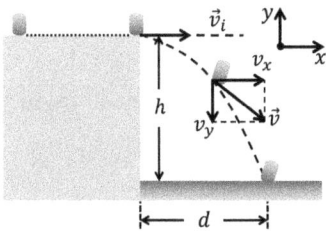

Figure A.11: Illustration of the trajectory of the mug.

(a) At any instant of time, the velocity vector has two components, v_x and v_y:

$$\mathbf{v} = v_x\mathbf{i} + v_y\mathbf{j} \tag{A.342}$$

where $v_x = v_i$ is constant and only the motion along the y-axis is with acceleration $a_y = -g = -9.80 \, \text{m/s}^2$. Let us denote by t the time passed until the mug strikes the floor. Then the displacement along the x-axis is given by

$$d = x_i + v_x t = 0 + v_i t. \tag{A.343}$$

Therefore, the initial speed is

$$v_i = d/t. \tag{A.344}$$

At any instant of time the position of the mug along the y directions is

$$y = y_i + v_i t + a_y t^2/2 = 0 + (d/t)t + (-g)t^2/2 = d - gt^2/2. \tag{A.345}$$

At the moment when the mug strikes the floor $y = -h$, therefore

$$-h = d - gt^2/2. \tag{A.346}$$

Solving eq. (A.346) for t, you get the physical solution $(t > 0)$ as

$$t = \sqrt{\frac{2(d+h)}{g}}. \tag{A.347}$$

Finally, you get the initial speed of the mug:

$$v_i = \sqrt{\frac{gd^2}{2(d+h)}}. \tag{A.348}$$

(b) The y component of the velocity at the moment the mug just strikes the floor is

$$v_y = v_{yi} + a_y t \tag{A.349}$$

where $v_{yi} = 0$. Therefore,

$$v_y = -\sqrt{2g(d+h)} \tag{A.350}$$

and the other component is

$$v_x = \sqrt{\frac{gd^2}{2(d+h)}}. \tag{A.351}$$

Thus, the x component is along the positive direction of the x-axis and y component is along the negative direction of the y-axis (downward).

4.9. In Fig. A.12, we have indicated all the parameters of the problem. The initial values of the components of the velocity are calculated as

$$v_{xi} = v_i \cos \alpha = (25.0 \text{ m/s}) \cos 70° = 8.60 \text{ m/s} \qquad \text{(A.352)}$$
$$v_{yi} = v_i \sin \alpha = (25.0 \text{ m/s}) \sin 70° = 23.5 \text{ m/s}. \qquad \text{(A.353)}$$

To find the time t the snowball reaches the highest point, we use this equation

$$v_{yf} = v_{yi} + a_y t \qquad \text{(A.354)}$$

where $a_y = -g = -9.80 \text{ m/s}^2$ and $v_{yf} = 0$, then

$$t_{1/2} = (23.5 \text{ m/s})/(9.80 \text{ m/s}^2) = 2.40 \text{ s}. \qquad \text{(A.355)}$$

This is the half of the time of the snowball flight because of the symmetry, the total time is

$$t = 2t_{1/2} = 4.80 \text{ s}. \qquad \text{(A.356)}$$

The distance traveled along the x-axis is

$$x_f = v_{xi} t = (8.60 \text{ m/s})(4.80 \text{ s}) \approx 41.3 \text{ m}. \qquad \text{(A.357)}$$

(a) Suppose that the second ball is thrown $\Delta t = 2.00 \text{ s}$ later. Then, for the second snowball to land at the same time and point as the first one, it should be that it travels the same horizontal distance for a time $t' = t - \Delta t = 4.80 \text{ s} - 2.00 \text{ s} = 2.80 \text{ s}$. Therefore,

$$x_f' = x_f = v_{xi}' t' \qquad \text{(A.358)}$$

where

$$v_{xi}' = v_i \cos \beta = (25.0 \text{ m/s}) \cos \beta \qquad \text{(A.359)}$$

since both snowballs are thrown with the same initial speed. β is the initial angle with which the second snowball is thrown. Then, you can write

$$41.3 \text{ m} = [(25.0 \text{ m/s}) \cos \beta](2.80 \text{ s}). \qquad \text{(A.360)}$$

Solving it for $\cos \beta$, you obtain

$$\cos \beta = \frac{41.3 \text{ m}}{(25.0 \text{ m/s})(2.80 \text{ s})} = 0.59. \qquad \text{(A.361)}$$

Thus,

$$\beta \approx 53.8° \qquad \text{(A.362)}$$

which, as expected, is smaller than 70°, the angle that the first snow ball is thrown.

(b) The time it takes to the first snowball to land as a function of the angle and initial speed is

$$t_1 = 2\frac{v_i \sin \alpha}{g}.$$

(A.363)

If the second snowball is thrown with an angle $\beta < \alpha$ some later time, let us say Δt seconds, the time it takes to the second snowball to land is

$$t_2 = \Delta t + 2\frac{v_i \sin \beta}{g}.$$

(A.364)

Requiring that these two times to be the same means that

$$t_1 = t_2 + \Delta t$$

(A.365)

or

$$2\frac{v_i \sin \alpha}{g} = \Delta t + 2\frac{v_i \sin \beta}{g}.$$

(A.366)

From this

$$\Delta t = \frac{2v_i}{g}(\sin \alpha - \sin \beta).$$

(A.367)

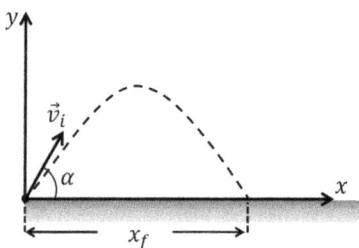

Figure A.12: Illustration of the trajectory of the snowball.

4.10. To illustrate the problem, you should refer to Fig. A.13. The components of the initial velocity in terms of the initial speed (which is unknown) are

$$v_{xi} = v_i \cos \alpha = v_i \cos 3.00° = 0.999v_i \text{ m/s}$$

(A.368)

$$v_{yi} = v_i \sin \alpha = v_i \sin 3.00° = 0.0523v_i \text{ m/s}.$$

(A.369)

The ball should reach the highest point $h = 0.330$ m in order to clear the net, which is related to the initial speed according to this equation

$$v_y^2 - v_{yi}^2 = 2a_y h$$

(A.370)

where the acceleration $a_y = -g = -9.8 \, \text{m/s}^2$, and $v_y = 0$. From this,

$$-0.002735 v_i^2 = 2(-9.8)(0.330) \tag{A.371}$$

or,

$$v_i^2 = 2364.899 \, \text{m}^2/\text{s}^2. \tag{A.372}$$

Solving it for v_i gives

$$v_i = \pm 48.6 \, \text{m/s}. \tag{A.373}$$

Since the direction of the velocity is along the positive y direction, the real physical solution is $v_i = +48.6 \, \text{m/s}$.

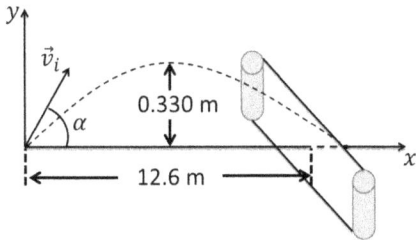

Figure A.13: Illustration of the trajectory of the racket ball.

4.11. Graphically the problem is illustrated in Fig. A.14. The components of the initial velocity are given by

$$v_{xi} = v_i \cos \alpha = (300 \, \text{m/s}) \cos 55° \approx 172.1 \, \text{m/s} \tag{A.374}$$

$$v_{yi} = v_i \sin \alpha = (300 \, \text{m/s}) \sin 55° \approx 245.7 \, \text{m/s}. \tag{A.375}$$

The motion along the x-axis is with constant velocity; therefore, the displacement along the x direction is

$$x = v_{xi} t = (172.1 \, \text{m/s})(42.0 \, \text{s}) \approx 7228 \, \text{m}. \tag{A.376}$$

The motion along the y-axis is accelerated with acceleration $a_y = -9.8 \, \text{m/s}^2$.

First, we find the time that the shell reaches the highest point using the equation

$$v_y = v_{yi} + a_y t_1. \tag{A.377}$$

From this,

$$t_1 = -\frac{v_{yi}}{a_y} = -\frac{245.7 \, \text{m/s}}{-9.80 \, \text{m/s}^2} = 25.071 \, \text{s} \approx 25.1 \, \text{s}. \tag{A.378}$$

The y position at the highest point is calculated from the equation:

$$v_y^2 - v_{yi}^2 = 2a_y h. \tag{A.379}$$

Solving it for h, you get

$$h = \frac{-v_{yi}^2}{2a_y} = \frac{-(245.7\,\text{m/s})^2}{2(-9.80\,\text{m/s}^2)} = 3080.025\,\text{m} \approx 3.08 \times 10^3\,\text{m}. \tag{A.380}$$

Next, the shell moves downward from the highest point with initial velocity $v_{yi} = 0$ until it explodes. The time left is $t_2 = t - t_1 = 42.0\,\text{s} - 25.1\,\text{s} = 16.9\,\text{s}$. The displacement along the y-axis is calculated as

$$y = h + v_{yi}t + \frac{1}{2}a_y t_2^2. \tag{A.381}$$

Replacing the numerical values, you get

$$y = h + 0 + \frac{1}{2}(-9.80\,\text{m/s}^2)(16.9\,\text{s})^2 \tag{A.382}$$

$$= 1680.511\,\text{m} \approx 1.68 \times 10^3\,\text{m} = 1.68\,\text{km}.$$

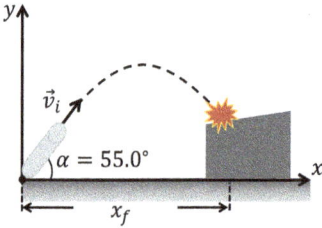

Figure A.14: Illustration of the trajectory of the artillery fire.

4.12. To find the acceleration, you can use this equation, assuming the y-axis positive direction is upward:

$$v_{yf}^2 - v_{yi}^2 = 2a_y h \tag{A.383}$$

where $h = 15.0\,\text{m}$, $v_{yf} = 0$, and $v_{yi} = 3.00\,\text{m/s}$. Replacing the numerical values, you get

$$a_y = \frac{v_{yf}^2 - v_{yi}^2}{2h} \tag{A.384}$$

$$= \frac{(0)^2 - (3.00)^2}{30.0} = -0.300\,\text{m/s}^2.$$

The minus sign means that the direction of the acceleration is along the negative y-axis.

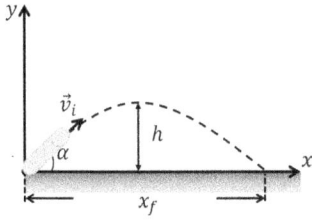

Figure A.15: Illustration of the trajectory of the projectile.

4.13. Graphically, the trajectory of the projectile is shown in Fig. A.15.
The components of the initial velocity along the axes are

$$v_{xi} = v_i \cos \alpha \tag{A.385}$$
$$v_{yi} = v_i \sin \alpha. \tag{A.386}$$

The maximum height is calculated using the following equation:

$$v_y^2 - v_{yi}^2 = 2a_y h \tag{A.387}$$

where $v_y = 0$ is the velocity at the maximum height h and $a_y = -g$ is the acceleration. Solving it for h, you get

$$h = \frac{v_i^2 \sin^2 \alpha}{2g}. \tag{A.388}$$

To find the time that the projectile needs to reach the maximum height, you can use this equation:

$$v_y = v_{yi} + a_y t_{1/2}. \tag{A.389}$$

Solving it for time, you get

$$t_{1/2} = \frac{v_i \sin \alpha}{g}. \tag{A.390}$$

The total time of the trajectory is

$$t = 2t_{1/2} = \frac{2v_i \sin \alpha}{g}. \tag{A.391}$$

The distance of the projectile along the x-axis is

$$x_f = v_{xi} t = \frac{v_i^2 \sin(2\alpha)}{g}. \tag{A.392}$$

Requiring that $x_f = 3h$, you can write

$$3\frac{v_i^2 \sin^2 \alpha}{2g} = \frac{v_i^2 \sin(2\alpha)}{g}. \tag{A.393}$$

Solving this equation for α, you get

$$\tan \alpha = \frac{4.00}{3.00} \tag{A.394}$$

or

$$\alpha \approx 53.1°. \tag{A.395}$$

4.14. Graphically the problem is illustrated in Fig. A.16.

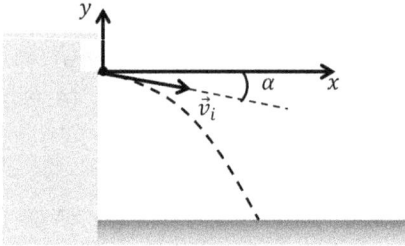

Figure A.16: Illustration of the trajectory of the ball thrown from the window.

(a) From Fig. A.16, the components of the initial velocity according to the axes of a Cartesian coordinate system are

$$v_{xi} = v_i \cos \alpha = 8.00 \cos 20° = 7.52 \, \text{m/s} \tag{A.396}$$
$$v_{yi} = -v_i \sin \alpha = -8.00 \sin 20° = -2.74 \, \text{m/s}. \tag{A.397}$$

The sign minus indicates that the direction of the velocity is along the negative y direction. The motion along the x-axis is with constant velocity; therefore, for the horizontal displacement you can get

$$x_f = v_{xi}t = (7.52 \, \text{m/s})(3.00 \, \text{s}) \approx 22.6 \, \text{m}. \tag{A.398}$$

(b) The displacement along the y direction is given by

$$y_f - y_i = v_{yi}t + a_y t^2/2 \tag{A.399}$$

where $y_f = -h$, $y_i = 0$ and $a_y = -9.80 \, \text{m/s}^2$. Solving this equation for h, you obtain

$$h = (2.74 \, \text{m/s})(3.00 \, \text{s}) + (9.80 \, \text{m/s}^2)(3.00 \, \text{s})^2/2 = 52.3 \, \text{m}. \tag{A.400}$$

(c) Using eq. (A.399) for the y coordinate at any time t, you find

$$y = v_{yi}t + a_y t^2/2. \tag{A.401}$$

After replacing the numerical values, you get

$$-10 \, \text{m} = (-2.74 \, \text{m/s})t + (-9.80 \, \text{m/s}^2)t^2/2 \tag{A.402}$$

or,

$$4.9t^2 + 2.74t - 10 = 0. \tag{A.403}$$

Solving it for t, the physical solution is

$$t = 1.18\,\text{s}. \tag{A.404}$$

4.15. Graphically the problem is illustrated in Fig. A.17.

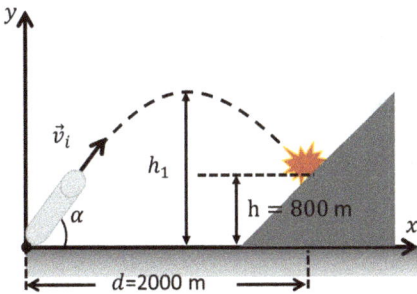

Figure A.17: Illustration of the trajectory of the cannon fire.

First, let us write the components of the initial velocity according to the x- and y-axes:

$$v_{xi} = v_i \cos \alpha = (1000\,\text{m/s}) \cos \alpha \tag{A.405}$$
$$v_{yi} = v_i \sin \alpha = (1000\,\text{m/s}) \sin \alpha. \tag{A.406}$$

The time needed to reach the target can be found using the displacement along the x direction:

$$x_f = v_{xi}t \tag{A.407}$$

from this

$$t = \frac{x_f}{v_{xi}} = \frac{d}{v_i \cos \alpha}. \tag{A.408}$$

The time to reach the maximum height can be calculated using this equation:

$$v_y = v_{yi} + a_y t_{1/2} \tag{A.409}$$

where $v_y = 0$ is the velocity at the maximum height, and $a_y = -g = -9.8\,\text{m/s}^2$ is the acceleration. You can calculate $t_{1/2}$ as

$$t_{1/2} = -\frac{v_{yi}}{a_y} = -\frac{v_i \sin \alpha}{-g} = \frac{v_i \sin \alpha}{g}. \tag{A.410}$$

Then, the time from the maximum height to the target is

$$\Delta t = t - t_{1/2} = \frac{d}{v_i \cos \alpha} - \frac{v_i \sin \alpha}{g}. \tag{A.411}$$

The maximum height is calculated as

$$v_y^2 - v_{yi}^2 = 2a_y h_1 \tag{A.412}$$

where $v_y = 0$, and hence

$$h_1 = \frac{-v_{yi}^2}{2a_y} = -\frac{v_i^2 \sin^2 \alpha}{-2g} = \frac{v_i^2 \sin^2 \alpha}{2g}. \tag{A.413}$$

The displacement from the maximum height to the target is found from

$$h - h_1 = v_y \Delta t + a_y (\Delta t)^2 / 2. \tag{A.414}$$

Replacing the expression for Δt, a_y and $v_y = 0$, you get

$$h - \frac{v_i^2 \sin^2 \alpha}{2g} = -\frac{g}{2} \left[\frac{d}{v_i \cos \alpha} - \frac{v_i \sin \alpha}{g} \right]^2. \tag{A.415}$$

After simplifying it, you find

$$h - \frac{v_i^2 \sin^2 \alpha}{2g} = -\frac{gd^2}{2v_i^2 \cos^2 \alpha} + \frac{d \sin \alpha}{\cos \alpha} - \frac{v_i^2 \sin^2 \alpha}{2g}. \tag{A.416}$$

Using $1 + \tan^2 \alpha = 1/\cos^2 \alpha$, you obtain

$$h = -\frac{gd^2}{2v_i^2}(1 + \tan^2 \alpha) + d \tan \alpha. \tag{A.417}$$

Replacing the numerical values, you get

$$800 = -19.6(1 + \tan^2 \alpha) + 2000 \tan \alpha \tag{A.418}$$

or, approximately, the following equation for $\tan \alpha$:

$$\tan^2 \alpha - 102 \tan \alpha + 41.8 = 0. \tag{A.419}$$

Solving it for $\tan \alpha$, you get

$$\tan \alpha = 51 \pm \sqrt{(51)^2 - 41.8} \approx 51 \pm 50.6. \tag{A.420}$$

Thus, solving it for α

$$\alpha_1 \approx 89.4°; \quad \alpha_2 \approx 21.8°. \tag{A.421}$$

This result indicates that the cannon can fire with two possible initial angles with the horizontal direction. For the first initial angle, the cannon fire will fly higher and reach the target longer in time. While for the second angle (which is smaller), the cannon fire will drift lower, and hence reach the same target earlier.

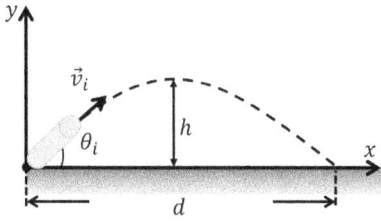

Figure A.18: Illustration of the trajectory of the projectile.

4.16. Graphically the problem is illustrated in Fig. A.18.

The components of the initial velocity according to the axes of a Cartesian coordinate system are

$$v_{xi} = v_i \cos \theta_i \tag{A.422}$$
$$v_{yi} = v_i \sin \theta_i. \tag{A.423}$$

The time passes until the projectile reaches the maximum height is calculated according to this equation:

$$v_y = v_{yi} + a_y t_{1/2} \tag{A.424}$$

where $a_y = -g$ is the acceleration, and $v_y = 0$. From this

$$t_{1/2} = \frac{v_i \sin \theta_i}{g}. \tag{A.425}$$

The motion along the x direction is with constant velocity; therefore, the x coordinate after $t_{1/2}$ is

$$x = v_{xi}t_{1/2} = v_i \cos \theta_i \frac{v_i \sin \theta_i}{g} = \frac{v_i^2 \sin(2\theta_i)}{2g}. \tag{A.426}$$

Since the trajectory is symmetric with this coordinate (because the trajectory is a parabola), the total distance along the horizontal axis is

$$d = 2x = \frac{v_i^2 \sin(2\theta_i)}{g}. \tag{A.427}$$

4.17. Graphically the problem is illustrated in Fig. A.19.

(a) The components of the initial velocity are given as

$$v_{xi} = v_i \cos \alpha = (20.0 \text{ m/s}) \cos 53° \approx 12.0 \text{ m/s} \tag{A.428}$$
$$v_{yi} = v_i \sin \alpha = (20.0 \text{ m/s}) \sin 53° \approx 16.0 \text{ m/s}. \tag{A.429}$$

First let us determine the time passed until the ball reaches the maximum height using this equation

$$v_y = v_{yi} + a_y t_{1/2} \tag{A.430}$$

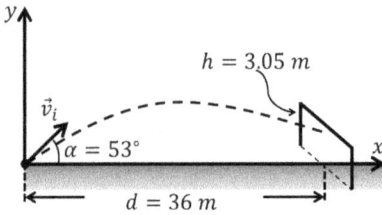

Figure A.19: Illustration of the trajectory of the ball.

where $v_y = 0$ and $a_y = -9.8\,\text{m/s}^2$ is the acceleration. Replacing the numerical values, you get

$$t_{1/2} = -(16.0\,\text{m/s})/(-9.80\,\text{m/s}^2) = 1.63\,\text{s}. \qquad (\text{A.431})$$

The maximum height traveled by the ball is given by the equation

$$v_y^2 - v_{yi}^2 = 2a_y h_{max}. \qquad (\text{A.432})$$

From this,

$$h_{max} = -\frac{v_{yi}^2}{2a_y} = \frac{256}{19.6} \approx 13.1\,\text{m}. \qquad (\text{A.433})$$

To find at what height the ball reaches the cross bar, you can use this equation

$$y_f - y_i = v_y t + a_y t^2/2 \qquad (\text{A.434})$$

where y_f is the height of the ball at the cross bar, y_i is the h_{max}, and t is the time from the maximum height until when the ball reaches the cross bar. This time is calculated using the formula

$$t = \frac{d}{v_{xi}} - t_{1/2} = \frac{36.0\,\text{m}}{12.0\,\text{m/s}} - 1.63\,\text{s} = 1.37\,\text{s}. \qquad (\text{A.435})$$

Replacing these values in eq. (A.434), you get

$$y_f = 13.1\,\text{m} + 0 - (9.80\,\text{m/s}^2)(1.37\,\text{s})^2/2 \qquad (\text{A.436})$$
$$= 13.1\,\text{m} - 9.20\,\text{m} = 3.90\,\text{m}.$$

Since $y_f > h$, you can say that the ball clears the cross bar.

(b) Because of the symmetry of the trajectory (it is a parabola), the total time traveled by the ball is

$$t = 2t_{1/2} = 3.27\,\text{s}. \qquad (\text{A.437})$$

The total horizontal distance covered by the ball is

$$x = v_{xi} t = (12.0\,\text{m/s})(3.27\,\text{s}) \approx 39.2\,\text{m}. \qquad (\text{A.438})$$

As you can see, this distance is longer than the distance from the point when the ball is kicked to the goal. Therefore, the ball approaches the cross bar while falling.

4.18. The components of the initial velocity are

$$v_{xi} = (40.0 \, \text{m/s}) \cos 30.0° = 34.6 \, \text{m/s} \qquad (\text{A.439})$$
$$v_{yi} = (40.0 \, \text{m/s}) \sin 30.0° = 20.0 \, \text{m/s.} \qquad (\text{A.440})$$

First, we determine the time passed until the water stream reaches the maximum height using this equation

$$v_y = v_{yi} + a_y t_{1/2} \qquad (\text{A.441})$$

where $v_y = 0$ and $a_y = -9.8 \, \text{m/s}^2$ is the acceleration. Replacing the numerical values, you get

$$t_{1/2} = -(20.0 \, \text{m/s})/(-9.80 \, \text{m/s}^2) = 2.04 \, \text{s.} \qquad (\text{A.442})$$

To find the height the water stream strikes the building, you can use this equation

$$y_f - y_i = v_y t + a_y t^2/2 \qquad (\text{A.443})$$

where y_f is the height the water stream strikes the building, y_i is the initial height, and t is the time from y_i until when the water stream strikes the building. This time is calculated using this formula

$$t = \frac{d}{v_{xi}} - t_{1/2} = \frac{50.0 \, \text{m}}{34.6 \, \text{m/s}} - 2.04 \, \text{s} = -0.600 \, \text{s} < 0. \qquad (\text{A.444})$$

Since this difference is less than zero, it means that the water stream strikes the building while still rising. Therefore, $y_i = 0$, and t is

$$t = \frac{d}{v_{xi}} = \frac{50.0 \, \text{m}}{34.6 \, \text{m/s}} = 1.44 \, \text{s.} \qquad (\text{A.445})$$

Replacing these values into eq. (A.443), you get

$$y_f = 0 + (20.0 \, \text{m/s})(1.44 \, \text{s}) - (9.80 \, \text{m/s}^2)(1.44 \, \text{s})^2/2 \qquad (\text{A.446})$$
$$= 28.8 \, \text{m} - 10.2 \, \text{m}$$
$$= 18.6 \, \text{m.}$$

4.19. Graphically the problem is illustrated in Fig. A.20.

Let us denote with d the horizontal distance kick point to the point when the ball strike the water. The total distance is

$$\ell = \sqrt{d^2 + h^2}. \qquad (\text{A.447})$$

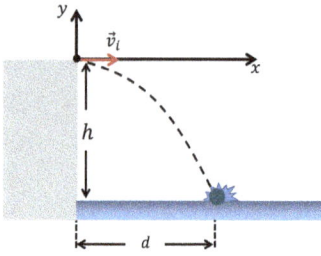

Figure A.20: Illustration of the trajectory of the ball.

If you denote with t the time the ball needs to hit the pool, then $d = v_i t$. Replacing this in eq. (A.447), you get

$$\ell = \sqrt{t^2 v_i^2 + h^2}. \tag{A.448}$$

On the other hand, the sound travels this distance in 3.00 s with speed 343 m/s; therefore,

$$\ell = (343\,\text{m/s})(3.00\,\text{s} - t) = 1029\,\text{m} - 343t \approx 1.03 \times 10^3\,\text{m} - 343t. \tag{A.449}$$

In addition, to find the time t, you can use the equation for the vertical displacement:

$$y_f - y_i = v_{yi}t + a_y t^2/2 \tag{A.450}$$

where $a_y = -g = -9.80\,\text{m/s}^2$ is the acceleration, and $y_i = 0$. Solving it for t, you get

$$t = \sqrt{2h/g} = \sqrt{80.0\,\text{m}/(9.80\,\text{m/s}^2)} = 2.86\,\text{s}. \tag{A.451}$$

Using eq. (A.449), you find that

$$\ell = 1.03 \times 10^3\,\text{m} - (343\,\text{m/s})(2.86\,\text{s}) = 48.0\,\text{m}. \tag{A.452}$$

Using eq. (A.447), you obtain d:

$$d = \sqrt{\ell^2 - h^2} = \sqrt{(48.0)^2 - (40.0)^2} \approx 26.5\,\text{m}. \tag{A.453}$$

The initial speed is

$$v_i = \frac{d}{t} = \frac{26.5\,\text{m}}{2.86\,\text{s}} \approx 9.27\,\text{m/s}. \tag{A.454}$$

4.20.

(a) The period is

$$T = 27.3\,\text{days}. \tag{A.455}$$

The angular speed is:

$$\omega = \frac{2\pi}{T} = 0.230 \, \text{radians/days}. \tag{A.456}$$

You can calculate the linear speed using the formula

$$v = \omega r \approx 88.4 \times 10^6 \, \text{m/s}. \tag{A.457}$$

(b) To calculate the centripetal acceleration, you can use the formula

$$a_r = \frac{v^2}{r} = 2034.12 \times 10^4 \, \text{m/days}^2 \approx 2.03 \times 10^7 \, \text{m/days}^2. \tag{A.458}$$

4.21. You can calculate the maximum radial (or centripetal) acceleration as

$$a_r = \frac{v^2}{r}. \tag{A.459}$$

After replacing the numerical values, you get

$$a_r = \frac{400 \, \text{m}^2/\text{s}^2}{1.06 \, \text{m}} \approx 377 \, \text{m/s}^2. \tag{A.460}$$

4.22. Assuming in one revolution, the stone will travel a distance which is equal to the circumference of the circle of $2\pi r$, we obtain the speed as

$$v = (200 \, \text{rev/min})(2\pi r) \approx 628 \, \text{m/s}. \tag{A.461}$$

The radial acceleration can then be calculated as

$$a_r = \frac{v^2}{r} = 790 \times 10^3 \, \text{m/s}^2. \tag{A.462}$$

4.23. First, you can calculate the speed from the formula

$$a_r = \frac{v^2}{r}. \tag{A.463}$$

From this,

$$v = \sqrt{1.40 \, \text{gr}} = 11.7 \, \text{m/s}. \tag{A.464}$$

Since one revolution is equal to $2\pi r$, you can calculate the rotation rate as

$$\text{rate} = v/(2\pi r) = 0.186 \, \text{rev/s}. \tag{A.465}$$

4.24.

(a) You can calculate the speed as

$$v = \text{rate} \times (2\pi r). \tag{A.466}$$

Replacing the numerical values for each case, you get

$$v_1 = 8.00(2\pi 0.600) \approx 30.2\,\text{m/s} \tag{A.467}$$
$$v_2 = 6.00(2\pi 0.900) \approx 33.9\,\text{m/s}. \tag{A.468}$$

As you can see, the second case provides a greater speed.

(b) You can calculate the centripetal acceleration at 8.00 rev/s with $r = 0.6$ m as

$$a_r = \frac{v_1^2}{r} = 1516.04\,\text{m/s}^2 \approx 1.52 \times 10^3\,\text{m/s}^2. \tag{A.469}$$

(c) You can calculate the centripetal acceleration at 6.00 rev/s with $r = 0.900$ m as

$$a_r = \frac{v_2^2}{r} = 1279.16\,\text{m/s}^2 \approx 1.28 \times 10^3\,\text{m/s}^2. \tag{A.470}$$

4.25. You can calculate the distance of the satellite from the center of the Earth as

$$r = 600\,\text{km} + 6400\,\text{km} = 7000\,\text{km}. \tag{A.471}$$

The radial acceleration is $a_r = 8.21\,\text{m/s}^2$. You can then determine the speed of the satellite as

$$v^2 = a_r r = (8.21 \times 10^{-3}\,\text{km/s}^2)(7000\,\text{km}) \approx 57.5\,\text{km}^2/\text{s}^2. \tag{A.472}$$

Solving it for v, you find that the speed is

$$v = 7.58\,\text{km/s}. \tag{A.473}$$

You calculate the time required to complete one orbit around the Earth as

$$T = \frac{2\pi r}{v} = 5802.41\,\text{s} \approx 1.61\,\text{h}. \tag{A.474}$$

4.26. The acceleration is calculated as

$$a = \sqrt{a_r^2 + a_t^2} \tag{A.475}$$

where a_r and a_t are the radial (centripetal) and tangential acceleration, respectively. The tangential acceleration is

$$a_t = \frac{50.0\,\text{km/h} - 90.0\,\text{km/h}}{15.0\,\text{s}} = -\frac{11.11\,\text{m/s}}{15.0\,\text{s}} = -0.741\,\text{m/s}^2. \tag{A.476}$$

The radial acceleration is

$$a_r = \frac{v^2}{r} \approx 1.29 \, \text{m/s}^2. \tag{A.477}$$

Then you can replace these numerical values into eq. (A.475) to get the total acceleration

$$a = \sqrt{(1.29 \, \text{m/s}^2)^2 + (-0.741 \, \text{m/s}^2)^2} = 1.48 \, \text{m/s}^2. \tag{A.478}$$

4.27.

(a) The tangential acceleration is

$$a_t = \frac{dv}{dt} = 0.600 \, \text{m/s}^2. \tag{A.479}$$

(b) You can calculate the radial acceleration component as

$$a_r = \frac{v^2}{r} = 0.800 \, \text{m/s}^2. \tag{A.480}$$

(c) The magnitude of the acceleration is calculated from

$$a = \sqrt{a_t^2 + a_r^2} = 1 \, \text{m/s}^2. \tag{A.481}$$

To find the direction of total acceleration vector, you can calculate the angle between the acceleration vector and the radius vector (see also Fig. A.21). From Fig. A.21, you get

$$\cos \alpha = \frac{a_r}{a} = 0.800 \tag{A.482}$$

or

$$\alpha \approx 36.9°. \tag{A.483}$$

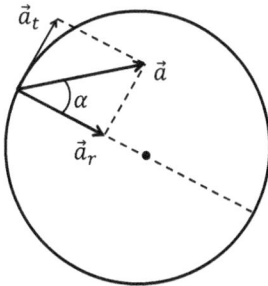

Figure A.21: Direction of the total acceleration vector.

4.28.

(a) Using Fig. 4.13, you can find the radial acceleration as:

$$a_r = a \cos 30° \approx 13.0 \, \text{m/s}^2. \qquad (A.484)$$

(b) The speed of the particle can be obtained from the following relation:

$$a_r = \frac{v^2}{r}. \qquad (A.485)$$

Hence,

$$v = \sqrt{r a_r} = \sqrt{(2.50 \, \text{m})(13.0 \, \text{m/s}^2)} = 5.70 \, \text{m/s}. \qquad (A.486)$$

(c) The tangential acceleration can be determined from Fig. 4.13 as

$$a_t = a \sin 30° = 7.50 \, \text{m/s}^2. \qquad (A.487)$$

4.29. Figure A.22 describes graphically the problem. Note that in our problem, we are ignoring the radius of the ball. We are also ignoring any gravitational forces upon the ball.

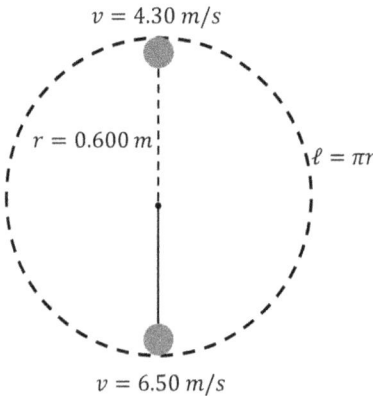

Figure A.22: Motion of a ball attached to a string.

(a) The radial acceleration when the ball is at its highest point is

$$a_r = \frac{v^2}{r} = \frac{(4.30 \, \text{m/s})^2}{0.600 \, \text{m}} \approx 30.8 \, \text{m/s}^2. \qquad (A.488)$$

The tangential acceleration can be found using the following relation:

$$v_2^2 - v_1^2 = 2 a_t \ell. \qquad (A.489)$$

Considering the motion is anticlockwise, you get

$$(6.50 \text{ m/s})^2 - (4.30 \text{ m/s})^2 = 2a_t(\pi 0.600 \text{ m}) \tag{A.490}$$

or

$$a_t = 6.30 \text{ m/s}^2. \tag{A.491}$$

The total acceleration is

$$a = \sqrt{a_t^2 + a_r^2} = 31.5 \text{ m/s}^2. \tag{A.492}$$

(b) The radial acceleration when the ball is at its lowest point is

$$a_r = \frac{v^2}{r} = \frac{(6.50 \text{ m/s})^2}{0.600 \text{ m}} \approx 70.4 \text{ m/s}^2. \tag{A.493}$$

The total acceleration is

$$a = \sqrt{a_t^2 + a_r^2} = 70.7 \text{ m/s}^2. \tag{A.494}$$

4.30. Figure A.23 illustrates graphically the problem. Note that in our problem, we are ignoring the radius of the ball and any gravitational forces upon the ball.

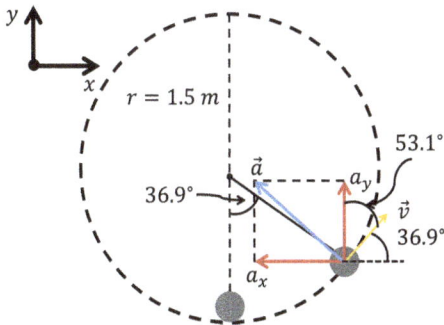

Figure A.23: Motion of a ball attached to a string.

(a) A sketch of the components of the acceleration is shown in Fig. A.23.
(b) First you can find the coordinates of the unit vector along the radius at 36.9° pointing to the center of the circle from the figure:

$$r_{0x} = -\sin 36.9° = -0.600 \tag{A.495}$$
$$r_{0y} = \cos 36.9° = 0.800. \tag{A.496}$$

Hence,

$$\mathbf{r}_0 = -0.600\mathbf{i} + 0.800\mathbf{j}. \tag{A.497}$$

Then you can find the radial acceleration as

$$a_r = \mathbf{a} \cdot \mathbf{r}_0 = (-0.6)(-22.5) + (0.8)(20.2) \approx 29.7 \, \text{m/s}^2. \tag{A.498}$$

(c) The speed can be calculated from

$$a_r = 29.7 \, \text{m/s}^2 = \frac{v^2}{r} = \frac{v^2}{1.50 \, \text{m}}. \tag{A.499}$$

Solving it for v, you get

$$v = 6.67 \, \text{m/s}. \tag{A.500}$$

Using Fig. A.23, you can find

$$v_x = v \cos 36.9° = 5.34 \, \text{m/s} \tag{A.501}$$
$$v_y = v \sin 36.9° = 4.00 \, \text{m/s}. \tag{A.502}$$

Hence

$$\mathbf{v} = (5.34 \, \text{m/s})\mathbf{i} + (4.00 \, \text{m/s})\mathbf{j}. \tag{A.503}$$

4.31.
(a) Let us find the velocities of the cars at $t = 5.00$ s.

$$\mathbf{v}_1 = \mathbf{v}_{1i} + \mathbf{a}_1 t = (15.0\mathbf{i} - 10.0\mathbf{j}) \, \text{m/s} \tag{A.504}$$
$$\mathbf{v}_2 = \mathbf{v}_{2i} + \mathbf{a}_2 t = (5.00\mathbf{i} + 15.0\mathbf{j}) \, \text{m/s}. \tag{A.505}$$

Then the relative velocity of car 1 with respect to car 2 is

$$\mathbf{v}_{rel} = \mathbf{v}_1 - \mathbf{v}_2 = (10.0\mathbf{i} - 25.0\mathbf{j}) \, \text{m/s}. \tag{A.506}$$

Furthermore, the relative speed is

$$v_{rel} = |\mathbf{v}_{rel}| = \sqrt{(10.0)^2 + (-25.0)^2} \approx 26.9 \, \text{m/s}. \tag{A.507}$$

(b) Let us first determine the distance traveled by car 1 during 5.00 s:

$$v_1^2 - v_{1i}^2 = 2a_1 \Delta r_1. \tag{A.508}$$

Replacing the numerical values, you get

$$325 = 2\Delta r_1 3.61 \tag{A.509}$$

or

$$\Delta r_1 \approx 45.1 \, \text{m}. \tag{A.510}$$

Now, you can determine the distance traveled by car 2 as

$$v_2^2 - v_{2i}^2 = 2a_2\Delta r_2. \tag{A.511}$$

Replacing the numerical values, you get

$$250 = 2\Delta r_2 3.16 \tag{A.512}$$

or

$$\Delta r_2 \approx 39.5 \, \text{m}. \tag{A.513}$$

Therefore, they relative distance is

$$\Delta r = 5.54 \, \text{m}. \tag{A.514}$$

(c) The relative acceleration of the car 1 with respect to car 2 is

$$\mathbf{a}_{\text{rel}} = \mathbf{a}_1 - \mathbf{a}_2 = (2.00\mathbf{i} - 5.00\mathbf{j}) \, \text{m/s}^2. \tag{A.515}$$

You can calculate its magnitude as

$$a_{\text{rel}} = \sqrt{(2.00)^2 + (-5.00)^2} \approx 5.39 \, \text{m/s}^2. \tag{A.516}$$

4.32. When the student swims upstream, the relative speed is

$$v_{\text{rel}} = 1.20 \, \text{m/s} - 0.500 \, \text{m/s} = 0.700 \, \text{m/s}. \tag{A.517}$$

Thus, the time to swim 1 km can be calculated as

$$t_{\text{upstream}} = (1.00 \, \text{km})/(0.700 \, \text{m/s}) = 1428.57 \, \text{s} \approx 1.43 \times 10^3 \, \text{s}. \tag{A.518}$$

The relative speed swimming downstream is

$$v_{\text{rel}} = 1.20 \, \text{m/s} + 0.500 \, \text{m/s} = 1.70 \, \text{m/s}. \tag{A.519}$$

The time it takes to swim downstream 1 km is

$$t_{\text{downstream}} = (1.00 \, \text{km})/(1.70 \, \text{m/s}) = 588.24 \, \text{s} \approx 0.588 \times 10^3 \, \text{s}. \tag{A.520}$$

Thus, the total time can be calculated as

$$t = t_{\text{upstream}} + t_{\text{downstream}} = 2016.81 \, \text{s} \approx 2.02 \times 10^3 \, \text{s}. \tag{A.521}$$

If the water would be still water, then the total time to swim 2.00 km with speed 1.20 m/s is

$$t = (2.00 \, \text{km})/(1.20 \, \text{m/s}) = 1666.67 \, \text{s} \approx 1.67 \times 10^3 \, \text{s}. \tag{A.522}$$

These results indicate, as expected, that swimming in still water is faster.

4.33. First you have to find the relative speed of the car on the left with respect to the car on the right as

$$v_{rel} = 60.0 \, km/h - 40.0 \, km/h = 20.0 \, km/h. \tag{A.523}$$

Since the distance between the cars is 100 m, the time the car on the left needs to reach the car on the right is

$$t = (100 \, m)/(20.0 \, km/h) = 0.005 \, h = 18.0 \, s. \tag{A.524}$$

4.34. Figure A.24 illustrates the motion of the airplane relative to the wind. The relative speed of the airplane with respect to air is

$$\mathbf{v}_{rel} = \mathbf{v}_{airplane} - \mathbf{v}_{wind}. \tag{A.525}$$

Using Fig. A.24, you can get

$$v_{rel}^2 = v_{airplane}^2 + v_{wind}^2 \tag{A.526}$$

where $v_{airplane}$ is the relative speed of the airplane relative to the ground. Therefore,

$$v_{airplane} = \sqrt{v_{rel}^2 - v_{wind}^2} = 147 \, km/h. \tag{A.527}$$

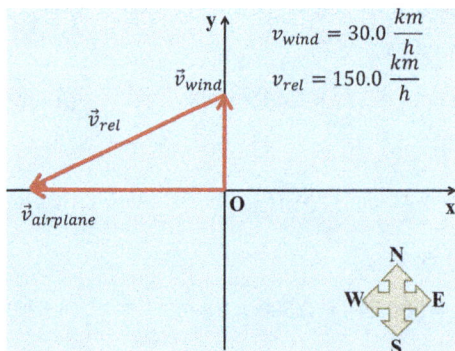

Figure A.24: Motion of the airplane relative to the wind.

4.35. Figure A.25 shows the motion of the boat relative to the fixed system of the shore and moving system of water. From Fig. A.25, the relative speed of the boat with respect to water when they move in the same direction is

$$\mathbf{v}_{boat} = \mathbf{v}_{rel} + \mathbf{v}_{wat} \tag{A.528}$$

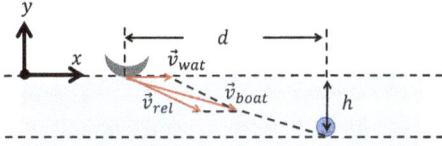

Figure A.25: Motion of the boat relative to the shore and water.

where $\mathbf{v}_{boat}$ is the velocity of the boar with respect to shore, $\mathbf{v}_{rel}$ is the velocity of the boat relative to the water stream and $\mathbf{v}_{wat}$ is the velocity of water stream (see Fig. A.25). Using Fig. A.25, direction of the $\mathbf{v}_{boat}$ relative to the shore direction is

$$\tan \alpha = h/d = 0.600/0.800 = 0.750 \qquad (A.529)$$

or

$$\alpha \approx 36.9°. \qquad (A.530)$$

In addition, you can write

$$v_{wat}^2 = v_{rel}^2 + v_{boat}^2 - 2v_{rel}v_{boat} \cos \beta \qquad (A.531)$$

where β is the angle between the two vectors $\mathbf{v}_{rel}$ and $\mathbf{v}_{boat}$. After replacing the values, you get

$$6.25 = 400 + v_{boat}^2 - 2(v_{boat})(20.0) \cos \beta. \qquad (A.532)$$

Using Fig. A.25, you can also write that

$$v_{rel}^2 = v_{boat}^2 + v_{wat}^2 - 2v_{wat}v_{boat} \cos 36.9° \qquad (A.533)$$

or

$$v_{boat}^2 - 4v_{boat} - 393.75 = 0. \qquad (A.534)$$

Solving it for the speed of the boat, you see that the physical solution is

$$v_{boat} = 21.9 \, \text{m/s}. \qquad (A.535)$$

Replacing this value into eq. (A.532), you obtain

$$\beta = 4.31°. \qquad (A.536)$$

Therefore, the angle between the shore direction and the heading is

$$4.31° + 36.9° = 41.2°. \qquad (A.537)$$

4.36.

(a) The acceleration of the bolt relative to the train is the free-fall acceleration $g = 9.80 \, \text{m/s}^2$ with direction towards the center of the Earth.

(b) To find the acceleration with respect to the Earth, you can use the formula

$$\mathbf{a} = \mathbf{a'} + \mathbf{g} \tag{A.538}$$

where $\mathbf{a'}$ is the acceleration vector of the train, which is perpendicular to free-fall acceleration vector $\mathbf{g}$. Therefore,

$$a^2 = (a')^2 + g^2. \tag{A.539}$$

Solving it for a, you obtain the magnitude of the acceleration of the bolt relative to the Earth

$$a = 10.1 \, \text{m/s}^2. \tag{A.540}$$

4.37. A graphical illustration of the problem is shown in Fig. A.26. From the figure, the velocity relative to the ground of the ball is

$$\mathbf{v}_{rel} = \mathbf{v} - \mathbf{v}_{train} \tag{A.541}$$

where $\mathbf{v}$ is the relative velocity of the ball to the train, and $\mathbf{v}_{train}$ is the velocity of the train.

It can be found that

$$v_{rel} = v_{train} \tan 60° \approx 17.3 \, \text{m/s} \tag{A.542}$$

which is the initial velocity of a ball thrown upward from the professor reference system. The acceleration, then, will be equal to the fall-free acceleration: $a_y = -g = -9.80 \, \text{m/s}^2$, where the sign minus indicates that the acceleration has negative y direction. You can get how high does the ball rise from the professor reference system using the following formula:

$$v_{yf}^2 - v_{rel}^2 = 2a_y h \tag{A.543}$$

where the velocity at maximum height is $v_{yf} = 0$. Therefore,

$$h = -\frac{v_{rel}^2}{2a_y}. \tag{A.544}$$

Replacing the numerical values, you get

$$h = -\frac{(17.3 \, \text{m/s})^2}{2(-9.8 \, \text{m/s}^2)} = 5.10 \, \text{m}. \tag{A.545}$$

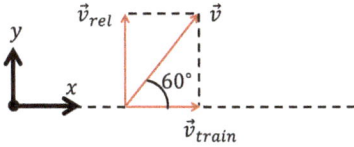

Figure A.26: Graphical illustration of the problem.

Solutions Chapter 5

5.1.

(a) Using Newton's second law, you can write

$$m_1\mathbf{a}_1 = \mathbf{F} \tag{A.546}$$
$$m_2\mathbf{a}_2 = \mathbf{F}. \tag{A.547}$$

Assuming that the force acts along the x direction, then the projections of the above equations along the x-axis give

$$m_1 a_1 = F_x \tag{A.548}$$

and

$$m_2 a_2 = F_x. \tag{A.549}$$

Dividing side by side eq. (A.548) and eq. (A.549), you get

$$\frac{m_1}{m_2}\frac{a_1}{a_2} = 1 \tag{A.550}$$

or

$$\frac{m_1}{m_2} = \frac{a_2}{a_1} = \frac{1}{3}. \tag{A.551}$$

(b) For the object of the combined masses, the total mass is $m_1 + m_2$, then you can write Newton's second law under the same force as

$$(m_1 + m_2)a_x = F_x. \tag{A.552}$$

Using eq. (A.548) and eq. (A.549), you can write

$$m_1 + m_2 = F_x\left(\frac{1}{a_1} + \frac{1}{a_2}\right). \tag{A.553}$$

Combining the last two equations, you get

$$a_x = \frac{a_1 a_2}{a_1 + a_2} = \frac{3}{4} = 0.750 \text{ m/s}^2. \tag{A.554}$$

5.2.

(a) You can find the body's acceleration using Newton's second law:

$$ma = F. \tag{A.555}$$

Therefore,

$$a = \frac{F}{m} = 5.00\,\text{m/s}^2. \tag{A.556}$$

(b) The weight of the body can be calculated using the formula

$$W = mg. \tag{A.557}$$

Here, $g = 9.80\,\text{m/s}^2$ is the free-fall acceleration. Replacing the numerical values, you get

$$W = 19.6\,\text{N}. \tag{A.558}$$

(c) If the force is doubled (i. e., $F = 20.0\,\text{N}$), using the second law of Newton you get

$$a = \frac{F}{m} = \frac{20.0\,\text{N}}{2.00\,\text{kg}} = 10.0\,\text{m/s}^2. \tag{A.559}$$

Hence, you can conclude that if the force doubles, for the same mass, the acceleration doubles as well.

5.3. You can get the resultant force using Newton's second law as

$$\sum_i \mathbf{F}_i = m\mathbf{a} = (6.00\,\text{N})\mathbf{i} + (15.0\,\text{N})\mathbf{j}. \tag{A.560}$$

The magnitude of the net force can be calculated by

$$\left| \sum_i \mathbf{F}_i \right| = \sqrt{(6.00\,\text{N})^2 + (15.0\,\text{N})^2} \tag{A.561}$$

$$= 16.16\,\text{N} \approx 16.2\,\text{N}. \tag{A.562}$$

5.4. First, you have to calculate the acceleration using Newton's second law as

$$a = \frac{F}{m} = \frac{(750000\,\text{N})}{(15000 \times 100\,\text{kg})} = 0.500\,\text{m/s}^2. \tag{A.563}$$

Then the time for increasing the speed from 0 to 80.0 km/h can be found from the formula

$$v_f = v_i + at \tag{A.564}$$

that gives

$$t = \frac{v_f}{a} = \frac{(80.0\,\text{km/h})}{(0.5\,\text{m/s}^2)} = \frac{80.0\frac{1000}{3600}\,\text{m/s}}{0.5\,\text{m/s}^2} = 44.44\,\text{s} \approx 44.4\,\text{s}. \tag{A.565}$$

5.5. Assuming that the motion of the bullet traveling down the barrel of the rifle starts from rest, $v_i = 0$, and the final speed is $v_f = 320$ m/s, you can write

$$v_f^2 - v_i^2 = 2a\ell \qquad (A.566)$$

where a is the unknown acceleration, and ℓ is the length of the barrel of the rifle. From eq. (A.566), you get

$$a = \frac{(320\,\text{m/s})^2}{2(0.820\,\text{m})} = 62.44\,\text{km/s}^2 \approx 62.4\,\text{km/s}^2. \qquad (A.567)$$

Then, the force exerted by the gases upon the bullet is calculated, based on the second law of Newton, as

$$F = ma \approx 312\,\text{N}. \qquad (A.568)$$

5.6.
(a) To find the distance the ball will accelerate before its release, you can use the formula

$$\mathbf{v}_f = \mathbf{v}_i + \mathbf{a}t \qquad (A.569)$$

where $v_i = 0$ is the initial speed, $\mathbf{v}_f = v\mathbf{i}$ is the final velocity, and $\mathbf{a}$ is the acceleration. Using eq. (A.569), you can determine the acceleration as

$$\mathbf{a} = \left(\frac{v}{t}\right)\mathbf{i}. \qquad (A.570)$$

Then the displacement vector of the ball before the release is

$$\mathbf{r} - \mathbf{r}_i = \mathbf{v}_i t + \frac{\mathbf{a}t^2}{2} \qquad (A.571)$$

where $\mathbf{r}_i = 0$ is the initial position. You can then determine the distance vector as

$$\mathbf{r} = \left(\frac{vt}{2}\right)\mathbf{i}. \qquad (A.572)$$

The magnitude of this vector is

$$r = \frac{vt}{2}. \qquad (A.573)$$

(b) Since the weight is $-F_g\mathbf{j}$, you can calculate the mass as

$$m = \frac{F_g}{g} \qquad (A.574)$$

where g is the free-fall acceleration and F_g is the magnitude of the weight vector. Then, using the second law of Newton, the force can be calculated as

$$F = ma = \frac{F_g}{g}\frac{v}{t} = \frac{F_g v}{gt}. \qquad (A.575)$$

5.7. From the information of the problem

$$1\,\text{pound} = mg \tag{A.576}$$

where g is the free-fall acceleration. You can then calculate

$$1\,\text{pound} = (0.45359237\,\text{kg})(32.174\,\text{ft/s}^2) \tag{A.577}$$
$$= (0.45359237\,\text{kg})(32.174 \times 0.3048\,\text{m/s}^2)$$
$$= 4.45\,\text{N}.$$

5.8.

(a) First, you have to determine the acceleration according to the formula

$$\mathbf{v}_f = \mathbf{v}_i + \mathbf{a}t \tag{A.578}$$

where $\mathbf{v}_f = (8.00\mathbf{i} + 10.0\mathbf{j})$ m/s is the final velocity, $\mathbf{v}_i = 3.00\mathbf{i}$ m/s is the initial velocity, and $t = 8\,\text{s}$ is the time. Using eq. (A.578), you get

$$\mathbf{a} = \frac{\mathbf{v}_f - \mathbf{v}_i}{t} = \frac{5.00\mathbf{i} + 10.0\mathbf{j}}{8\,\text{s}} \tag{A.579}$$
$$= (0.625\mathbf{i} + 1.25\mathbf{j})\,\text{m/s}^2.$$

The force is then calculated using the second law of Newton as

$$\mathbf{F} = m\mathbf{a} = (4.00\,\text{kg})[(0.625\mathbf{i} + 1.25\mathbf{j})\,\text{m/s}^2] \tag{A.580}$$
$$= (2.50\mathbf{i} + 5.00\mathbf{j})\,\text{N}.$$

Hence, the components of the force are

$$F_x = 2.50\,\text{N} \tag{A.581}$$
$$F_y = 5.00\,\text{N}. \tag{A.582}$$

(b) The magnitude of the force is calculated as

$$F = \sqrt{F_x^2 + F_y^2} = \sqrt{(2.50\,\text{N})^2 + (5.00\,\text{N})^2} = 5.59\,\text{N}. \tag{A.583}$$

5.9.

(a) Assume that the direction of the movement is along the positive x-axis. The average acceleration is calculated according to

$$\bar{\mathbf{a}} = \frac{\mathbf{v}_f - \mathbf{v}_i}{\Delta t} \tag{A.584}$$

where the final velocity is $\mathbf{v}_f = -6.70 \times 10^2 \mathbf{i}$ m/s and the initial velocity is $\mathbf{v}_i = 6.70 \times 10^2 \mathbf{i}$ m/s. The time interval is $\Delta t = 3.00 \times 10^{-13}$ s. After replacing the numerical values, you get

$$\bar{\mathbf{a}} = \frac{-13.4 \times 10^2 \mathbf{i}\,\text{m/s}}{3.00 \times 10^{-13}\,\text{s}} = (-4.47 \times 10^{15}\,\text{m/s}^2)\mathbf{i}. \tag{A.585}$$

The minus sign means that the average acceleration is along the negative x-axis direction.

(b) The average force acting upon the molecule can be calculated using the second law of Newton as

$$\mathbf{F} = m\mathbf{a} = (-20.9 \times 10^{-11}\mathbf{i})\,\text{N}. \tag{A.586}$$

Then the force the molecule exerts on the wall can be calculated as a reaction force using Newton's third law:

$$\mathbf{F}_R = -\mathbf{F} = (20.9 \times 10^{-11}\mathbf{i})\,\text{N}. \tag{A.587}$$

5.10.

(a) First, you have to determine the acceleration of the electron using the following formula, assuming the direction of the displacement, $x_f - x_i = 5.00$ cm, is along the positive x direction:

$$v_{xf}^2 - v_{xi}^2 = 2a_x(x_f - x_i). \tag{A.588}$$

Here $v_{xf} = 7.00 \times 10^5$ m/s is the final speed, and $v_{xi} = 3.00 \times 10^5$ m/s is the initial speed. Then you get the acceleration a_x as

$$a_x = \frac{(7.00 \times 10^5\,\text{m/s})^2 - (3.00 \times 10^5\,\text{m/s})^2}{2(5.00 \times 10^2\,\text{m})} = 4.00 \times 10^8\,\text{m/s}^2. \tag{A.589}$$

You can then determine the force acting on the electron using the second law of Newton:

$$F_x = ma_x = (9.11 \times 10^{-31}\,\text{kg})(4.00 \times 10^8\,\text{m/s}^2) \tag{A.590}$$
$$\approx 36.4 \times 10^{-23}\,\text{N}.$$

(b) The weight of the electron is determined as

$$W = mg = (9.11 \times 10^{-31}\,\text{kg})(9.8\,\text{m/s}^2) \tag{A.591}$$
$$\approx 89.3 \times 10^{-31}\,\text{N}.$$

As one can see, the weight of electron is much smaller than the force exerted upon it. Exactly, it is

$$\frac{W}{F_x} = 2.45 \times 10^{-8} \tag{A.592}$$

times smaller.

5.11. First, you have to convert the mass from pound to kilogram, as

$$m = (120 \text{ lb})(0.45359237 \text{ kg}) = 54.43 \text{ kg}. \tag{A.593}$$

Then you can determine the weight as

$$W = mg = (54.43 \text{ kg})(9.8 \text{ m/s}^2) \approx 533.4 \text{ N}. \tag{A.594}$$

5.12. First, you have to determine the mass of that person using the weight on Earth as

$$m = \frac{W_{\text{Earth}}}{g} = \frac{(900 \text{ N})}{(9.8 \text{ m/s}^2)} \approx 91.8 \text{ kg}. \tag{A.595}$$

Then, the Jupiter's weight is:

$$W_{\text{Jupiter}} = (91.8 \text{ kg})(25.9 \text{ m/s}^2) = 2378.656 \text{ N} \approx 2.38 \times 10^3 \text{ N}. \tag{A.596}$$

As you can see, the weight is larger on Jupiter compared to the Earth. Exactly, it is

$$\frac{W_{\text{Jupiter}}}{W_{\text{Earth}}} = \frac{(2.38 \times 10^3 \text{ N})}{(900 \text{ N})} = 2.64 \tag{A.597}$$

times larger.

5.13. To determine the acceleration in each case, you have to use the second law of Newton in the form

$$\mathbf{F}_1 + \mathbf{F}_2 = m\mathbf{a}. \tag{A.598}$$

If you consider the coordinate system as in Fig. 5.11, then, for the case (a), you get

$$\mathbf{F}_1 = F_1\mathbf{i} = (20.0 \text{ N})\mathbf{i} \tag{A.599}$$
$$\mathbf{F}_2 = F_2\mathbf{j} = (15.0 \text{ N})\mathbf{j}.$$

You find the acceleration as

$$\mathbf{a} = \frac{(20.0 \text{ N})\mathbf{i} + (15.0 \text{ N})\mathbf{j}}{5.00 \text{ kg}} \tag{A.600}$$
$$= (4.00\mathbf{i} + 3.00\mathbf{j}) \text{ m/s}^2.$$

The magnitude of the acceleration is

$$a = \sqrt{(4.00)^2 + (3.00)^2} = \sqrt{16.0 + 9.00} = 5.00 \text{ m/s}^2. \tag{A.601}$$

For the case (b), you determine the forces as

$$\mathbf{F}_1 = F_1\mathbf{i} = (20.0 \text{ N})\mathbf{i} \tag{A.602}$$

$$\mathbf{F}_2 = F_2 \cos 60.0°\mathbf{i} + F_2 \sin 60.0°\mathbf{j} = (7.5\,\text{N})\mathbf{i} + (13.0\,\text{N})\mathbf{j}. \tag{A.603}$$

Then you can determine the acceleration as

$$\mathbf{a} = \frac{(27.5\,\text{N})\mathbf{i} + (13.0\,\text{N})\mathbf{j}}{5.00\,\text{kg}} \tag{A.604}$$
$$= (5.50\mathbf{i} + 2.60\mathbf{j})\,\text{m/s}^2.$$

The magnitude of the acceleration is

$$a = \sqrt{(5.50)^2 + (2.60)^2} = \sqrt{30.3 + 6.76} = 6.08\,\text{m/s}^2. \tag{A.605}$$

5.14. A graphical representation of the forces exerted upon the mass $m = 2.80\,\text{kg}$ is shown in Fig. A.27. To calculate the extra force, we first determine the acceleration from the following formula:

$$\mathbf{r}_f - \mathbf{r}_i = \mathbf{v}_i t + \frac{at^2}{2} \tag{A.606}$$

where the initial velocity is $\mathbf{v}_i = 0$. Then you can get the acceleration as

$$\mathbf{a} = \frac{2}{t^2}(\mathbf{r}_f - \mathbf{r}_i) \tag{A.607}$$
$$= \frac{2}{(1.20\,\text{s})^2}((4.20\,\text{m})\mathbf{i} - (3.30\,\text{m})\mathbf{j})$$
$$= (5.83\,\text{m/s}^2)\mathbf{i} - (4.58\,\text{m/s}^2)\mathbf{j}.$$

The resultant force is

$$\mathbf{F}_R = m\mathbf{a} = (2.80\,\text{kg})((5.83\,\text{m/s}^2)\mathbf{i} - (4.58\,\text{m/s}^2)\mathbf{j}) \tag{A.608}$$
$$= (16.3\,\text{N})\mathbf{i} + (-12.8\,\text{N})\mathbf{j}.$$

The forces acting on the mass can be written in terms of the components along the x and y directions as

$$\mathbf{w} = -mg\mathbf{j} = -(2.80\,\text{kg})(9.80\,\text{m/s}^2) \approx (-27.4\,\text{N})\mathbf{j} \tag{A.609}$$
$$\mathbf{F} = F_x\mathbf{i} + F_y\mathbf{j}. \tag{A.610}$$

Since the resultant force is

$$\mathbf{F}_R = \mathbf{w} + \mathbf{F} = (16.3\,\text{N})\mathbf{i} + (-12.8\,\text{N})\mathbf{j} \tag{A.611}$$

we have

$$\mathbf{F} = \mathbf{F}_R - \mathbf{w} \tag{A.612}$$
$$= (16.3\,\text{N})\mathbf{i} + (-12.8\,\text{N})\mathbf{j} - (-27.4\,\text{N})\mathbf{j}$$
$$= (16.3\,\text{N})\mathbf{i} + (14.6\,\text{N})\mathbf{j}.$$

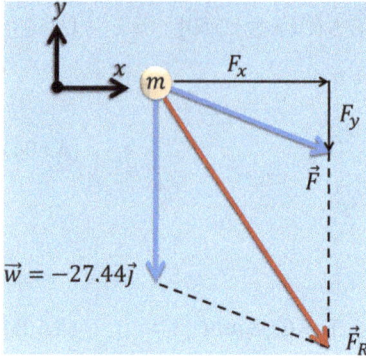

Figure A.27: A graphical representation of the forces acting upon mass m = 2.80 kg.

$\vec{w} = -27.44\vec{j}$

5.15. A graphical representation of all forces acting on the mass m = 4.00 kg is shown in Fig. A.28. To obtain the object's acceleration, you can use Newton's second law as

$$m\mathbf{a} = \mathbf{F}_R \tag{A.613}$$

where $\mathbf{F}_R$ is the resultant force acting on the object, that is,

$$\mathbf{F}_R = \mathbf{F}_1 + \mathbf{F}_2 + \mathbf{F}_3 \tag{A.614}$$

where $\mathbf{F}_1 = 10.0\mathbf{j}$ N, $\mathbf{F}_2 = 20.0\mathbf{i}$ N and $\mathbf{F}_3 = -15.0\mathbf{j}$ N. Therefore, the net force vector is

$$\mathbf{F}_R = (20.0\mathbf{i} - 5.00\mathbf{j}) \text{ N.} \tag{A.615}$$

The magnitude of $\mathbf{F}_R$ is calculated as

$$F_R = \sqrt{(20.0)^2 + (-5.00)^2} \approx 20.6 \text{ N.} \tag{A.616}$$

Using Newton's second law, you get the acceleration vector as

$$\mathbf{a} = \frac{\mathbf{F}_R}{m} \tag{A.617}$$
$$= \frac{(20.0\mathbf{i} - 5.00\mathbf{j}) \text{ N}}{4.00 \text{ kg}}$$
$$= (5.00 \text{ m/s}^2)\mathbf{i} + (-1.25 \text{ m/s}^2)\mathbf{j}.$$

To determine the magnitude of the acceleration, you can calculate the magnitude of a vector as

$$a = \sqrt{a_x^2 + a_y^2} = \sqrt{(5.00)^2 + (-1.25)^2} = 5.15 \text{ m/s}^2. \tag{A.618}$$

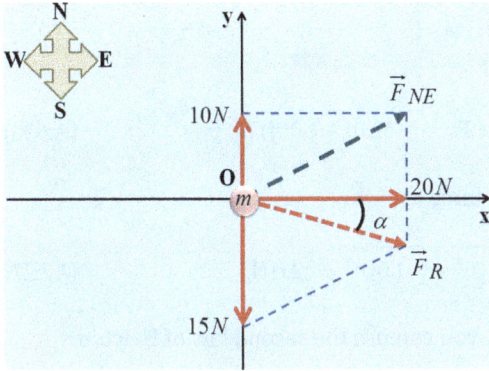

Figure A.28: A graphical representation of the forces acting upon mass $m = 4.00$ kg.

5.16.

(a) The resultant force is

$$\mathbf{F}_R = \mathbf{F}_1 + \mathbf{F}_2 \tag{A.619}$$

where $\mathbf{F}_1$ is the push force and $\mathbf{F}_2$ is the resistance force. Assuming the direction of the displacement of the boat is along the positive x-axis direction, you can project eq. (A.619) along the x-axis:

$$F_R = F_1 - F_2 = 2000\,\text{N} - 1800\,\text{N} = 200\,\text{N}. \tag{A.620}$$

You can then use Newton's second law to determine the accelerations as

$$a = F/m = 200\,\text{N}/1000\,\text{kg} = 0.2\,\text{m/s}^2. \tag{A.621}$$

(b) To find the displacement in 10.0 s knowing that the initial speed $v_{xi} = 0$, you can use the following equation:

$$x_f - x_i = v_{xi}t + at^2/2. \tag{A.622}$$

Replacing the numerical values, you find

$$x_f - x_i = 0 + (0.2\,\text{m/s}^2)(10.0\,\text{s})^2/2 = 10.0\,\text{m}. \tag{A.623}$$

(c) To determine the speed at the end of the 10.0 s time, you can use the expression

$$v_{xf} = v_{xi} + at. \tag{A.624}$$

After replacing the numerical values, you obtain

$$v_{xf} = (0.2\,\text{m/s}^2)(10.0\,\text{s}) = 2.00\,\text{m/s}. \tag{A.625}$$

5.17.

(a) First, you find the resultant force as

$$\mathbf{F}_R = \mathbf{F}_1 + \mathbf{F}_2 + \mathbf{F}_3 = (-42.0\mathbf{i} - 1.00\mathbf{j})\,\text{N}. \tag{A.626}$$

Next, you get the magnitude of the force as

$$F_R = \sqrt{(-42.0)^2 + (-1.00)^2} \approx 42.0\,\text{N}. \tag{A.627}$$

To calculate the mass of the object, you can use the second law of Newton:

$$m = F_R/a = (42.0\,\text{N})/(3.75\,\text{m/s}^2) \approx 11.2\,\text{kg}. \tag{A.628}$$

You can then calculate the acceleration vector using the second law of Newton

$$\mathbf{a} = \frac{\mathbf{F}_R}{m} = ((-3.75)\mathbf{i} + (-0.0890)\mathbf{j})\,\text{m/s}^2. \tag{A.629}$$

(b) The mass of the object is

$$m = 11.2\,\text{kg}. \tag{A.630}$$

(c) To get the speed of the object after 10.0 s, assuming that initially it was at rest, you can use the formula

$$\mathbf{v}_f = \mathbf{v}_i + \mathbf{a}t \tag{A.631}$$

where the initial velocity is $\mathbf{v}_i = 0$. From eq. (A.631), you obtain

$$\begin{aligned} \mathbf{v}_f &= (-3.75)(10.0)\mathbf{i} + (-0.089)(10.0)\mathbf{j} \\ &= (-37.5\,\text{m/s})\mathbf{i} + (-0.89\,\text{m/s})\mathbf{j}. \end{aligned} \tag{A.632}$$

The speed then can be calculated as

$$v_f = \sqrt{(-37.5)^2 + (-0.89)^2} = 37.5\,\text{m/s}. \tag{A.633}$$

(d) The velocity components are

$$v_x = -37.5\,\text{m/s} \tag{A.634}$$
$$v_y = -0.890\,\text{m/s} \tag{A.635}$$

where the minus signs indicate that the components of the velocity are along the negative directions of the corresponding axes.

5.18. First, you have to determine the components of the velocity as

$$v_x = \frac{dx}{dt} = 10.0t \tag{A.636}$$

$$v_y = \frac{dy}{dt} = 9.00t^2. \tag{A.637}$$

Then you can determine the initial velocity components at $t = 0$:

$$v_{xi} = 0 \tag{A.638}$$
$$v_{yi} = 0 \tag{A.639}$$

and the final velocity components at $t = 2.00$ s:

$$v_{xf} = 20.0 \text{ m/s} \tag{A.640}$$
$$v_{yf} = 36.0 \text{ m/s}. \tag{A.641}$$

The initial velocity vector is then $\mathbf{v}_i = 0$ and the final velocity vector is

$$\mathbf{v}_f = (20.0\mathbf{i} + 36.0\mathbf{j}) \text{ m/s}. \tag{A.642}$$

You can calculate the acceleration from the formula

$$\mathbf{v}_f = \mathbf{v}_i + \mathbf{a}t. \tag{A.643}$$

From eq. (A.643), you write

$$
\begin{aligned}
\mathbf{a} &= \frac{\mathbf{v}_f}{t} \\
&= \frac{(20.0\mathbf{i} + 36.0\mathbf{j}) \text{ m/s}}{2.00 \text{ s}} \\
&= (10.0\mathbf{i} + 18.0\mathbf{j}) \text{ m/s}^2.
\end{aligned}
\tag{A.644}
$$

The force can be determined using the second law of Newton:

$$
\begin{aligned}
\mathbf{F} &= m\mathbf{a} \\
&= (3.00 \text{ kg})((10.0\mathbf{i} + 18.0\mathbf{j}) \text{ m/s}^2) \\
&= (30.0\mathbf{i} + 54.0\mathbf{j}) \text{ N}.
\end{aligned}
\tag{A.645}
$$

Then the magnitude of the force is

$$F = \sqrt{(30.0)^2 + (54.0)^2} \approx 61.8 \text{ N}. \tag{A.646}$$

5.19. Figure 5.4 illustrates the force acting from the bird to the telephone wire and the tension produced by the bird. It can be seen that the weight of the bird causes the telephone wire to bend, which can be calculated as

$$\mathbf{w} = m\mathbf{g}. \tag{A.647}$$

Projecting along the y-axis, you get

$$w = -mg = -9.80 \text{ N}. \tag{A.648}$$

Here, the sign minus indicates that the direction of the weight is along the negative y direction. Based on the third law of Newton, the wire will react with force in the opposite direction, but with the same magnitude, which is equal to the resultant of the tensions produced on the telephone wire:

$$\mathbf{T}_1 + \mathbf{T}_2 = -m\mathbf{g} \tag{A.649}$$

where $T_1 = T_2$. From Fig. A.29, you can write

$$T_1^2 = T_1^2 + (mg)^2 - 2(mg)T_1 \cos\alpha \tag{A.650}$$

where

$$\cos\alpha = \frac{0.200}{\sqrt{(0.200)^2 + (25.0)^2}} = 0.0800. \tag{A.651}$$

Then

$$mg = 2T_1 0.0800 \tag{A.652}$$

or

$$T_1 = T_2 = \frac{mg}{0.160} \approx 61.3 \text{ N}. \tag{A.653}$$

The total tension is $T_1 + T_2 \approx 123$ N.

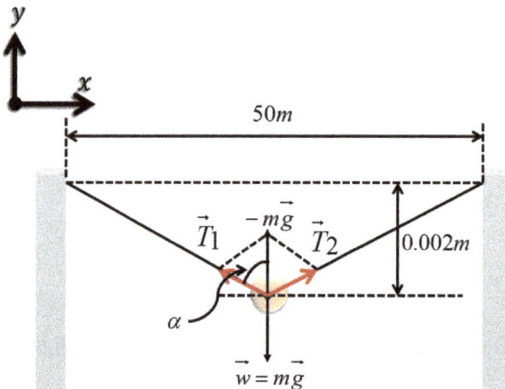

Figure A.29: A graphical representation of the forces acting upon telephone wire.

5.20. A graphical representation of the forces acting upon the bag of cement is shown in Fig. 5.12.

Since the system is in equilibrium, you can write

$$\mathbf{T}_1 + \mathbf{T}_2 + \mathbf{T}_3 = 0. \tag{A.654}$$

Projections along the vertical and horizontal directions will give

$$T_1 \sin \theta_1 + T_2 \sin \theta_2 - T_3 = 0 \tag{A.655}$$
$$-T_1 \cos \theta_1 + T_2 \cos \theta_2 = 0 \tag{A.656}$$

where

$$T_3 = F_g. \tag{A.657}$$

Using eq. (A.656), you get

$$T_2 = T_1 \frac{\cos \theta_1}{\cos \theta_2}. \tag{A.658}$$

Replacing this expression into eq. (A.655), you get

$$T_1 \sin \theta_1 + T_1 \sin \theta_2 \frac{\cos \theta_1}{\cos \theta_2} = F_g. \tag{A.659}$$

Using the relation

$$\sin(\theta_1 + \theta_2) = \sin \theta_1 \cos \theta_2 + \sin \theta_2 \cos \theta_1 \tag{A.660}$$

you get

$$T_1 = \frac{F_g \cos \theta_2}{\sin(\theta_1 + \theta_2)}. \tag{A.661}$$

5.21. First, convert units into SI units:

$$x = 2.0 \text{ cm} = 2.0 \times 10^{-2} \text{ m}. \tag{A.662}$$

Method 1: We find the magnitude of elastic force in the spring using equation

$$F_l = kx \tag{A.663}$$

which has a direction upward. Replacing the numerical values, we get

$$F_l = (1.0 \times 10^3 \text{ N/m})(2.0 \times 10^{-2} \text{ m}) = 20 \text{ N}. \tag{A.664}$$

Therefore, the elastic force vector is $\mathbf{F}_l = (20\mathbf{j})$ N, where $\mathbf{j}$ is a unit vector along the positive y-axis.

The other force acting on the object is the gravity force, $\mathbf{F}_g$, downward:

$$\mathbf{F}_g = m\mathbf{g} = -mg\mathbf{j}. \tag{A.665}$$

Projecting the resulting force, $\mathbf{F}_R = \mathbf{F}_l + \mathbf{F}_g$, acting on the object along the vertical axis, we get

$$F_R = F_l - F_g \tag{A.666}$$

which, based on Newton's second law, equals ma, where a is the acceleration. Since the object is at rest, $a = 0$. Thus,

$$F_l - F_g = 0 \tag{A.667}$$

or

$$20\,\text{N} - mg = 0. \tag{A.668}$$

Solving it for m, knowing that $g = -9.8\,\text{m/s}^2$, we get

$$m \approx 2.0\,\text{kg}. \tag{A.669}$$

Method 2: We can also think about the problem in this way: The object acts on the spring with the force which has a magnitude equal to its weight, $W = mg$, with direction downward. On the other hand, since the object is at rest, the spring acts on the object with a resistance force which is the elastic force in the spring, F_l. Based on Newton's third law, $W = F_l$; therefore,

$$20\,\text{N} = mg \tag{A.670}$$

or

$$m = \frac{20\,\text{N}}{g} \approx 2.0\,\text{kg}. \tag{A.671}$$

5.22. The terminal speed is

$$y = \frac{mg}{v_t}. \tag{A.672}$$

Replacing the numerical values, you get

$$y = \frac{(2.00\,\text{g})(980\,\text{cm/s}^2)}{5.00\,\text{cm/s}} = 392\,\text{g/s}. \tag{A.673}$$

Therefore, the time constant τ is

$$\tau = \frac{m}{y} = \frac{2.00\,\text{g}}{392\,\text{g/s}} = 5.10 \times 10^{-3}\,\text{s}. \tag{A.674}$$

The speed of the sphere is

$$v(t) = \frac{mg}{\gamma}(1 - e^{-\gamma t/m}) = v_t(1 - e^{-t/\tau}). \tag{A.675}$$

To find time t for which $v = 0.900v_t$, we have

$$0.900v_t = v_t(1 - e^{-t/\tau}) \tag{A.676}$$

and solve it for t

$$t = -\tau \ln(0.100) = 2.30\tau = 11.7 \times 10^{-3}\,\text{s} = 11.7\,\text{ms}. \tag{A.677}$$

Solutions Chapter 6

6.1. The forces, which equal the force **T** exerted by the cord into the ball, cause a centripetal acceleration:

$$T = \sum_i F_{ri} = ma_r \tag{A.678}$$

or

$$T = m\frac{v^2}{r}. \tag{A.679}$$

Now, you can solve it for v, and you get

$$v = \sqrt{\frac{rT}{m}}. \tag{A.680}$$

The maximum speed is obtained for the maximum value of T as

$$v_{max} = \sqrt{\frac{rT_{max}}{m}} = \sqrt{\frac{(1.50\,\text{m})(50.0\,\text{N})}{(0.500\,\text{kg})}} = 12.2\,\text{m/s}. \tag{A.681}$$

6.2. In this example, the force causing the centripetal acceleration is the tension force **T** exerted by the cord on the sphere.
(a) Using Newton's second law, we have

$$\sum_i F_{ri} \equiv T = ma_r \tag{A.682}$$

where $a_r = v^2/r$, hence

$$T = m\frac{v^2}{r}. \tag{A.683}$$

From eq. (A.683), you find

$$v = \sqrt{\frac{rT}{m}}.$$

(A.684)

This expression indicates that v increases with increasing T and decreases with increasing m. For a given v, a large mass m requires large tension T, and a small mass m needs only a small tension T. The maximum, speed the sphere can have corresponds to the maximum tension. Hence, we find

$$v_{max} = \sqrt{\frac{rT_{max}}{m}} = \sqrt{\frac{(50.0\ \mathrm{N})(1.50\ \mathrm{m})}{0.500\ \mathrm{kg}}} = 12.2\ \mathrm{m/s}.$$

(A.685)

(b) The tension in the cord can be calculated as

$$T = m\frac{v^2}{r} = (0.500\ \mathrm{kg})\frac{(5.00\ \mathrm{m/s})^2}{1.50\ \mathrm{m}} = 8.33\ \mathrm{kg}\frac{\mathrm{m}}{\mathrm{s}^2} = 8.33\ \mathrm{N}$$

(A.686)

given that

$$1\ \mathrm{N} = 1\ \mathrm{kg}\frac{\mathrm{m}}{\mathrm{s}^2}.$$

(A.687)

6.3. Denote θ the angle between string and vertical. In the free-body diagram shown in Fig. 6.7, the force **T** exerted by the string is resolved into the vertical and horizontal components:

$$T_y = T\cos\theta$$

(A.688)

$$T_x = T\sin\theta$$

(A.689)

acting toward the center of revolution. Since the acceleration of the object $a_y = 0$

$$\sum_i F_{yi} = ma_y = 0$$

(A.690)

and the upward vertical component of **T** must balance the downward force of gravity. Therefore,

$$T\cos\theta = mg.$$

(A.691)

In this example, the force giving the centripetal acceleration is the component $T\sin\theta$, and thus we can use Newton's second law

$$\sum_i F_{ri} = T\sin\theta = ma_r = mv^2/r.$$

(A.692)

Combining eq. (A.691) and eq. (A.692), you find

$$\tan\theta = v^2/(rg)$$

(A.693)

or

$$v = \sqrt{rg \tan \theta}. \tag{A.694}$$

From the geometry $r = L \sin \theta$, hence

$$v = \sqrt{Lg \sin \theta \tan \theta}. \tag{A.695}$$

6.4. Here, the speed is not constant because, at most points along the path, a tangential component of acceleration arises from the gravitational force exerted on the sphere. From the free-body diagram, we see that the only forces acting on the sphere are the gravitational force

$$\mathbf{F}_g = m\mathbf{g} \tag{A.696}$$

exerted by Earth, and force **T** exerted by the cord. The magnitudes of the components of $\mathbf{F}_g$ according to radial and tangential directions are

$$F_t = mg \sin \theta \tag{A.697}$$
$$F_r = mg \cos \theta. \tag{A.698}$$

Applying Newton's second law to the forces acting on the sphere along the tangential direction, we get

$$\sum_i F_{ti} = mg \sin \theta = ma_t \tag{A.699}$$

or

$$a_t = g \sin \theta. \tag{A.700}$$

Applying Newton's second law to the forces acting on the sphere along the radial direction, we get

$$\sum_i F_{ri} = T - mg \cos \theta = mv^2/r \tag{A.701}$$

or

$$T = m\left(\frac{v^2}{R} + g \cos \theta\right). \tag{A.702}$$

6.5. The mass of Earth is approximately $M_{\text{Earth}} = 5.96 \times 10^{24}$ kg, therefore, the mass of Mars is

$$M_{\text{Mars}} = 0.11 \times (5.96 \times 10^{24} \text{ kg}) = 6.56 \times 10^{23} \text{ kg}. \tag{A.703}$$

The centripetal force is equal to

$$F_c = M_{\text{Mars}} \frac{v^2}{R} \tag{A.704}$$

where $R = 228000000 \text{ km} = 2.28 \times 10^8 \text{ km} = 2.28 \times 10^{11} \text{ m}$. The velocity in SI units is

$$v = 24.1 \text{ km/s} = 24.1 \times 10^3 \text{ m/s}. \tag{A.705}$$

Replacing the numerical values, we get

$$F_c = (6.56 \times 10^{23} \text{ kg}) \frac{(24.1 \times 10^3 \text{ m/s})^2}{2.28 \times 10^{11} \text{ m}} = 1.68 \times 10^{21} \text{ N}. \tag{A.706}$$

6.6. A strict answer is no. The centripetal force is a name for a force acting in the role of a force that causes circular motion. Hence, when you draw a force diagram, adding a new vector to indicate the centripetal force is a mistake.

When the Earth moves around the Sun, the gravity force plays the role of the centripetal force. Also, for an object sitting on a rotating turntable, the centripetal force is friction. For a rock rotating at the end of a string, the centripetal force is the force of tension in the string.

In general, the centripetal force could also be a combination of two or more forces.

6.7. The radial component of the acceleration has a magnitude defined from:

$$a_r = \frac{v^2}{r}$$

where $v = 2.0 \text{ m/s}$ and $r = 1.0 \text{ m}$. Therefore,

$$a_r = \frac{(2.0 \text{ m/s})^2}{1.0 \text{ m}} = 4.0 \text{ m/s}^2.$$

The tangential acceleration is

$$a_t = g \sin(30°) = (9.8 \text{ m/s}^2)(0.5) = 4.9 \text{ m/s}^2.$$

The magnitude of the total acceleration is given as

$$a = \sqrt{a_r^2 + a_t^2} = \sqrt{(4.0 \text{ m/s}^2)^2 + (4.9 \text{ m/s}^2)^2} \approx 6.3 \text{ m/s}^2.$$

6.8. The centripetal acceleration is given as

$$a_r = \frac{v^2}{R}$$

where $R = 3.8 \times 10^5 \text{ km} = 3.8 \times 10^8 \text{ m}$, and the speed of the Moon is

$$v = \frac{2\pi R}{27.3 \text{ days}} = \frac{2\pi(3.8 \times 10^8 \text{ m})}{27.3(24)(3600 \text{ s})} \approx 1.01 \times 10^3 \text{ m/s}.$$

Thus, the centripetal acceleration is

$$a_r = \frac{(1.01 \times 10^3 \text{ m/s})^2}{3.8 \times 10^8 \text{ m}} \approx 0.0027 \text{ m/s}^2.$$

6.9. The centripetal acceleration is given as

$$a_r = \frac{v^2}{r}$$

where $r = 1.00$ m and $v = 20.0$ m/s. Therefore,

$$a_r = \frac{(20.0 \text{ m/s})^2}{1.00 \text{ m}} = 400.0 \text{ m/s}^2.$$

6.10. One revolution corresponds to the circumference of the circle; that is, $2\pi r$, and hence the speed is

$$v = 200 \text{ rev/min} = 200 \frac{2\pi (0.500 \text{ m})}{60 \text{ s}} \approx 10.5 \text{ m/s}.$$

The centripetal acceleration is given as

$$a_r = \frac{v^2}{r}$$

where $r = 0.500$ m and $v = 10.5$ m/s. Therefore,

$$a_r = \frac{(10.5 \text{ m/s})^2}{0.500 \text{ m}} \approx 219 \text{ m/s}^2.$$

Solutions Chapter 7

7.1. Using the definition of the work given as

$$W = Fd \cos \theta \qquad\qquad\qquad\qquad (A.707)$$
$$= (50.0 \text{ N})(\cos 30.0°)(3.00 \text{ m})$$
$$= 130 \text{ J}.$$

The normal force $\mathbf{n}$, the force of gravity $\mathbf{F}_g = m\mathbf{g}$, and the vertical component of the applied force $(50.0 \text{ N}) (\sin 30.0°)$ do not do any work on the vacuum cleaner as it moves because these forces are perpendicular to its displacement, and so $\cos 90° = 0$.

7.2. The work done by applied force is

$$W = \mathbf{F} \cdot \mathbf{d} = Fd \cos \theta \qquad\qquad\qquad (A.708)$$

where $F = 35.0$ N, $d = 50.0$ m and $\theta = 25.0°$. Then

$$W = (35.0 \text{ N})(50.0 \text{ m}) \cos 25.0° = 1585.5 \text{ J} \approx 1.59 \text{ kJ}. \qquad (A.709)$$

7.3. The work done by the force is equal to the area under the curve from $x_A = 0$ to $x_C = 6.0$ m. The net area equals the area of the rectangular section from A to B plus the area of the triangular section from B to C. The area of the rectangle is

$$(4.0)(5.0)\,\text{N} \cdot \text{m} = 20\,\text{J}, \tag{A.710}$$

and the area of the triangle is

$$\frac{1}{2}(2.0)(5.0)\,\text{N} \cdot \text{m} = 5.0\,\text{J}. \tag{A.711}$$

Therefore, the total work done is

$$W = 20\,\text{J} + 5.0\,\text{J} = 25\,\text{J}. \tag{A.712}$$

7.4. *Method 1*: We could apply the equations of kinematics to find speed; however, we could also use the energy approach as follows. The normal force balances the force of gravity on the block, and neither of these vertically acting forces does work on the block because the displacement is horizontal. Because there is no friction, the net external force acting on the block is 12 N. Then the work done by this force is

$$W = Fd = (12\,\text{N})(3.0\,\text{m}) = 36\,\text{J}. \tag{A.713}$$

Applying the work–kinetic energy theorem and noting that the initial kinetic energy is zero, we obtain

$$W = K_f - K_i = \frac{1}{2}mv_f^2 - 0 \tag{A.714}$$

or

$$v_f^2 = \frac{2W}{m} = \frac{2(36\,\text{J})}{6.0\,\text{kg}} = 12\,\text{m}^2/\text{s}^2. \tag{A.715}$$

Solving it for v_f, we obtain the final speed as

$$v_f \approx 3.5\,\text{m/s}. \tag{A.716}$$

Method 2: To find the final speed we can use the following kinematic equation:

$$v_f^2 - v_i^2 = 2a(x_f - x_i) \tag{A.717}$$

where

$$v_i = 0, \quad x_f - x_i = 3.0\,\text{m}. \tag{A.718}$$

Using the second law of Newton:

$$F = ma \tag{A.719}$$

we obtain the acceleration as

$$a = \frac{F}{m} = \frac{12\,\text{N}}{6.0\,\text{kg}} = 2.0\,\text{m/s}^2. \tag{A.720}$$

Thus,

$$v_f^2 = 2(2.0\,\text{m/s}^2)(3.0\,\text{m}) = 12\,\text{m}^2/\text{s}^2 \tag{A.721}$$

or, the speed is

$$v_f \approx 3.5\,\text{m/s}. \tag{A.722}$$

7.5. The work done by the force is

$$W = Fd = (12\,\text{N})(3.0\,\text{m}) = 36\,\text{J}. \tag{A.723}$$

First calculate the magnitude of the friction force

$$f_k = \mu_k mg = (0.15)(6.0\,\text{kg})(9.80\,\text{m/s}^2) \approx 8.8\,\text{N}. \tag{A.724}$$

Next, the change in kinetic energy due to friction force is given as

$$\Delta K_{\text{friction}} = -f_k d = -(8.8\,\text{N})(3.0\,\text{m}) \approx -26\,\text{J}. \tag{A.725}$$

Applying the work–kinetic energy theorem and taking zero the initial kinetic energy, we obtain

$$W - f_k d = K_f - K_i = \frac{1}{2}mv_f^2 - 0 \tag{A.726}$$

or

$$\begin{aligned}
v_f^2 &= \frac{2W}{m} - \frac{2f_k d}{m} \\
&= \frac{2(36\,\text{J})}{6.0\,\text{kg}} - \frac{2(26\,\text{J})}{6.0\,\text{kg}} \\
&\approx 3.3\,\text{m}^2/\text{s}^2.
\end{aligned} \tag{A.727}$$

Then, solving it for the speed v_f, we get

$$v_f \approx 1.8\,\text{m/s}. \tag{A.728}$$

7.6. In this situation, the block starts with $v_i = 0$ at $x_i = -2.0$ cm, and we want to find v_f at $x_f = 0$. We find the work done by the spring with $x_{\text{max}} = x_i = -2.0\,\text{cm} = -2.0{\times}10^{-2}\,\text{m}$:

$$W_s = \frac{1}{2}kx_{\text{max}}^2 = \frac{1}{2}(1.0 \times 10^3\,\text{N/m})(-2.0 \times 10^{-2}\,\text{m})^2 = 0.20\,\text{J}. \tag{A.729}$$

We can use the work–kinetic energy theorem with $v_i = 0$. Therefore, we obtain the change in kinetic energy of the block due to the work done on it by the spring:

$$W_s = \frac{1}{2}mv_f^2 - \frac{1}{2}mv_i^2. \tag{A.730}$$

Replacing the numerical values, we obtain

$$0.20\,\text{J} = \frac{1}{2}(1.6\,\text{kg})v_f^2 - 0. \tag{A.731}$$

Solving it for the speed v_f, we get

$$v_f^2 = \frac{0.40\,\text{J}}{1.6\,\text{kg}} = 0.25\,\text{m}^2/\text{s}^2 \tag{A.732}$$

or,

$$v_f = 0.50\,\text{m/s}. \tag{A.733}$$

7.7. If **F** is the engine force that propels the car, using the second law of Newton, we write

$$\sum_i \mathbf{F}_i = \mathbf{F} + \mathbf{f}_t + \mathbf{F}_g \tag{A.734}$$

where $\mathbf{F}_g = m\mathbf{g}$ is the gravity force. Projecting along the direction of motion, we obtain

$$\sum F_x = F - f_t - mg\sin\theta = ma \tag{A.735}$$

or,

$$F = ma + mg\sin\theta + f_t = ma + mg\sin\theta + (218 + 0.70v^2) \tag{A.736}$$

where θ is the angle with the horizontal level.

Thus, the power required to move the car forward is

$$P = Fv = mav + mgv\sin\theta + 218v + 0.70v^3 \tag{A.737}$$

where
-   mav is the power that engine must deliver to accelerate the car. If the car moves with constant speed ($a = 0$), this term is zero.
-   $mgv\sin\theta$ is the power required to provide a force to balance a component of the gravity force as the car moves up the incline. This term would be zero if the motion is in the horizontal surface.
-   $218v$ is the power required to provide a force balance road friction.
-   $0.70v^3$ is the power needed to do work on the air.

A compact car with a mass of 800 kg, and its efficiency rated at 18 %, that is, only 18 % of the available fuel energy is delivered to the wheels. What is the amount of gasoline needed to accelerate the car from zero to 27 m/s (or 60 mi/h)? Note that the energy equivalent of 1 gal of gasoline is 1.3×10^8 J.

The energy needed to accelerate the car from the speed zero to a speed v is its kinetic energy

$$K = \frac{1}{2}mv^2 = \frac{1}{2}(800 \text{ kg})(27 \text{ m/s})^2 \approx 2.9 \times 10^5 \text{ J}. \tag{A.738}$$

If the engine was 100 % efficient, each gallon of gasoline would supply

$$1.3 \times 19^8 \text{ J} \tag{A.739}$$

energy. Since the engine is only 18 % efficient, each gallon delivers only

$$(0.18)(1.3 \times 10^8 \text{ J}) = 2.3 \times 10^7 \text{ J}. \tag{A.740}$$

Therefore, the number of gallons used to accelerate the car is

$$\text{Number of gallons} = \frac{2.9 \times 10^5 \text{ J}}{2.3 \times 10^7 \text{ J/gal}} = 0.013 \text{ gal}. \tag{A.741}$$

7.8. The motor must supply the force of magnitude T that pulls the elevator car upward. Since the speed is constant, from Newton's second law:

$$\sum_i F_{yi} = T - f - Mg = 0. \tag{A.742}$$

M denotes the total mass of the system including the mass of passengers, equal to 1800 kg. Therefore,

$$T = f + Mg = 4.00 \times 10^3 \text{ N} + (1.80 \times 10^3 \text{ kg})(9.80 \text{ m/s}^2) \tag{A.743}$$
$$= 2.16 \times 10^4 \text{ N}.$$

Using the fact that **T** is in the same direction as **v**, we find that

$$P = \mathbf{Tv} = Tv = (2.16 \times 10^4 \text{ N})(3.00 \text{ m/s}) = 6.48 \times 10^4 \text{ W}. \tag{A.744}$$

7.9. Note, the block is 10 meters along an inclined plane, so we must find the vertical h first:

$$h = 10.0 \cdot \sin(30°) = 5.00 \text{ m}. \tag{A.745}$$

Then the potential energy of the block is

$$U = mgh = (100 \text{ kg}) \cdot (9.80 \text{ m/s}^2) \cdot (5.00 \text{ m}) = 4900 \text{ J} = 4.90 \text{ kJ}. \tag{A.746}$$

7.10. First convert units from cm to m:

$$x = 10.0 \text{ cm} = 0.100 \text{ m}. \tag{A.747}$$

The potential energy is calculated as

$$U = \frac{1}{2}kx^2. \tag{A.748}$$

From this,

$$U = \frac{1}{2}(800 \text{ N/m})(0.100 \text{ m})^2 = 4.00 \text{ J}. \tag{A.749}$$

7.11. Since the friction is neglected, the forces are conservative. Therefore, the elastic potential energy of the spring equals the kinetic energy of the steel ball. Thus, we can write

$$\frac{1}{2}kx^2 = \frac{1}{2}mv^2. \tag{A.750}$$

Replacing the values in SI units as given by the problem, we get

$$\frac{1}{2}(20.0 \times 10^3 \text{ N/m})(10.0 \times 10^{-2} \text{ m})^2 = \frac{1}{2}(20.0 \times 10^{-3} \text{ kg})v^2. \tag{A.751}$$

Solving it for v, we get

$$v = 100 \text{ m/s}. \tag{A.752}$$

7.12. We can chose a vertical y-axis, as shown in Fig. 7.12, and $\mathbf{j}$ is a unit vector along the y-axis. Therefore, we can write

$$\mathbf{F} = F\mathbf{j} \tag{A.753}$$
$$m\mathbf{g} = -mg\mathbf{j}. \tag{A.754}$$

Let $\mathbf{d}$ be the displacement vector of the mass m, thus

$$\mathbf{d} = h\mathbf{j}. \tag{A.755}$$

(a) The work done by the applied force $\mathbf{F}$ is

$$W_F = \mathbf{F} \cdot \mathbf{d} = (F\mathbf{j}) \cdot (h\mathbf{j}) = Fh. \tag{A.756}$$

(b) The work done by the gravity force is

$$W_g = m\mathbf{g} \cdot \mathbf{d} = (-mg\mathbf{j}) \cdot (h\mathbf{j}) = -mgh. \tag{A.757}$$

Since the person is holding the block at the height h, the total work is zero. That is,

$$W_g + W_F = 0. \tag{A.758}$$

Thus,

$$W_F = -W_g = mgh. \tag{A.759}$$

7.13. We can chose a coordinate system, as shown in Fig. 7.13, where **i** and **j** are unit vectors along the x- and y-axis, respectively. Therefore, we can write

$$\mathbf{F} = F\mathbf{j} \tag{A.760}$$

and

$$m\mathbf{g} = -mg\cos(90° - \theta)\mathbf{i} - mg\sin(90° - \theta)\mathbf{j} \tag{A.761}$$
$$= -mg\sin(\theta)\mathbf{i} - mg\cos(\theta)\mathbf{j}.$$

Let **d** be the displacement vector of the mass m, thus

$$\mathbf{d} = d\mathbf{i}. \tag{A.762}$$

(a) The work done by the gravity force is

$$W_g = m\mathbf{g} \cdot \mathbf{d} \tag{A.763}$$
$$= (-mg\sin(\theta)\mathbf{i} - mg\cos(\theta)\mathbf{j}) \cdot (d\mathbf{i})$$
$$= -mgd\sin(\theta).$$

Using the triangle filled in yellow in Fig. 7.13, we have

$$h = d\sin(\theta). \tag{A.764}$$

Therefore,

$$W_g = -mgh. \tag{A.765}$$

(b) Since the person is holding the block at the height h, the total work is zero. That is,

$$W_g + W_F = 0. \tag{A.766}$$

Thus,

$$W_F = -W_g = mgh. \tag{A.767}$$

Solutions Chapter 8

8.1. The meaning of the "rest" is that the velocity is zero. Thus, the initial and the final velocity and kinetic energy are zero. In this situation, the system consists of the two blocks, the spring, and Earth. We need to consider two forms of potential energy:

gravitational and elastic. Because the initial and final kinetic energies of the system are zero, $\Delta K = 0$, we can write

$$\Delta E = \Delta U_g + \Delta U_s \tag{A.768}$$

where ΔU_g is the change in the gravitational potential energy, and ΔU_s is the change in the potential energy of the block–spring system. Here

$$\Delta U_g = U_{gf} - U_{gi} \tag{A.769}$$

and

$$\Delta U_s = U_{sf} - U_{si}. \tag{A.770}$$

Since the mass m_2 falls a distance h, the mass m_1 moves the same distance horizontally to the right. Thus, the loss in energy due to the friction between mass m_1 and the table is

$$\Delta E = -f_k h = -\mu_k m_1 g h. \tag{A.771}$$

The change in the gravitational potential energy of the system is associated with only the falling block because the vertical coordinate of the horizontally sliding block does not change. Therefore, we obtain

$$\Delta U_g = U_{gf} - U_{gi} = 0 - m_2 g h = -m_2 g h \tag{A.772}$$

where the coordinates have been measured from the lowest position of the falling block.

The change in the elastic potential energy stored in the spring is

$$\Delta U_s = U_{sf} - U_{si} = \frac{1}{2} k h^2 - 0 = \frac{1}{2} k h^2. \tag{A.773}$$

Then

$$\Delta E = -\mu_k m_1 g h = -m_2 g h + \frac{1}{2} k h^2. \tag{A.774}$$

Solving this equation for μ_k, you get

$$\mu_k = \frac{m_2 g - \frac{1}{2} k h}{m_1 g}. \tag{A.775}$$

8.2. The only force that does work on the sphere is the gravitational force, as indicated in Fig. 8.5.

Because the gravitational force is conservative, the total mechanical energy of the pendulum–Earth system is constant, and hence, this can be considered as an energy

conservation problem. As the pendulum swings, the continuous transformation between potential and kinetic energy occurs. At the instant, the pendulum is released
(Point A), the energy of the system is entirely potential energy. At point B, the pendulum has kinetic energy, but the system has lost some potential energy. At C, the system has regained its initial potential energy, and the kinetic energy of the pendulum
is again zero.

We denote with y the coordinates of the sphere from the center of rotation, Point P.
Then

$$y_A = y_C = -L \cos \theta_A, \quad y_B = -L \tag{A.776}$$

the minus sign indicates that position is in the negative direction of y-axis. The potential energies at each position are

$$U_A = mgy_A = -mgL \cos \theta_A, \quad U_B = mgy_B = -mgL. \tag{A.777}$$

Applying the principle of conservation of mechanical energy to the system gives

$$K_A + U_A = K_B + U_B \tag{A.778}$$

or

$$0 - mgL \cos \theta_A = \frac{1}{2} mv_B^2 - mgL. \tag{A.779}$$

Solving eq. (A.779) for v_B, you get the speed at Point B as

$$v_B = \sqrt{2gL(1 - \cos \theta_A)}. \tag{A.780}$$

Since the cord cannot be extended, the tension force T is doing no work, we cannot
determine the tension using the energy method. To find T_B, we can apply Newton's
second law to the radial direction. First, recall that the centripetal acceleration a_r of
a particle moving in a circle is equal to v^2/r directed toward the center of rotation.
Because $r = L$ in this example, we obtain

$$\sum F_r = T_B - mg = ma_r = m\frac{v_B^2}{L} \tag{A.781}$$

or

$$T_B = mg + m\frac{2gL(1 - \cos \theta_A)}{L} = mg(3 - 2\cos \theta_A). \tag{A.782}$$

8.3.

(a) Because the ball is in free fall ($v_i = 0$), the only force acting on it is the gravitational force. Therefore, we apply the principle of conservation of mechanical energy to the ball-Earth system. Initially, the system has potential energy but no kinetic energy. As the ball falls, the total mechanical energy remains constant and equal to the initial potential energy of the system.

At the instant the ball is released, its kinetic energy is $K_i = 0$ and the potential energy of the system is $U_i = mgh$.

When the ball is at a distance y above the ground, the kinetic energy is $K_f = mv_f^2/2$ and the potential energy relative to the ground is $U_f = mgy$. Applying conservation law of energy, you get

$$K_i + U_i = K_f + U_f \tag{A.783}$$

or

$$0 + mgh = \frac{mv_f^2}{2} + mgy. \tag{A.784}$$

Solving eq. (A.784) for v_f, you get

$$v_f = \pm\sqrt{2g(h - y)}. \tag{A.785}$$

The speed is always positive, thus the speed is

$$v_f = \sqrt{2g(h - y)}. \tag{A.786}$$

If we had been asked to find the ball's velocity, we would use the negative value of the square root as the y component to indicate the downward motion:

$$v_f = -\sqrt{2g(h - y)}. \tag{A.787}$$

(b) In this case, v_i is not zero, thus

$$K_i = \frac{1}{2}mv_i^2. \tag{A.788}$$

Using

$$K_i + U_i = K_f + U_f \tag{A.789}$$

you can write

$$\frac{1}{2}mv_i^2 + mgh = \frac{1}{2}mv_f^2 + mgy. \tag{A.790}$$

Solving eq. (A.790) for v_f, you get the final speed as

$$v_f = \sqrt{v_i^2 + 2g(h - y)}. \tag{A.791}$$

8.4. Because $v_i = 0$, the initial kinetic energy at the top of the ramp is zero

$$K_i = 0. \tag{A.792}$$

If the y coordinate is measured from the bottom of the ramp, the final position is $y_f = 0$, and the potential energy is zero

$$U_f = 0. \tag{A.793}$$

Considering the upward direction being positive, then $y_i = 0.500\,\text{m}$. Therefore, the total mechanical energy of the crate–Earth system at the top is all potential energy:

$$E_i = K_i + U_i = 0 + mgy_i = mgy_i \tag{A.794}$$
$$= (3.00\,\text{kg})(9.80\,\text{m/s}^2)(0.500\,\text{m}) = 14.7\,\text{J}.$$

When the crate reaches the bottom of the ramp, the potential energy of the system is zero because the elevation of the crate is $y_f = 0$. Therefore, the total mechanical energy of the system when the crate reaches the bottom is all kinetic energy:

$$E_f = K_f + U_f = \frac{1}{2}mv_f^2 + 0 = \frac{1}{2}mv_f^2. \tag{A.795}$$

We cannot say that $E_i = E_f$ because a non-conservative force reduces the mechanical energy of the system: the force of kinetic friction acting on the crate. In this case,

$$\Delta E = E_f - E_i = W_f \tag{A.796}$$

where W_f is the work done by the friction force, which is

$$W_f = -f_k d \tag{A.797}$$

where d is the displacement along the ramp. You can get

$$W_f = -(5.00\,\text{N})(1.00\,\text{m}) = -5.00\,\text{J}. \tag{A.798}$$

Then finally

$$\frac{1}{2}mv_f^2 - 14.7\,\text{J} = -5.00\,\text{J} \tag{A.799}$$

or

$$v_f^2 = \frac{19.4\,\text{J}}{3.00\,\text{kg}} = 6.47\,\text{m}^2/\text{s}^2. \tag{A.800}$$

Therefore, you get the speed as

$$v_f = 2.54\,\text{m/s}. \tag{A.801}$$

8.5. Initially, the object is at rest, therefore, it has only potential energy, which is equal to

$$U_i = mgh. \tag{A.802}$$

Hence, the total mechanical energy is

$$E_i = U_i + K_i = mgh + 0 = mgh. \tag{A.803}$$

When it reaches the ground, the potential energy becomes zero and it owns only kinetic energy, K_{f1}, hence, the total mechanical energy when it reaches the ground is

$$E_{f1} = U_{f1} + K_{f1} = 0 + K_{f1} = K_{f1}. \tag{A.804}$$

Using the conservation law of energy, we get

$$E_i = E_{f1} \tag{A.805}$$

or

$$K_{f1} = m\frac{v_1^2}{2} = mgh. \tag{A.806}$$

Solving eq. (A.806) for the speed v_1, we get

$$v_1 = \sqrt{2gh}. \tag{A.807}$$

When the objects enters the second circle, at any height l (at an angle θ with vertical axis; see also Fig. 8.8), the potential energy due to the gravity is

$$U_\theta = mgl = mgR(1 - \cos\theta). \tag{A.808}$$

From the conservation law of the energy, U_θ equals the initial kinetic energy K_{f1}:

$$mgR(1 - \cos\theta) = mgh. \tag{A.809}$$

Equation (A.809) gives

$$\frac{h}{R} = 1 - \cos\theta. \tag{A.810}$$

At point B, $\theta = \pi$, thus, we get

$$\frac{h}{R} = 1 - \cos\pi = 2 \tag{A.811}$$

and at point A, $\theta = 3\pi/2$, we get

$$\frac{h}{R} = 1 - \cos\left(\frac{3}{2}\pi\right) = 1. \tag{A.812}$$

8.6. First, let us find the acceleration of the system using the conservation law of the energy. When the mass m_2 is at rest at the height h_2, the mass m_1 is at rest at some distance h_1, then the potential energy of the system is

$$U_i = U_1 + U_2 = m_1 g h_1 + m_2 g h_2 \qquad (A.813)$$

and the total energy is

$$E_i = U_i. \qquad (A.814)$$

When the mass m_2 hits the ground, then $U_{2f} = 0$, and the mass m_1 is at the height $h_1 + h_2$, hence, we get

$$U_f = m_1 g(h_1 + h_2). \qquad (A.815)$$

In order to find the kinetic energy of the system, first we determine the velocities using the equation

$$v_i^2 - v_f^2 = 2\mathbf{a} \cdot \Delta \mathbf{S}. \qquad (A.816)$$

For mass m_1:

$$v_{1f}^2 = 2 a_y h_2. \qquad (A.817)$$

Similarly, for mass m_2, we get

$$v_{2f}^2 = 2 a_y h_2. \qquad (A.818)$$

Therefore, we get

$$K_f = m_1 \frac{v_{1f}^2}{2} + m_2 \frac{v_{2f}^2}{2} = m_1 a_y h_2 + m_2 a_y h_2 = (m_1 + m_2) a_y h_2 \qquad (A.819)$$

where a_y is the projection of the acceleration of the mass m_1 along the y vertical axis. The total final mechanical energy is

$$E_f = U_f + K_f = m_1 g(h_1 + h_2) + a_y h_2(m_1 + m_2). \qquad (A.820)$$

Using the conservation law of energy:

$$E_i = E_f \qquad (A.821)$$

or

$$m_1 g h_1 + m_2 g h_2 = m_1 g(h_1 + h_2) + a_y h_2(m_1 + m_2). \qquad (A.822)$$

Solving eq. (A.822) for a_y, we get the projection of the acceleration along the y-axis as

$$a_y = \frac{g(m_2 - m_1)}{m_1 + m_2}. \qquad (A.823)$$

It can seen that if $m_2 > m_1$, then the system rotates clockwise, and hence $a_y > 0$. While if $m_2 < m_1$, then it rotates counterclockwise, thus $a_y < 0$.

8.7. First, we convert in SI units the stretch of the spring:

$$x = 10.0 \, \text{cm} = 0.10 \, \text{m}. \tag{A.824}$$

Then the elastic potential energy of the spring is

$$U = \frac{1}{2}kx^2 = \frac{1}{2}(800 \, \text{N/m})(0.10 \, \text{m})^2 = 4.00 \, \text{J}. \tag{A.825}$$

8.8. Since the block moves 10 meters along an inclined plane, the vertical h is

$$h = 10.0 \, \text{m} \sin 30° = 5.00 \, \text{m}. \tag{A.826}$$

The potential energy is given by

$$U = mgh = (100 \, \text{kg})(9.80 \, \text{m/s}^2)(5.00 \, \text{m}) = 490 \, \text{J}. \tag{A.827}$$

8.9. The work along the path (1) of the gravitational force is calculates as product of the force $m\mathbf{g} = -mg\mathbf{j}$ and the displacement $\mathbf{s} = -h\mathbf{j}$:

$$W_1 = (-mg\mathbf{j}) \cdot (-h\mathbf{j}) = mgh \tag{A.828}$$

where $\mathbf{j}$ is a unit vector and $\mathbf{j} \cdot \mathbf{j} = 1$.

While along the path (2), the force is the same (gravitational force) and the displacement can be written $\mathbf{s} = -d \sin \theta \mathbf{j}$, and hence

$$W_2 = (-mg\mathbf{j}) \cdot (-d \sin \theta \mathbf{j}) = mgd \sin \theta. \tag{A.829}$$

From Fig. A.30, $h = d \sin \theta$, then you obtain the work along the second path as

$$W_2 = mgh. \tag{A.830}$$

Therefore, $W_1 = W_2$, indicating that the work done by the gravitational force does not depend on the path, as expected since it is a conservative force.

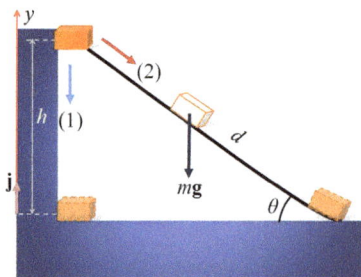

Figure A.30: Illustration of the movement of an object under the gravitational force $m\mathbf{g}$, which can either fall vertically or slide down an inclined frictionless surface.

8.10. Figure 8.10 illustrates the problem. The displacement vector is **d** and the gravitational force is *m***g**. The work done by the gravitational force is

$$W_g = (m\mathbf{g}) \cdot \mathbf{d} \tag{A.831}$$

where $\mathbf{g} = -g\mathbf{j}$, where **j** is a unit vector along the vertical direction, and the displacement vector can be written as

$$\mathbf{d} = \mathbf{h} + \mathbf{s} = h\mathbf{j} + s\mathbf{i} \tag{A.832}$$

where **i** is a unit vector along the positive x direction such that $\mathbf{i} \cdot \mathbf{j} = 0$. Therefore, W_g is

$$W_g = (-mg\mathbf{j}) \cdot (h\mathbf{j} + s\mathbf{i}) = -mgh. \tag{A.833}$$

8.11. Figure 8.11 illustrates the problem. The net work done on the object is

$$W = W_g + W_F. \tag{A.834}$$

Based on the work–kinetic energy theorem, you can write

$$W_g + W_F = \Delta K \tag{A.835}$$

where ΔK is the change in kinetic energy. Since the gravitational force is a conservative force,

$$W_g = -\Delta U. \tag{A.836}$$

Therefore, you obtain

$$W_F = \Delta E = E_f - E_i \tag{A.837}$$

where $E = U + K$ is the mechanical energy. The initial mechanical energy is zero $E_i = 0$, and the final mechanical energy is $E_f = mgh$ (only potential energy). Therefore,

$$W_F = mgh. \tag{A.838}$$

Solutions Chapter 9

9.1. You can denote with A the moment when the club first contacts the ball, B the moment when the club loses contact with the ball, and C when it is landing. Here, you can neglect air resistance. Assuming that the initial angle is $\theta_B = 45°$, then projection of the initial velocity along the horizontal axis is

$$v_{i,x} = v_B \cos \theta_B. \tag{A.839}$$

Then the distance traveled by the ball along the horizontal direction is

$$R = x_C - x_B = v_{i,x}t_{BC} = v_B \cos\theta_B t_{BC}. \tag{A.840}$$

To find the t_{BC}, we denote with D the point at maximum height along the vertical axis, where $v_y = 0$, since the motion along the y-axis is with acceleration $a_y = -g$, you can write

$$v_y = 0 = v_{i,y} + a_y t_{BD} \tag{A.841}$$

where

$$v_{i,y} = v_B \sin\theta_B \tag{A.842}$$

and $t_{BD} = t_{BC}/2$, thus

$$0 = v_B \sin\theta_B - gt_{BC}/2 \tag{A.843}$$

or

$$t_{BC} = \frac{2v_B \sin\theta_B}{g}. \tag{A.844}$$

Replacing it in the equation for R (eq. (A.840)), you get

$$R = \frac{v_B^2}{g}\sin(2\theta_B). \tag{A.845}$$

Solving eq. (A.845) for v_B, you get the speed at point B as

$$v_B = \sqrt{\frac{Rg}{\sin(2\theta_B)}} = \sqrt{(200\,\text{m})(9.80\,\text{m/s}^2)} = 44.3\,\text{m/s}. \tag{A.846}$$

If you consider as the interval of collision, the time between A and B, where, respectively,

$$v_i = v_A = 0; \quad v_f = v_B \tag{A.847}$$

for the ball, then you get the magnitude of the impulse as

$$I = \Delta p = mv_B - mv_A = (50 \times 10^{-3}\,\text{kg})(44.3\,\text{m/s}) - 0 = 2.22\,\text{kg} \cdot \text{m/s}. \tag{A.848}$$

9.2. You can calculate the impulse using the formula

$$\mathbf{I} = \Delta\mathbf{p} = \mathbf{p}_f - \mathbf{p}_i \tag{A.849}$$

where

$$\mathbf{p}_i = m\mathbf{v}_i = (1500 \text{ kg})(-15.0\mathbf{i} \text{ m/s}) = -2.25 \times 10^4 \mathbf{i} \, \frac{\text{kg} \cdot \text{m}}{\text{s}} \tag{A.850}$$

$$\mathbf{p}_f = m\mathbf{v}_i = (1500 \text{ kg})(2.60\mathbf{i} \text{ m/s}) = 0.39 \times 10^4 \mathbf{i} \, \frac{\text{kg} \cdot \text{m}}{\text{s}}. \tag{A.851}$$

Thus, the impulse is

$$\mathbf{I} = \Delta\mathbf{p} = 2.64 \times 10^4 \mathbf{i} \, \frac{\text{kg} \cdot \text{m}}{\text{s}}. \tag{A.852}$$

The average force acting upon the car is

$$\mathbf{F} = \frac{\Delta\mathbf{p}}{\Delta t} = \frac{2.64 \times 10^4 \mathbf{i} \, \frac{\text{kg} \cdot \text{m}}{\text{s}}}{0.150 \text{ s}} = 1.76 \times 10^5 \mathbf{i} \text{ N}. \tag{A.853}$$

9.3. From the problem, for the first car, you can write:

$$m_1 = 1800 \text{ kg}; \quad v_{i,1} = 0 \tag{A.854}$$

where $v_{i,1}$ is the initial velocity. Therefore, the initial momentum of the car 1 is

$$p_{i,1} = 0. \tag{A.855}$$

For the second car,

$$m_2 = 900 \text{ kg}; \quad v_{i,2} = 20.0 \text{ m/s} \tag{A.856}$$

and hence the initial momentum is

$$p_{i,2} = m_2 v_{i,2} = (900 \text{ kg})(20.0 \text{ m/s}) = 1.80 \times 10^4 \text{ kg} \cdot \text{m/s}. \tag{A.857}$$

After the collision, since the two cars become entangled, then the momentum of the system after collision is

$$p_f = (m_1 + m_2)v_f = (2700 \text{ kg})v_f. \tag{A.858}$$

Using the conservation law of the momentum:

$$p_{i,1} + p_{i,2} = p_f. \tag{A.859}$$

Replacing the numerical values, you get

$$0 + 1.80 \times 10^4 \text{ kg} \cdot \text{m/s} = 2700 v_f \tag{A.860}$$

or

$$v_f = \frac{1.80 \times 10^4 \text{ kg} \cdot \text{m/s}}{2.70 \times 10^2 \text{ kg}} = 6.67 \text{ m/s}. \tag{A.861}$$

9.4. We can label the masses of the particles, as shown in Fig. 9.10, with $m_1 = m_2 = 1.0$ kg and $m_3 = 2.0$ kg. Using the basic defining equations for the coordinates of the center of mass and noting that $z_{c.m.} = 0$, we obtain

$$x_{c.m.} = \frac{\sum_{i=1}^3 m_i x_i}{\sum_{i=1}^3 m_i} \tag{A.862}$$
$$= \frac{(1.0\,\text{kg})(1.0\,\text{m}) + (1.0\,\text{kg})(2.0\,\text{m}) + (2.0\,\text{kg})(0.0\,\text{m})}{1.0\,\text{kg} + 1.0\,\text{kg} + 2.0\,\text{kg}}$$
$$= \frac{3.0\,\text{kg} \cdot \text{m}}{4.0\,\text{kg}} = 0.75\,\text{m}$$

$$y_{c.m.} = \frac{\sum_{i=1}^3 m_i y_i}{\sum_{i=1}^3 m_i} \tag{A.863}$$
$$= \frac{(1.0\,\text{kg})(0.0\,\text{m}) + (1.0\,\text{kg})(0.0\,\text{m}) + (2.0\,\text{kg})(2.0\,\text{m})}{1.0\,\text{kg} + 1.0\,\text{kg} + 2.0\,\text{kg}}$$
$$= \frac{4.0\,\text{kg} \cdot \text{m}}{4.0\,\text{kg}} = 1.0\,\text{m}.$$

Therefore, the position vector of the center of mass measured with respect to the origin O is

$$\mathbf{r}_{c.m.} = x_{c.m.}\mathbf{i} + y_{c.m.}\mathbf{j} = (0.75\,\text{m})\mathbf{i} + (1.0\,\text{m})\mathbf{j}. \tag{A.864}$$

9.5. We will assume that the rod is aligned along the x-axis, hence, $y_{c.m.} = z_{c.m.} = 0$. Furthermore, if we denote the mass per unit length μ (this quantity is called the *linear mass density*), then

$$\mu = \frac{M}{L} \tag{A.865}$$

for the uniform rod as assumed in our case. If we divide the rod into elements of length dx, then the mass of each element is

$$dm = \mu dx. \tag{A.866}$$

The center of mass coordinate $x_{c.m.}$ is

$$x_{c.m.} = \frac{\int x\,dm}{\int dm} \tag{A.867}$$
$$= \frac{1}{M}\int_0^L x\mu\,dx$$
$$= \frac{1}{M}\mu\int_0^L x\,dx$$

$$= \frac{\mu L^2}{2M}.$$

Replacing μ, we get

$$x_{\text{c.m.}} = \frac{L}{2}. \tag{A.868}$$

9.6. Assume the total mass of the rocket is M. Then the mass of each fragment is $M/3$. Because the forces of the explosion are internal to the system and cannot affect its total momentum, the total momentum $\mathbf{p}_i$ of the rocket just before the blast must equal the total momentum $\mathbf{p}_f$ of the fragments right after the explosion.

Before the explosion

$$\mathbf{p}_i = M\mathbf{v}_i = (300M\mathbf{j}) \, \text{m/s}. \tag{A.869}$$

After the explosion

$$\mathbf{p}_f = \frac{M}{3}(240\mathbf{i}) \, \text{m/s} + \frac{M}{3}(450\mathbf{j}) \, \text{m/s} + \frac{M}{3}\mathbf{v}_f \tag{A.870}$$

where $\mathbf{v}_f$ is unknown velocity of the third fragment. Since $\mathbf{p}_i = \mathbf{p}_f$, we get

$$(300M\mathbf{j}) \, \text{m/s} = \frac{M}{3}(240\mathbf{i}) \, \text{m/s} + \frac{M}{3}(450\mathbf{j}) \, \text{m/s} + \frac{M}{3}\mathbf{v}_f. \tag{A.871}$$

Solving it for $\mathbf{v}_f$, we get

$$\mathbf{v}_f = (-240\mathbf{i} + 450\mathbf{j}) \, \text{m/s}. \tag{A.872}$$

9.7. First, we calculate the change on the momentum of the car and so of the passenger assuming the car moves in a straight line along the x-axis:

$$\Delta\mathbf{p} = \mathbf{p}_f - \mathbf{p}_i = \mathbf{p}_f = mv_f\mathbf{i} = (70.0 \, \text{kg})(5.20 \, \text{m/s})\mathbf{i} \tag{A.873}$$
$$= (364\mathbf{i}) \, \text{kg} \cdot \text{m/s}.$$

The linear impulse is

$$\mathbf{I} = \Delta\mathbf{p} = (364\mathbf{i}) \, \text{kg} \cdot \text{m/s}. \tag{A.874}$$

The average force exerted on the passenger is

$$\bar{\mathbf{F}} = \frac{\mathbf{I}}{\Delta t} = \frac{(364\mathbf{i}) \, \text{kg} \cdot \text{m/s}}{0.832 \, \text{s}} \approx 438 \, \text{N}. \tag{A.875}$$

9.8. Taking the horizontal direction as the x-axis and the position of the tennis as the origin of the axis, then the momentum of the ball before hit is

$$\mathbf{p}_i = m\mathbf{v}_i = (0.0600 \, \text{kg})(-50.0\mathbf{i} \, \text{m/s}) = -(3.00\mathbf{i}) \, \text{kg} \cdot \text{m/s} \tag{A.876}$$

the minus sign indicates that the ball is traveling initially in the negative direction of the x-axis. After the hit, the momentum of the ball is

$$\mathbf{p}_f = m\mathbf{v}_f = (0.0600\ \text{kg})(40.0\mathbf{i}\ \text{m/s}) = (2.40\mathbf{i})\ \text{kg} \cdot \text{m/s}. \tag{A.877}$$

The change on the momentum of the ball, which equals the impulse, is

$$\mathbf{I} = \Delta\mathbf{p} = \mathbf{p}_f - \mathbf{p}_i = (2.40\mathbf{i})\ \text{kg} \cdot \text{m/s} - (-(3.00\mathbf{i})\ \text{kg} \cdot \text{m/s}) \tag{A.878}$$
$$= (5.40\mathbf{i})\ \text{kg} \cdot \text{m/s}.$$

9.9. First, let us write the momentum of the ball before it strikes the wall:

$$\mathbf{p}_i = m\mathbf{v}_i = (mv_i \cos 60°\mathbf{i}) + (mv_i \sin 60°\mathbf{j}) \tag{A.879}$$
$$= (3.00\ \text{kg})\left(10.0\ \frac{\text{m}}{\text{s}}\right)(0.500)\mathbf{i} + (3.00\ \text{kg})\left(10.0\ \frac{\text{m}}{\text{s}}\right)(0.866)\mathbf{j}$$
$$= (15.0\mathbf{i} + 26.0\mathbf{j})\ \text{kg} \cdot \text{m/s}.$$

The momentum of the ball after it strikes the wall is

$$\mathbf{p}_f = m\mathbf{v}_f = (-mv_f \sin 60°\mathbf{i}) + (mv_f \cos 60°\mathbf{j}) \tag{A.880}$$
$$= -(3.00\ \text{kg})\left(10.0\ \frac{\text{m}}{\text{s}}\right)(0.866)\mathbf{i} + (3.00\ \text{kg})\left(10.0\ \frac{\text{m}}{\text{s}}\right)(0.500)\mathbf{j}$$
$$= (-26.0\mathbf{i} + 15.0\mathbf{j})\ \text{kg} \cdot \text{m/s}$$

where $v_i = v_f = 10.0\ \text{m/s}$ and the minus sign indicates that the ball is moving in the negative direction of the x-axis after it hits the wall.

The change on the momentum of the ball, which equals the impulse gained by the ball is

$$\mathbf{I} = \Delta\mathbf{p} = \mathbf{p}_f - \mathbf{p}_i = (-41.0\mathbf{i} - 11.0\mathbf{j})\ \text{kg} \cdot \text{m/s}. \tag{A.881}$$

The average force exerted on the ball by the wall is

$$\bar{\mathbf{F}} = \frac{\mathbf{I}}{\Delta t} = \frac{(-41.0\mathbf{i} - 11.0\mathbf{j})\ \text{kg} \cdot \text{m/s}}{0.200\ \text{s}} \tag{A.882}$$
$$= -(205\mathbf{i} + 55\mathbf{j})\ \text{N}$$

and its magnitude is

$$\bar{F} = \sqrt{(205)^2 + (55)^2} \approx 212\ \text{N}. \tag{A.883}$$

9.10. Let us first determine the total momentum before the collision of the system as

$$\mathbf{p}_i = \mathbf{p}_{1i} + \mathbf{p}_{2i} = m_1\mathbf{v}_{1i} \tag{A.884}$$

since the mass m_2 is at rest. Mass m_1 initially has a potential energy

$$U_{1i} = m_1gh = (5.00 \text{ kg})(9.80 \text{ m/s}^2)(5.00 \text{ m}) \qquad \text{(A.885)}$$
$$= 245 \text{ J}$$

which is transferred into kinetic energy completely at point B based on the conservation law of the energy, therefore,

$$U_{1i} = \frac{m_1v_{1i}^2}{2} \qquad \text{(A.886)}$$

solving it for v_{1i}, we get the initial speed of the mass m_1 as

$$v_{1i} \approx 9.90 \text{ m/s}. \qquad \text{(A.887)}$$

Thus, the initial momentum of the system is

$$\mathbf{p}_i = m_1v_{1i}\mathbf{i} \approx (49.5\mathbf{i}) \text{ kg} \cdot \text{m/s}. \qquad \text{(A.888)}$$

The kinetic energy before the collision is

$$K_i = \frac{m_1v_{1i}^2}{2} = 245 \text{ J}. \qquad \text{(A.889)}$$

After the collision, the total momentum is

$$\mathbf{p}_f = \mathbf{p}_{1f} + \mathbf{p}_{2f} = -m_1v_{1f}\mathbf{i} + m_2v_{2f}\mathbf{i} \qquad \text{(A.890)}$$

and the kinetic energy is

$$K_f = \frac{m_1v_{1f}^2}{2} + \frac{m_2v_{2f}^2}{2}. \qquad \text{(A.891)}$$

Since the collision is elastic, the conservation law of the kinetic energy and momentum gives

$$(49.5\mathbf{i}) \text{ kg} \cdot \text{m/s} = -m_1v_{1f}\mathbf{i} + m_2v_{2f}\mathbf{i} \qquad \text{(A.892)}$$

$$245 \text{ J} = \frac{m_1v_{1f}^2}{2} + \frac{m_2v_{2f}^2}{2} \qquad \text{(A.893)}$$

or

$$9.90 = -v_{1f} + 2.00v_{2f} \qquad \text{(A.894)}$$
$$98.0 = v_{1f}^2 + 2.00v_{2f}^2. \qquad \text{(A.895)}$$

Solution of these two equations for the speeds v_{1f} and v_{2f}, we get

$$v_{1f} \approx 3.30 \text{ m/s} \qquad \text{(A.896)}$$

$$v_{2f} \approx 6.60 \text{ m/s}. \tag{A.897}$$

The kinetic energy just after the collision of the first mass is

$$K_{1f} = \frac{m_1 v_{1f}^2}{2} = \frac{(5.00 \text{ kg})(3.3 \text{ m/s})^2}{2} \tag{A.898}$$

$$\approx 27.2 \text{ J}.$$

Based on the conservation law of the mechanical energy, when the mass m_1 reaches the highest point this kinetic energy is completely transferred into gravitational potential energy, $U = m_1 g h'$, hence

$$m_1 g h' = K_{1f}. \tag{A.899}$$

Solving it for h', we obtain

$$h' = \frac{K_{1f}}{m_1 g} = \frac{27.2 \text{ J}}{(5.00 \text{ kg})(9.80 \text{ m/s}^2)} \tag{A.900}$$

$$= 0.556 \text{ m}.$$

Solutions Chapter 10

10.1.

(a) You can use the following equation for angular displacement:

$$\theta_f - \theta_i = \omega_i t + \frac{1}{2} \alpha t^2. \tag{A.901}$$

Replacing the numerical values:

$$\theta_f - \theta_i = (2.00 \text{ rad/s})(2.00 \text{ s}) + (1/2)(3.50 \text{ rad/s}^2)(2.00 \text{ s})^2 \tag{A.902}$$

$$= 11.0 \text{ rad}.$$

(b) The angular speed at $t = 2.00$ s is calculated as:

$$\omega_f = \omega_i + \alpha t = 2.00 \text{ rad/s} + (3.50 \text{ rad/s}^2)(2.00 \text{ s}) \tag{A.903}$$

$$= 9.00 \text{ rad/s}.$$

10.2.

(a) Because each atom is at a distance $d/2$ from the z-axis, the moment of inertia about that axis is

$$I = \sum_i m_i r_i^2 = m(d/2)^2 + m(d/2)^2 = (1/2)md^2 \tag{A.904}$$

$$= (1/2)(2.66 \times 10^{-26} \text{ kg})(1.21 \times 10^{-10} \text{ m})^2$$

$$= 1.95 \times 10^{-46} \text{ kg} \cdot \text{m}^2.$$

(b) The kinetic energy is

$$K_R = \frac{1}{2}I\omega^2 \tag{A.905}$$

$$= \frac{1}{2}(1.95 \times 10^{-46}\ \text{kg} \cdot \text{m}^2)(4.60 \times 10^{12}\ \text{rad/s})^2$$

$$= 2.06 \times 10^{-21}\ \text{J}.$$

10.3. First, note that the two spheres of mass m, which lie on the y-axis, do not contribute to I_y (that is, $r_i = 0$ for these spheres about this axis). Then, the moment of inertia is

$$I_y = \sum_i m_i r_i^2 = Ma^2 + Ma^2 = 2Ma^2. \tag{A.906}$$

Therefore, the rotational kinetic energy about the y-axis is

$$K_r = I\omega^2/2 = Ma^2\omega^2. \tag{A.907}$$

10.4. Consider a shaded length element dx with a mass dm:

$$dm = \mu dx = \frac{M}{L}dx \tag{A.908}$$

where $\mu = M/L$ is the mass per unit length.
(a) The moment of inertia about the y-axis is

$$I_y = \int r^2 dm = \int_{-L/2}^{L/2} x^2 \frac{M}{L}dx \tag{A.909}$$

$$= \frac{M}{L}\left[\frac{x^3}{3}\right]_{-L/2}^{L/2}$$

$$= \frac{1}{12}ML^2.$$

(b) Because the distance between the center of mass axis and the y'-axis is $d = L/2$, the parallel-axis theorem gives

$$I_{y'} = I_y + Md^2 = \frac{1}{12}ML^2 + \frac{1}{4}ML^2 = \frac{1}{3}ML^2. \tag{A.910}$$

10.5. All mass elements dm are at the same distance $r = R$ from the axis; thus, we obtain the moment of inertia about the z-axis through O:

$$I_z = \int r^2 dm = R^2 \int dm = MR^2 \tag{A.911}$$

which is the same as that of a single particle of mass M located a distance R from the axis of rotation.

10.6. We can divide the cylinder into small cylindrical shells, with radius r, thickness dr, and length L, as shown in Fig. 10.15. The elementary volume dV of one shell like this is its cross-sectional area multiplied by its length:

$$dV = (2\pi r dr)L. \tag{A.912}$$

The mass per unit of volume is (where $V = \pi R^2 L$ is the cylinder's volume)

$$\rho = \frac{M}{V} = \frac{M}{\pi R^2 L}. \tag{A.913}$$

The moment of inertia is calculated as

$$I = \rho \int r^2 dV = \frac{M}{\pi R^2 L} \int_0^R r^2 2\pi r L dr \tag{A.914}$$

$$= \frac{2M}{R^2} \int_0^R r^3 dr$$

$$= \frac{2M}{R^2} \frac{R^4}{4} = \frac{1}{2} MR^2.$$

It is interesting to note that the result does not depend on L, the length of the cylinder. Therefore, it applies equally well to an infinitely long cylinder and a flat disc.

10.7. We use the law of conservation of the mechanical energy of the system. When the rod is horizontal, it has no rotational energy, but only potential energy due to gravitation, which is equal to

$$E_i = U_i = \frac{1}{2} MgL. \tag{A.915}$$

When the rod reaches its lowest position, the potential energy is zero, and so the mechanical energy is entirely rotational:

$$E_f = K_R = \frac{1}{2} I\omega^2 \tag{A.916}$$

where the moment of inertia I is

$$I = \frac{1}{3} ML^2 \tag{A.917}$$

given by eq. (A.910).
Therefore, we get

$$E_f = \frac{1}{6} ML^2 \omega^2. \tag{A.918}$$

From the conservation law of energy:

$$E_i = E_f \tag{A.919}$$

or

$$\frac{1}{2}MgL = \frac{1}{6}ML^2\omega^2. \tag{A.920}$$

Solving it for ω, we get the angular speed as

$$\omega = \sqrt{\frac{3g}{L}}. \tag{A.921}$$

10.8. We will use the law of conservation of the mechanical energy to solve the problem. The system is composed of two cylinders, pulley, and Earth. Initially, the mechanical energy of the system is

$$E_i = U_i + K_i \tag{A.922}$$

where $K_i = 0$, because the system is at rest, and

$$U_i = m_2gh. \tag{A.923}$$

We are neglecting the potential energy of the pulley because the pulley does not move and so its potential energy will be the same. Thus,

$$E_i = m_2gh. \tag{A.924}$$

The final mechanical energy of the system is

$$E_f = U_f + K_f \tag{A.925}$$

where

$$K_f = \frac{1}{2}m_1v_f^2 + \frac{1}{2}m_2v_f^2 + \frac{1}{2}I\omega_f^2 \tag{A.926}$$

where term 1 is the kinetic energy of the first cylinder, term 2 is the kinetic energy of the second cylinder, and the third term is rotational kinetic energy of the pulley. Here, v_f is the same for both cylinders and $v_f = R\omega_f$. Therefore, we get

$$K_f = \frac{1}{2}\left(m_1 + m_2 + \frac{I}{R^2}\right)v_f^2. \tag{A.927}$$

The potential energy is

$$U_f = m_1gh. \tag{A.928}$$

From the conservation law of energy, we obtain

$$m_2 gh = m_1 gh + \frac{1}{2}\left(m_1 + m_2 + \frac{I}{R^2}\right)v_f^2$$

(A.929)

which can be solved for v_f, to get the final linear speed as

$$v_f = \sqrt{\frac{(m_2 - m_1)gh}{\frac{1}{2}(m_1 + m_2 + \frac{I}{R^2})}}.$$

(A.930)

The angular velocity is

$$\omega_f = \frac{v_f}{R} = \frac{1}{R}\sqrt{\frac{(m_2 - m_1)gh}{\frac{1}{2}(m_1 + m_2 + \frac{I}{R^2})}}.$$

(A.931)

10.9.
(a) We use the formula

$$\omega_f = \omega_i + \alpha t$$

(A.932)

where $\omega_i = 0$, thus

$$\omega_f = \alpha t.$$

(A.933)

Solving it for α, we get the magnitude of angular acceleration

$$\alpha = \frac{\omega_f}{t} = \frac{12.0\,\text{rad/s}}{3.00\,\text{s}} = 4.00\,\text{rad/s}^2.$$

(A.934)

(b) To find the angular displacement, we use the following equation:

$$\theta_f - \theta_i = \omega_i t + \frac{1}{2}\alpha t^2.$$

(A.935)

Replacing the numerical values, we get

$$\theta_f - \theta_i = 0 + \frac{1}{2}(4.00\,\text{rad/s}^2)(3.00\,\text{s})^2 = 18.0\,\text{rad}.$$

(A.936)

10.10.
(a) The angular position at $t = 0$ is

$$\theta(t = 0) = (5.00 + 10.0 \cdot 0 + 2.00(0)^2)\,\text{rad} = 5.00\,\text{rad}.$$

(A.937)

The angular speed is

$$\omega = \frac{d\theta}{dt} = (10.0 + 4.00t)\,\text{rad/s}.$$

(A.938)

Hence, we get

$$w(t = 0) = 10.0 \, \text{rad/s}. \tag{A.939}$$

The angular acceleration is

$$\alpha = \frac{dw}{dt} = 4.00 \, \text{rad/s}^2. \tag{A.940}$$

(b) At $t = 3.00$ s, we get

$$\theta(t = 3.00 \, \text{s}) = (5.00 + 10.0 \cdot 3.00 + 2.00(3.00)^2) \, \text{rad} = 53.0 \, \text{rad} \tag{A.941}$$

$$w(t = 3.00 \, \text{s}) = (10.0 + 4.00(3.00)) \, \text{rad/s} = 22.0 \, \text{rad/s} \tag{A.942}$$

$$\alpha(t = 3.00 \, \text{s}) = 4.00 \, \text{rad/s}^2. \tag{A.943}$$

Therefore, we can say that this is a rotation motion with constant angular acceleration.

Solutions Chapter 11

11.1. Note that accelerated rolling motion is possible only if a frictional force is present between the sphere and the incline to produce a net torque about the center of mass. Despite the presence of friction, no loss of mechanical energy occurs because the contact point is at rest relative to the surface at any instant. On the other hand, if the sphere was to slip, mechanical energy would be lost as motion progressed due to the friction with the rough surface.

The system is composed of the Earth and the sphere.

Initially, the system's total mechanical energy is

$$E_i = K_i + U_i = U_i = Mgh \tag{A.944}$$

where K_i its initial kinetic energy, which is zero because the system is at rest, and Mgh is the potential energy of the system due to gravitational force.

The final mechanical energy of system is

$$E_f = K_f + U_f = K_f \tag{A.945}$$

because $U_f = 0$. The kinetic energy K_f is

$$K_f = \frac{1}{2}I_{\text{c. m.}}w^2 + \frac{1}{2}Mv_{\text{c. m.}}^2 \tag{A.946}$$

where $v_{\text{c. m.}} = Rw$, thus, we write the kinetic energy as

$$K_f = \frac{1}{2}\left(\frac{I_{\text{c. m.}}}{R^2} + M\right)v_{\text{c. m.}}^2 \tag{A.947}$$

where $I_{c.\,m.}$ is the moment of inertia of the sphere about the axis through the center of mass.

Applying the conservation law of energy:

$$E_i = E_f \tag{A.948}$$

we get

$$Mgh = \frac{1}{2}\left(\frac{I_{c.\,m.}}{R^2} + M\right)v_{c.\,m.}^2. \tag{A.949}$$

Solving eq. (A.949) for $v_{c.\,m.}$, we get

$$v_{c.\,m.} = \sqrt{\frac{Mgh}{\frac{1}{2}(\frac{I_{c.\,m.}}{R^2} + M)}} = \sqrt{\frac{2gh}{\frac{I_{c.\,m.}}{MR^2} + 1}}. \tag{A.950}$$

11.2.

(a) From the definition, the angular momentum **L** is given by

$$\mathbf{L} = \mathbf{r} \times \mathbf{p} \tag{A.951}$$

where linear momentum $\mathbf{p} = m\mathbf{v}$. Since **v** changes its direction with time, but not the magnitude, we expect angular momentum to change with time its direction, but the magnitude remains constant. Thus, the magnitude L is

$$L = mvr\sin 90° = mvr. \tag{A.952}$$

The direction of **L** can be found by applying the right-hand rule:

$$\mathbf{L} = (mvr)\mathbf{k}. \tag{A.953}$$

That is, if the particle was to move counterclockwise, **k** (and so **L**) would point upward (along the positive direction of z-axis) and into the page.

(b) Using the relation

$$v = r\omega \tag{A.954}$$

we get

$$L = mr^2\omega = I\omega \tag{A.955}$$

where I is the moment of inertia of particle about z-axis through O. Assuming the rotation is counterclockwise, the direction of ω is along the z-axis. The direction of **L** is the same as that of ω, and so we can write the angular momentum as

$$\mathbf{L} = I\boldsymbol{\omega} = I\omega\mathbf{k}. \tag{A.956}$$

11.3. If we consider the bowling ball as a solid sphere, then the moment of inertia about vertical axes through center of the sphere is

$$I = \frac{2}{5}mR^2 = \frac{2}{5}(6.0\,\text{kg})(0.12\,\text{m})^2 \approx 0.035\,\text{kg} \cdot \text{m}^2. \tag{A.957}$$

The magnitude of angular momentum is calculated as

$$L = I\omega \tag{A.958}$$

where angular velocity is

$$\omega = 10\,\text{rev/s} = (10\,\text{rev/s})(2\pi\,\text{rad/rev}) \approx 63\,\text{rad/s}. \tag{A.959}$$

Replacing the numerical values at expression for L, we get

$$L = (0.035\,\text{kg} \cdot \text{m}^2)(63\,\text{rad/s}) \approx 2.2\,\text{kg} \cdot \text{m}^2/\text{s}. \tag{A.960}$$

11.4.

(a) The rigid body is composed of three bodies: masses m_1 and m_2 and the rod. Therefore, the moment of inertia about the z-axis through O is

$$I = I_{\text{rod}} + I_1 + I_2 \tag{A.961}$$

where

$$I_{\text{rod}} = \frac{1}{12}Md^2; \quad I_1 = m_1\left(\frac{d}{2}\right)^2; \quad I_2 = m_2\left(\frac{d}{2}\right)^2. \tag{A.962}$$

Therefore, we get

$$I = \frac{1}{12}Md^2 + m_1\left(\frac{d}{2}\right)^2 + m_2\left(\frac{d}{2}\right)^2 = \frac{d^2}{4}\left(\frac{M}{3} + m_1 + m_2\right). \tag{A.963}$$

The magnitude of the angular momentum is

$$L = I\omega = \frac{d^2}{4}\left(\frac{M}{3} + m_1 + m_2\right)\omega. \tag{A.964}$$

(b) Upon two masses are acting the gravitational forces, with magnitudes, respectively, as

$$F_{1g} = m_1g; \quad F_{2g} = m_2g \tag{A.965}$$

which will cause the rod to rotate, since there is a net torque different from zero. Note that, if the masses of the two particles were equal, then the system will have no angular acceleration because the net torque on the system is zero when $m_1 = m_2$. Moreover, if initially the angle $\theta = \pm\pi/2$, then the rod will be at equilibrium. Let us calculate the net torque at any angle θ. Due to the force F_{1g}, the torque is

$$T_1 = +F_{1g}\frac{d}{2}\cos\theta = +m_1g\frac{d}{2}\cos\theta \qquad (A.966)$$

the sign plus indicates that this force is rotating the rod counterclockwise. The torque due to the force F_{2g} is

$$T_2 = -F_{2g}\frac{d}{2}\cos\theta = -m_2g\frac{d}{2}\cos\theta \qquad (A.967)$$

the sign minus indicates that this force is rotating the rod clockwise. The net torque on the system is then

$$\tau = T_1 + T_2 = (m_1 - m_2)g\frac{d}{2}\cos\theta. \qquad (A.968)$$

If $m_1 > m_2$, then the direction of τ is out of page, and if $m_2 > m_1$, it is into the page. To find the angular acceleration, we use this equation:

$$\tau = I\alpha \qquad (A.969)$$

or

$$\alpha = \frac{\tau}{I} = \frac{2(m_1 - m_2)g\cos\theta}{d(\frac{M}{3} + m_1 + m_3)}. \qquad (A.970)$$

If $\theta = \pm\pi/2$, then $\alpha = 0$, so the rod is at equilibrium. If $\theta = 0, \pi$, then α has maximum magnitude.

11.5. Let us first calculate the angular momentum about an axis that coincides with the axle of the pulley. At any instant of time the two objects have the same speed v, the angular momentum of the block m_1 is

$$L_1 = |\mathbf{r} \times \mathbf{p}_1| = m_1vR \qquad (A.971)$$

and that of the block m_2 is

$$L_2 = |\mathbf{r} \times \mathbf{p}_2| = m_2vR. \qquad (A.972)$$

The angular momentum of the pulley is

$$L_3 = I\omega = Iv/R. \qquad (A.973)$$

The total angular momentum of the system is

$$L = L_1 + L_2 + L_3 = m_1vR + m_2vR + I\frac{v}{R}. \qquad (A.974)$$

Now let us determine the net external torque acting on the system about the pulley axle. Forces on the system that contribute to the net external torque are the force exerted by the axle on the pulley, which does not contribute since the moment arm is zero; gravity force acting on the mass m_1, $\mathbf{F}_{1g} = m_1\mathbf{g}$, which rotates the pulley counterclockwise, thus its torque is

$$\tau_1 = m_1 g R \tag{A.975}$$

where R is the moment arm; the normal force $\mathbf{N}$ acting on the block m_2 is balanced by the force of gravity $\mathbf{F}_{2g} = m_2\mathbf{g}$, and so these forces do not contribute to the torque.
Therefore, the net external torque is

$$\tau = \tau_1 = m_1 g R. \tag{A.976}$$

Using the equation

$$\tau = \frac{dL}{dt} \tag{A.977}$$

we get

$$m_1 g R = m_1 R \frac{dv}{dt} + m_2 R \frac{dv}{dt} + \frac{I}{R}\frac{dv}{dt} = \frac{dv}{dt}\left(m_1 R + m_2 R + \frac{I}{R}\right). \tag{A.978}$$

Solving it for $a = dv/dt$, which is the linear acceleration, we get

$$a = \frac{dv}{dt} = \frac{m_1 g}{(m_1 + m_2) + \frac{I}{R^2}}. \tag{A.979}$$

11.6. We assume that during the collapse of the stellar core, no torque acts on it, it remains spherical, and it does not change its mass. Let T be the period, such that T_i is the initial period of the star and T_f is the period of the neutron star. The angular speed of the star is

$$\omega = \frac{2\pi}{T} \tag{A.980}$$

and the moment of inertia is

$$I = \frac{2}{5}mr^2 \tag{A.981}$$

where m is the mass and r the radius of the star. Using the conservation law of the angular momentum, since there is no torque, we get

$$L_i = L_f \tag{A.982}$$

where

$$L_i = I\omega_i = \frac{2}{5}mr_i^2\frac{2\pi}{T_i}; \quad L_f = I\omega_f = \frac{2}{5}mr_f^2\frac{2\pi}{T_f}. \tag{A.983}$$

Therefore, we get

$$\frac{r_i^2}{T_i} = \frac{r_f^2}{T_f} \tag{A.984}$$

or

$$T_f = \left(\frac{r_f}{r_i}\right)^2 T_i = \left(\frac{3.0\,\text{km}}{1.0 \times 10^4\,\text{km}}\right)^2 (30\,\text{days}) \tag{A.985}$$

$$= 27 \times 10^{-7}\,\text{days} = 0.23\,\text{s}.$$

11.7. We know that the angular momentum is

$$\mathbf{L} = \mathbf{r} \times \mathbf{p} = \det \begin{pmatrix} \mathbf{i} & \mathbf{j} & \mathbf{k} \\ x & y & z \\ mv_x & mv_y & mv_z \end{pmatrix} = \det \begin{pmatrix} \mathbf{i} & \mathbf{j} & \mathbf{k} \\ 1.50 & 2.20 & 0.00 \\ 6.30 & -5.40 & 0.00 \end{pmatrix} \tag{A.986}$$

$$= (-1.50 \cdot 5.40 - 2.20 \cdot 6.30)\mathbf{k}$$

$$\approx -(22.0\mathbf{k})\,\text{kg} \cdot \text{m}^2/\text{s}.$$

Here, the minus sign indicates that the direction of $\mathbf{L}$ is along the negative direction of z-axis.

11.8. We know that the angular momentum is

$$\mathbf{L} = \mathbf{r} \times \mathbf{p} = m\mathbf{r} \times \mathbf{v} \tag{A.987}$$

where the velocity $\mathbf{v}$ is

$$\mathbf{v} = \frac{d\mathbf{r}}{dt} = (5.00\mathbf{j})\,\text{m/s}. \tag{A.988}$$

Then the angular momentum is

$$\mathbf{L} = \mathbf{r} \times \mathbf{p} = \det \begin{pmatrix} \mathbf{i} & \mathbf{j} & \mathbf{k} \\ x & y & z \\ mv_x & mv_y & mv_z \end{pmatrix} \tag{A.989}$$

$$= (2.00\,\text{kg}) \cdot \det \begin{pmatrix} \mathbf{i} & \mathbf{j} & \mathbf{k} \\ 6.00 & 5.00t & 0.00 \\ 0.00 & 5.00 & 0.00 \end{pmatrix}$$

$$= (60.0\mathbf{k})\,\text{kg} \cdot \text{m}^2/\text{s}$$

which does not depend on time t, and thus it is constant.

11.9. The angular momentum about the point O is

$$\mathbf{L}_O = \mathbf{r} \times \mathbf{p} \tag{A.990}$$

where $\mathbf{p} = m\mathbf{v}$, and the direction of $\mathbf{v}$ changes with time as particle moves along the circle. At an angle θ, we have

$$\mathbf{r} = r\cos\theta\mathbf{i} + r\sin\theta\mathbf{j} \tag{A.991}$$

and

$$\mathbf{v} = -v\sin\theta\mathbf{i} + v\cos\theta\mathbf{j}. \tag{A.992}$$

Therefore, we get

$$\mathbf{L}_O = \det\begin{pmatrix} \mathbf{i} & \mathbf{j} & \mathbf{k} \\ x & y & z \\ mv_x & mv_y & mv_z \end{pmatrix} = m\cdot\det\begin{pmatrix} \mathbf{i} & \mathbf{j} & \mathbf{k} \\ r\cos\theta & r\sin\theta & 0 \\ -v\sin\theta & v\cos\theta & 0 \end{pmatrix} \tag{A.993}$$

$$= mvr\mathbf{k}.$$

To find the angular momentum about point P, we first express $\mathbf{r}$ with respect to point P:

$$\mathbf{r}_P = r\mathbf{i} + \mathbf{r} = r(1 + \cos\theta)\mathbf{i} + r\sin\theta\mathbf{j}. \tag{A.994}$$

Then the angular momentum about P is

$$\mathbf{L}_P = \det\begin{pmatrix} \mathbf{i} & \mathbf{j} & \mathbf{k} \\ x & y & z \\ mv_x & mv_y & mv_z \end{pmatrix} = m\cdot\det\begin{pmatrix} \mathbf{i} & \mathbf{j} & \mathbf{k} \\ r(1 + \cos\theta) & r\sin\theta & 0 \\ -v\sin\theta & v\cos\theta & 0 \end{pmatrix} \tag{A.995}$$

$$= mrv(\cos\theta + \cos^2\theta + \sin^2\theta)\mathbf{k}$$

$$= mvr(1 + \cos\theta)\mathbf{k}$$

which depends on the angle θ, and so the position of the pass on the circle. Moreover, this result indicates that angular momentum depends on the origin.

11.10. The direction of the angular momentum is shown in Fig. A.31. The magnitude of the angular momentum is

$$L = I\omega \tag{A.996}$$

where I is the moment of inertia about the z-axis through the center of mass, which is

$$I = \frac{2}{5}mr^2 = \frac{2}{5}(15.0\,\text{kg})(0.500\,\text{m})^2 = 1.50\,\text{kg}\cdot\text{m}^2. \tag{A.997}$$

Then for L we get

$$L = (1.50\,\text{kg}\cdot\text{m}^2)(3.00\,\text{rad/s}) = 4.50\,\text{kg}\cdot\text{m}^2/\text{s}. \tag{A.998}$$

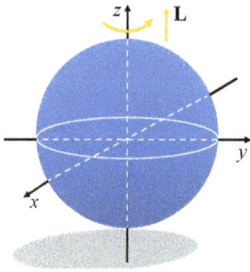

Figure A.31: A rotating solid sphere.

Solutions Chapter 12

12.1. We have two unknowns T_1 and T_2, therefore we need two equations. These two equations are provided by taking the components of T_1 and T_2 along the x and the y directions. We use first the condition for translational equilibrium:

$$\sum F_x = 0; \quad \sum F_y = 0 \tag{A.999}$$

or

$$T_1 \cos \theta - T_2 \cos \phi = 0 \tag{A.1000}$$
$$T_1 \sin \theta + T_2 \sin \phi - mg = 0. \tag{A.1001}$$

Using simple algebra, we get

$$T_2 = \frac{mg}{\cos \phi \tan \theta + \sin \phi} \tag{A.1002}$$

and

$$T_1 = \frac{mg \cos \phi}{\cos \phi \sin \theta + \sin \phi \cos \theta}. \tag{A.1003}$$

Using the relation: $\cos \phi \sin \theta + \sin \phi \cos \theta = \sin(\theta + \phi)$, we get

$$T_1 = \frac{mg \cos \phi}{\sin(\theta + \phi)}. \tag{A.1004}$$

12.2. The translational equilibrium condition gives

$$F_1 + F_2 + F_3 = 0 \tag{A.1005}$$

or

$$5\mathbf{i} - 2\mathbf{j} + 9\mathbf{i} + \mathbf{j} + F_{3x}\mathbf{i} + F_{3y}\mathbf{j} = 0 \tag{A.1006}$$

which can further be simplified as

$$F_{3x}\mathbf{i} + F_{3y}\mathbf{j} = -14\mathbf{i} + \mathbf{j}. \tag{A.1007}$$

From eq. (A.1007), we get

$$F_{3x} = -14; \quad F_{3y} = 1. \tag{A.1008}$$

Therefore, we can write

$$\mathbf{F}_3 = -14\mathbf{i} + \mathbf{j}. \tag{A.1009}$$

12.3. To understand the problem, we have shown a bridge and a man standing at one end, at equilibrium, in Fig. A.32. The conditions of the static equilibrium for the bridge can be written as

$$\sum_i \mathbf{F}_i = \mathbf{N}_1 + \mathbf{W} + M\mathbf{g} + \mathbf{N}_2 = 0 \tag{A.1010}$$

$$\sum_i \boldsymbol{\tau}_i = \mathbf{r}_1 \times \mathbf{N}_1 + \mathbf{r}_2 \times \mathbf{W} + \mathbf{r}_3 \times (M\mathbf{g}) + \mathbf{r}_4 \times \mathbf{N}_2 = 0. \tag{A.1011}$$

The axis of rotation is perpendicular to the page through the point O. From Fig. A.32, the moment arm for the gravity force $M\mathbf{g}$ exerted on the bridge is zero; therefore, its contribution to the net torque is zero. $\mathbf{W}$ is the weight of the man standing on the bridge, which produces a force ($= mg$) exerted on the bridge.

We can project the first equation along the vertical axis, for example the y-axis:

$$N_1 - mg - Mg + N_2 = 0 \tag{A.1012}$$

$$-2.0N_1 + 0.5mg + 0 + 2.0N_2 = 0. \tag{A.1013}$$

In the second equation, the sign minus indicates that the force creates a rotation clockwise, and the plus sign indicates that the force creates a rotation counterclockwise. We have two equations and two unknowns (N_1 and N_2), we can solve this system of equations for N_1 and N_2. The solution gives

$$N_1 = \frac{g}{4}(2.5m + 2.0M) \tag{A.1014}$$

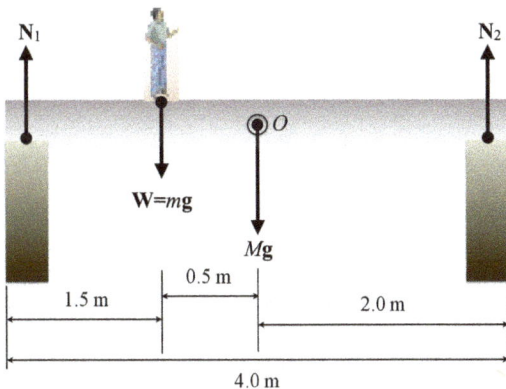

Figure A.32: Bridge system at equilibrium.

$$N_2 = \frac{g}{4}(1.5m + 2.0M).$$ (A.1015)

Replacing the numerical values, we get

$$N_1 = 2.94 \times 10^3 \, N$$ (A.1016)

$$N_2 = 2.74 \times 10^3 \, N.$$ (A.1017)

12.4. From Fig. 12.7, the forces exerted on the ladder are the reaction forces on the floor $\mathbf{N_1} = \mathbf{f} + \mathbf{R}$ and wall $\mathbf{N_2}$, the gravity forces on the ladder $M\mathbf{g}$ and the person's weight $\mathbf{W} = m\mathbf{g}$. We are going to assume that the person's center of gravity is at the same height as the center of gravity of the ladder. Moreover, the rotation axis passes through the point O. The friction force is

$$\mathbf{f} = \mu_s R \mathbf{i}.$$ (A.1018)

Using the static equilibrium conditions for translational motion, we can write

$$\sum_i F_{ix} = f - N_2 = 0$$ (A.1019)

$$\sum_i F_{iy} = R - mg - Mg = 0.$$ (A.1020)

From eq. (A.1020), we get

$$R = (m + M)g.$$ (A.1021)

Therefore, the magnitude of the friction force is

$$f = \mu_s R = \mu_s(m + M)g.$$ (A.1022)

From eq. (A.1019), the magnitude N_2 is

$$N_2 = f = \mu_s(m + M)g.$$ (A.1023)

The magnitude N_1 is calculated as

$$N_1 = \sqrt{R^2 + f^2} = \sqrt{1 + \mu_s^2}(m + M)g.$$ (A.1024)

Then the vectors $\mathbf{N_1}$ and $\mathbf{N_2}$ are

$$\mathbf{N_1} = \mu_s(m + M)g\mathbf{i} + (m + M)g\mathbf{j}$$ (A.1025)

$$\mathbf{N_2} = -\mu_s(m + M)g\mathbf{i}.$$ (A.1026)

To find θ_{min}, we must use the second condition for equilibrium. When we calculate the net torque about an axis through the origin O at the bottom of the ladder, we have

$$\sum_i \tau_i = N_2 l \sin\theta - Mg\frac{l}{2}\cos\theta - mg\frac{l}{2}\cos\theta = 0$$ (A.1027)

or

$$\tan \theta = \frac{(m + M)g}{2N_2} = \frac{1}{2\mu_s}$$ (A.1028)

or

$$\theta_{min} = \tan^{-1}\left(\frac{1}{2\mu_s}\right).$$ (A.1029)

12.5. To understand the problem, we have shown it graphically in Fig. A.33, along with a free-body diagram.

The external forces acting on the beam include the 200 N force of gravity, the force **T** exerted by the cable, the force **N** exerted by the wall at the axis of rotation, and the 600 N force that the person exerts on the beam. These forces are also indicated in the free-body diagram for the beam. We now apply the condition for the equilibrium of translational motion:

$$\sum_i F_{ix} = N \cos \theta - T \cos 53.0° = 0$$ (A.1030)

$$\sum_i F_{iy} = N \sin \theta - 600\,N - 200\,N + T \sin 53.0° = 0.$$ (A.1031)

From the condition for rotational equilibrium, we get

$$\sum_i \tau_i = -(600\,N)(2.00\,m) - (200\,N)(4.00\,m)$$ (A.1032)

$$+ (T \sin 53.0°)(8.00\,m) = 0.$$

Solving it for T, we get

$$T \approx \frac{2000\,Nm}{6.38908\,m} \approx 313\,N$$ (A.1033)

as a vector

$$\mathbf{T} = T_x\mathbf{i} + T_y\mathbf{j} = (-188\mathbf{i} + 250\mathbf{j})\,N.$$ (A.1034)

Moreover, from eq. (A.1030) and eq. (A.1031), we get

$$N_x \equiv N \cos \theta = T \cos 53.0° = 188\,N$$ (A.1035)
$$N_y \equiv N \sin \theta = 600\,N + 200\,N - T \sin 53.0° = 550\,N$$ (A.1036)

or

$$\tan \theta = 2.93.$$ (A.1037)

Thus,

$$\theta = \tan^{-1}(2.93) = 71.2°. \tag{A.1038}$$

The vector **N** is

$$\mathbf{N} = N_x\mathbf{i} + N_y\mathbf{j} = (188\mathbf{i} + 550\mathbf{j})\,\text{N} \tag{A.1039}$$

and its magnitude is

$$N = \sqrt{N_x^2 + N_y^2} = \sqrt{(188)^2 + (550)^2} \approx 581\,\text{N}. \tag{A.1040}$$

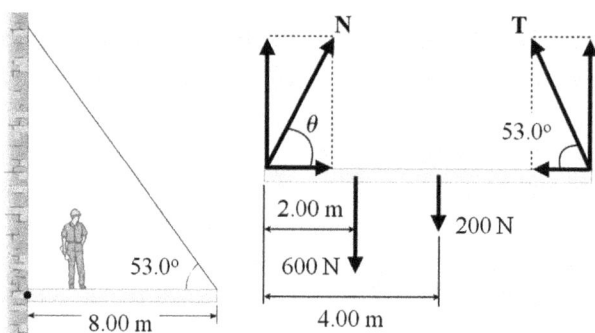

Figure A.33: A uniform beam supported by a cable and a free-body diagram.

12.6. From the definition of Young's modulus:

$$E = \frac{\sigma_n}{\varepsilon_n} = \frac{F_n/A}{\Delta l/l} \tag{A.1041}$$

or

$$A = \frac{F_n l}{E \Delta l} = \frac{(940\,\text{N})(10.0\,\text{m})}{(20.0 \times 10^{10}\,\text{N/m}^2)(0.500 \times 10^{-2})\,\text{m}} \tag{A.1042}$$
$$= 9.40 \times 10^{-6}\,\text{m}^2.$$

The radius is found from the relation: $A = \pi R^2$ as

$$R = \sqrt{A/\pi} \approx 1.73 \times 10^{-3}\,\text{m} = 1.73\,\text{mm} \tag{A.1043}$$

and the diameter is

$$D = 2R = 3.46\,\text{mm}. \tag{A.1044}$$

12.7. From the definition of the bulk modulus, we can write

$$B = -\frac{\Delta P}{\Delta V / V_i} \tag{A.1045}$$

or

$$\Delta V = -\frac{V_i \Delta P}{B}. \tag{A.1046}$$

Assuming that the final pressure is much greater than the initial pressure, we can neglect the initial pressure and hence

$$\Delta P = P_f - P_i \approx P_f = 2.0 \times 10^7 \text{ N/m}^2. \tag{A.1047}$$

Therefore, we obtain

$$\Delta V = -\frac{(0.50 \text{ m}^3)(2.0 \times 10^7 \text{ N/m}^2)}{6.1 \times 10^{10} \text{ N/m}^2} = -1.6 \times 10^{-4} \text{ m}^2 \tag{A.1048}$$

where the negative sign indicates a decrease in volume.

12.8. The cross-sectional surface area is

$$A = \pi R^2 = \pi (D/2)^2 = 4.91 \times 10^{-4} \text{ m}^2. \tag{A.1049}$$

Using the normal stress definition:

$$\sigma_n = \frac{F_n}{A} \tag{A.1050}$$

or

$$F_n = \sigma_n A = (1.50 \times 10^8 \text{ N/m}^2)(4.91 \times 10^{-4} \text{ m}^2) \approx 2.46 \times 10^4 \text{ N}. \tag{A.1051}$$

From the definition of Young's modulus:

$$E = \frac{\sigma_n}{\varepsilon_n} = \frac{F_n/A}{\Delta l/l} \tag{A.1052}$$

we get

$$\Delta l = \frac{F_n/A}{E/l} = \frac{1.50 \times 10^8 \text{ N/m}^2}{(1.50 \times 10^{10} \text{ N/m}^2)/(0.250 \text{ m})} \tag{A.1053}$$
$$= 0.250 \times 10^{-2} \text{ m} = 0.250 \text{ cm}.$$

12.9. The normal force is $F_n = mg$, where $g = 9.80 \text{ m/s}^2$.
From the definition of Young's modulus:

$$E = \frac{\sigma_n}{\varepsilon_n} = \frac{F_n/A}{\Delta l/l} \tag{A.1054}$$

we get

$$\Delta l = \frac{F_n/A}{E/l} = \frac{(200\ \text{kg} \cdot 9.80\ \text{m/s}^2)/(0.200 \times 10^{-4}\ \text{m}^2)}{(8.00 \times 10^{10}\ \text{N/m}^2)/(4.00\ \text{m})} \tag{A.1055}$$

$$= 4900 \times 10^{-6}\ \text{m} = 0.490\ \text{cm}.$$

12.10. From the definition of the bulk modulus, we can write

$$B = +\frac{\Delta P}{\Delta V/V_i}. \tag{A.1056}$$

The plus sign indicates that the volume increases.

From this equation we obtain

$$\Delta P = +B\frac{\Delta V}{V_i} = +(2.00 \times 10^9\ \text{N/m}^2)(9.00\,\%) \tag{A.1057}$$

$$= +0.180 \times 10^9\ \text{N/m}^2$$

where the plus sign indicates that the pressure increases.

Solutions Chapter 13

13.1. We can compare the above equation with the general form of the equation of the simple harmonic motion:

$$x(t) = A\cos(\omega t + \phi). \tag{A.1058}$$

(a) The amplitude is $A = 4.00$ m, the angular frequency is $\omega = \pi$, the frequency f is

$$f = \frac{\omega}{2\pi} = \frac{\pi}{2\pi} = 0.500\ \text{Hz}. \tag{A.1059}$$

The period is

$$T = \frac{1}{f} = 2.00\ \text{s}. \tag{A.1060}$$

(b) The velocity at any time t is

$$v(t) = \frac{dx}{dt} = -(4.00\ \text{m})(\pi)\sin\left(\pi t + \frac{\pi}{4}\right) \tag{A.1061}$$

$$= -(12.6\ \text{m/s})\sin\left(\pi t + \frac{\pi}{4}\right).$$

The acceleration is

$$a = \frac{dv}{dt} = -(12.6\pi\ \text{m/s})\cos\left(\pi t + \frac{\pi}{4}\right) \tag{A.1062}$$

$$= -(39.5\ \text{m/s}^2)\cos\left(\pi t + \frac{\pi}{4}\right).$$

(c) The position at $t = 1.00$ s is

$$x(t = 1.00\,\text{s}) = (4.00\,\text{m})\cos(\pi(1.00\,\text{s}) + \pi/4) \tag{A.1063}$$

$$= (4.00\,\text{m})\cos\left(\frac{5\pi}{4}\right) \approx -2.83\,\text{m}.$$

The velocity at $t = 1.00$ s is

$$v(t) = -(12.6\,\text{m/s})\sin\left(\pi + \frac{\pi}{4}\right) = 8.89\,\text{m/s} \tag{A.1064}$$

and the acceleration is

$$a = -(39.5\,\text{m/s}^2)\cos\left(\pi + \frac{\pi}{4}\right) = 27.9\,\text{m/s}^2. \tag{A.1065}$$

(d) The maximum speed is

$$v_{max} = 12.6\,\text{m/s} \tag{A.1066}$$

and maximum acceleration is

$$a_{max} = 39.5\,\text{m/s}^2. \tag{A.1067}$$

(e) To determine the displacement of the object between $t = 0$ and $t = 1.00$ s, we calculate $x_i = x(t = 0)$ and $x_f = x(t = 1.00$ s$)$, and then the displacement is

$$\Delta x = x_f - x_i. \tag{A.1068}$$

At $t = 0$

$$x_i = (4.00\,\text{m})\cos\left(0 + \frac{\pi}{4}\right) \approx 2.83\,\text{m}. \tag{A.1069}$$

Then Δx is

$$\Delta x = x_f - x_i = (-2.83\,\text{m}) - (2.83\,\text{m}) = -5.66\,\text{m}. \tag{A.1070}$$

13.2. The total mass is

$$M = m_{car} + m_{2\text{-people}} = 1460\,\text{kg}. \tag{A.1071}$$

Assuming that this mass is uniformly distributed, the mass supported by each spring is

$$M_{1/4} = \frac{M}{4} = \frac{1300\,\text{kg}}{4} = 365\,\text{kg}. \tag{A.1072}$$

The frequency of vibration is then

$$f = \frac{1}{2\pi}\sqrt{\frac{k}{M_{1/4}}} = \frac{1}{2\pi}\sqrt{\frac{20000\,\text{N/m}}{365\,\text{kg}}} = 1.18\,\text{Hz}. \tag{A.1073}$$

13.3.

(a) The initial velocity is zero, since the block is released from the rest: $v_0 = 0$ and the amplitude $A = 5.00$ cm $= 0.0500$ m. The angular frequency is

$$\omega = \sqrt{\frac{k}{m}} = \sqrt{\frac{5.00 \text{ N/m}}{200 \times 10^{-3} \text{ kg}}} = 5.00 \text{ rad/s} \tag{A.1074}$$

and the period is given by

$$T = \frac{2\pi}{\omega} = \frac{2\pi}{5.00 \text{ rad/s}} = 1.26 \text{ s.} \tag{A.1075}$$

(b) The maximum speed is

$$v_{max} = A\omega = (0.0500 \text{ m})(5.00 \text{ rad/s}) = 0.250 \text{ m/s.} \tag{A.1076}$$

(c) The maximum acceleration is

$$a_{max} = A\omega^2 = (0.0500 \text{ m})(5.00 \text{ rad/s})^2 = 1.25 \text{ m/s}^2. \tag{A.1077}$$

(d) The displacement is

$$x(t) = A \cos(\omega t + \phi) \tag{A.1078}$$

where $x_0 = x(t = 0) = A \cos \phi = 0.0500$ m, from this

$$\cos \phi = 1 \tag{A.1079}$$

or

$$\phi = 0 \tag{A.1080}$$

therefore, we get

$$x(t) = (0.0500 \text{ m}) \cos(5.00t). \tag{A.1081}$$

The speed is

$$v(t) = \frac{dx}{dt} = -(0.250 \text{ m/s}) \sin(5.00t) \tag{A.1082}$$

and the acceleration

$$a(t) = \frac{dv}{dt} = -(1.25 \text{ m/s}^2) \cos(5.00t). \tag{A.1083}$$

13.4.

(a) The total mechanical energy is

$$E = K + U = \frac{1}{2}kA^2. \tag{A.1084}$$

Replacing the numerical values, we get

$$E = \frac{1}{2}(20.0\,\text{N/m})(3.00 \times 10^{-2}\,\text{m})^2 = 9.00 \times 10^{-3}\,\text{J}. \tag{A.1085}$$

The maximum speed is at $x = 0$, where $U = 0$ and $E = K$, thus

$$E = K = \frac{1}{2}mv_{max}^2 \tag{A.1086}$$

or

$$v_{max} = \sqrt{\frac{2E}{m}} = \sqrt{\frac{2(9.00 \times 10^{-3}\,\text{J})}{0.500\,\text{kg}}} = 0.190\,\text{m/s}. \tag{A.1087}$$

(b) Using the equation

$$v = \pm\sqrt{\frac{k}{m}(A^2 - x^2)} \tag{A.1088}$$

we get the velocity of the mass as

$$v = \pm\sqrt{\frac{20.0\,\text{N/m}}{0.500\,\text{kg}}((0.0300\,\text{m})^2 - (0.0200\,\text{m})^2)} = \pm 0.141\,\text{m/s}. \tag{A.1089}$$

The positive and negative signs indicate that the object is moving to either the right or the left at this instant.

(c) The kinetic energy is

$$K = \frac{1}{2}mv^2 = \frac{1}{2}(0.500\,\text{kg})(0.141\,\text{m/s})^2 = 5.00 \times 10^{-3}\,\text{J} \tag{A.1090}$$

and the potential energy is

$$U = \frac{1}{2}kx^2 = \frac{1}{2}(20.0\,\text{N})(0.0200\,\text{m})^2 = 4.00 \times 10^{-3}\,\text{J}. \tag{A.1091}$$

13.5. Using the formula of the period

$$T = 2\pi\sqrt{\frac{L}{g}} \tag{A.1092}$$

and solving it for L, we get

$$L = \frac{T^2 g}{4\pi^2} = \frac{(1\,\text{s})^2(9.80\,\text{m/s}^2)}{4\pi^2} = 0.248\,\text{m}. \tag{A.1093}$$

Thus, the meter's length would be slightly less than one-fourth of its current length.

13.6. To find the period, we can use the equation

$$T = 2\pi \sqrt{\frac{I}{Mg\frac{L}{2}}} \qquad \text{(A.1094)}$$

where the moment of inertia of the rod at point O is

$$I_O = I_{\text{c.m.}} + M\left(\frac{L}{2}\right)^2 = \frac{1}{12}ML^2 + \frac{1}{4}ML^2 = \frac{1}{3}ML^2. \qquad \text{(A.1095)}$$

Then the period is

$$T = 2\pi \sqrt{\frac{\frac{1}{3}ML^2}{Mg\frac{L}{2}}} = 2\pi \sqrt{\frac{2L}{3g}}. \qquad \text{(A.1096)}$$

13.7. The period is calculated using the formula

$$T = \frac{2\pi}{\omega} = 2\pi \sqrt{\frac{m}{k}}. \qquad \text{(A.1097)}$$

To find k, we use the third law of Newton:

$$mg = kx \qquad \text{(A.1098)}$$

where $m = 10.0$ g and $x = 3.90$ cm. Therefore, we get

$$k = \frac{mg}{x} = \frac{(10.0 \times 10^{-3}\,\text{kg})(9.80\,\text{m/s}^2)}{3.90 \times 10^{-2}\,\text{m}} \approx 2.51\,\text{N/m}. \qquad \text{(A.1099)}$$

Then the period is

$$T = 2\pi \sqrt{\frac{25.0 \times 10^{-3}\,\text{kg}}{2.51\,\text{N/m}}} \approx 0.627\,\text{s}. \qquad \text{(A.1100)}$$

13.8. To find ϕ, we express x as follows:

$$x = -A\cos \omega t = A\cos(\omega t + \pi).$$

By comparing this expression with $x = A\cos(\omega t + \phi)$, we determine

$$\phi = \pi.$$

13.9. Using

$$x = (4.00\,\text{m})\cos(3.00\pi t + \pi)$$

we find the following.

(a) The angular frequency is

$$\omega = 3.00\pi \approx 9.42\,\text{rad/s}.$$

The linear frequency is

$$f = \frac{\omega}{2\pi} = \frac{9.42\,\text{rad/s}}{2\pi} = 1.5\,\text{Hz}.$$

(b) The period is

$$T = \frac{1}{f} = \frac{1}{1.5\,\text{Hz}} \approx 0.667\,\text{s}.$$

(c) The amplitude is

$$A = 4.00\,\text{m}.$$

(d) The phase constant is

$$\phi = \pi\,\text{rad} \approx 3.14\,\text{rad}.$$

(e) The displacement $x(t = 0.250\,\text{s})$ is

$$x(0.250\,\text{s}) = (4.00\,\text{m})\cos(3.00\pi(0.250\,\text{s}) + \pi) \approx 2.83\,\text{m}.$$

13.10. Based on the data, the angular frequency is

$$\omega = 2\pi \approx 6.28\,\text{rad/s}.$$

Therefore, the period is

$$T = \frac{2\pi}{\omega} = \frac{2\pi}{2\pi} = 1.00\,\text{s}.$$

Using the relationship

$$T = 2\pi\sqrt{\frac{L}{g}}$$

where L is the length of the pendulum and $g = 9.80\,\text{m/s}^2$ is the gravitational acceleration. Thus, we obtain

$$L = g\left(\frac{T}{2\pi}\right)^2 = (9.80\,\text{m/s}^2)\left(\frac{1.00\,\text{s}}{2\pi}\right)^2 \approx 0.248\,\text{m} = 24.8\,\text{cm}.$$

Solutions Chapter 14

14.1. The radius of Mars in SI units is

$$R = 3396\,\text{km} = 3396 \times 10^3\,\text{m}.$$

Then, using eq. (14.6), we get

$$g = \left(6.67384 \times 10^{-11}\,\text{N} \cdot \frac{\text{m}^2}{\text{kg}^2}\right) \frac{6.4 \times 10^{23}\,\text{kg}}{(3396 \times 10^3\,\text{m})^2} \qquad (\text{A.1101})$$
$$\approx 3.70 \times 10^{-11+23-12}\,\text{m/s}^2.$$

Thus,

$$g \approx 3.70\,\text{m/s}^2.$$

14.2. The radius of Jupiter in SI units is

$$R = 71492\,\text{km} = 71.492 \times 10^6\,\text{m}.$$

Using eq. (14.6), we get

$$g = \left(6.67384 \times 10^{-11}\,\text{N} \cdot \frac{\text{m}^2}{\text{kg}^2}\right) \frac{1898.19 \times 10^{24}\,\text{kg}}{(71.492 \times 10^6\,\text{m})^2} \qquad (\text{A.1102})$$
$$\approx 2.4786 \times 10^{-11+24-12}\,\text{m/s}^2.$$

Thus,

$$g \approx 24.79\,\text{m/s}^2.$$

14.3. The radius of Mercury in SI units is

$$R = 2439.7\,\text{km} = 2.4397 \times 10^6\,\text{m}.$$

Using eq. (14.6), we find that

$$g = \left(6.67384 \times 10^{-11}\,\text{N} \cdot \frac{\text{m}^2}{\text{kg}^2}\right) \frac{0.33011 \times 10^{24}\,\text{kg}}{(2.4397 \times 10^6\,\text{m})^2} \qquad (\text{A.1103})$$
$$\approx 0.370 \times 10^{-11+24-12}\,\text{m/s}^2.$$

Thus,

$$g \approx 3.70\,\text{m/s}^2.$$

14.4. The equatorial and polar radii of Earth in SI units are

$$R_e = 6378.1 \, \text{km} = 6.3781 \times 10^6 \, \text{m} \tag{A.1104}$$
$$R_p = 6356.8 \, \text{km} = 6.3568 \times 10^6 \, \text{m}.$$

Using eq. (14.6), we get the gravitational acceleration on equator as

$$g_e = \left(6.67384 \times 10^{-11} \, \text{N} \cdot \frac{\text{m}^2}{\text{kg}^2} \right) \frac{5.9724 \times 10^{24} \, \text{kg}}{(6.3781 \times 10^6 \, \text{m})^2} \tag{A.1105}$$
$$\approx 0.9798 \times 10^{-11+24-12} \, \text{m/s}^2.$$

Thus,

$$g_e \approx 9.80 \, \text{m/s}^2.$$

For the gravitational acceleration on poles, we obtain

$$g_p = \left(6.67384 \times 10^{-11} \, \text{N} \cdot \frac{\text{m}^2}{\text{kg}^2} \right) \frac{5.9724 \times 10^{24} \, \text{kg}}{(6.3568 \times 10^6 \, \text{m})^2} \tag{A.1106}$$
$$\approx 0.986 \times 10^{-11+24-12} \, \text{m/s}^2.$$

Thus,

$$g_p \approx 9.86 \, \text{m/s}^2.$$

14.5. Here, AU stands for the astronomical unit, which is the distance between the Earth and the Sun:

$$1 \, \text{AU} = 1.5 \times 10^{11} \, \text{m}.$$

Therefore,

$$r = 19 \times 1.5 \times 10^{11} \, \text{m} = 2.85 \times 10^{12} \, \text{m}.$$

Using eq. (14.44), we obtain the speed of Uranus as

$$v = \sqrt{ \left(6.67384 \times 10^{-11} \, \text{N} \cdot \frac{\text{m}^2}{\text{kg}^2} \right) \frac{1.99 \times 10^{30} \, \text{kg}}{2.85 \times 10^{12} \, \text{m}} } \approx 6.8 \times 10^3 \, \text{m/s}.$$

Thus,

$$v \approx 6.8 \, \text{km/s}.$$

14.6. The acceleration of an object on Earth is

$$g = 9.80 \text{ m/s}^2.$$

At a distance r from the center of the Earth, g is given by eq. (14.22), which is written as

$$g = G\frac{M}{(R_E + h)^2}$$

where R_E is the radius of the Earth, taken according to the equator: $R_E = 6.3781 \times 10^6$ m. Mass of Earth is $M = 5.9724 \times 10^{24}$ kg. Therefore,

$$\frac{9.80}{2} \text{ m/s}^2 = \left(6.67384 \times 10^{-11} \text{ N} \cdot \frac{\text{m}^2}{\text{kg}^2}\right)\frac{5.9724 \times 10^{24} \text{ kg}}{(6.3781 \times 10^6 \text{ m} + h)^2}.$$

Solving it for h, we find

$$h \approx 2.64 \times 10^6 \text{ m}.$$

14.7. Using eq. (14.1), we find the magnitude of the force as

$$F = \left(6.67384 \times 10^{-11} \text{ N} \cdot \frac{\text{m}^2}{\text{kg}^2}\right)\frac{(100.0 \text{ kg})^2}{(50.0 \times 10^{-2} \text{ m})^2}.$$

After the calculations, we find

$$F \approx 26.7 \times 10^{-7} \text{ N}.$$

Note that this magnitude is practically tiny in value.

14.8. Using eq. (14.39), we obtain

$$T = 2\pi(228 \times 10^9 \text{ m})^{3/2}\sqrt{\frac{1}{(6.67384 \times 10^{-11} \text{ N} \cdot \frac{\text{m}^2}{\text{kg}^2})(1.99 \times 10^{30} \text{ kg})}}.$$

Thus,

$$T_{\text{Mars}} \approx 59373505 \text{ s}.$$

14.9. Using eq. (14.50), we obtain

$$V_{\text{escape}} = \sqrt{\frac{2GM}{R}}$$

$$= \sqrt{\frac{2(6.67384 \times 10^{-11} \text{ N} \cdot \frac{\text{m}^2}{\text{kg}^2})(5.98 \times 10^{24} \text{ kg})}{6.37 \times 10^6 \text{ m}}}$$

$$\approx 1.1 \times 10^4 \text{ m/s}.$$

The kinetic energy is

$$K = \frac{1}{2}mv_{\text{escape}}^2 = \frac{1}{2}(6000 \text{ kg})(1.1 \times 10^4 \text{ m/s})^2 \approx 3.6 \times 10^8 \text{ kJ}.$$

14.10. First we calculate the total gravitational potential energy of the three-body system:

$$U_{total} = U_{12} + U_{13} + U_{23} = -G\left(\frac{m_1 m_2}{r_{12}} + \frac{m_1 m_3}{r_{13}} + \frac{m_2 m_3}{r_{23}}\right).$$

Then the total work necessary to separate the three masses equals U_{total}:

$$W = -G\left(\frac{m_1 m_2}{r_{12}} + \frac{m_1 m_3}{r_{13}} + \frac{m_2 m_3}{r_{23}}\right).$$

Bibliography

[1] D. Holliday, R. Resnick, and J. Walker. *Fundamentals of Physics*. John Wiley & Sons, (2011).

[2] M. Sternheim and J. Kane. *General Physics*. John Wiley & Sons, (1991).

[3] H. Goldstein, C. Poole, and J. Safko. *Classical Mechanics. Third Edition*. Addison Wesley, (2002).

[4] A. Fazely. *Foundation of Physics for Scientists and Engineers. Volume I: Mechanics, Heat and Sound*. ISBN 978-87-403-1002-3. Bookboon.com, (2015).

https://doi.org/10.1515/9783110755824-016

Index